T0073841

VOLUME ONE HUNDRED AND TWENTY TWO

ADVANCES IN
COMPUTERS

Hardware Accelerator Systems
for Artificial Intelligence
and Machine Learning

VOLUME ONE HUNDRED AND TWENTY TWO

Advances in
COMPUTERS

Hardware Accelerator Systems
for Artificial Intelligence
and Machine Learning

Edited by

SHIHO KIM
*School of Integrated Technology, Yonsei University,
Seoul, South Korea*

GANESH CHANDRA DEKA
*Ministry of Skill Development and Entrepreneurship,
New Delhi, India*

ACADEMIC PRESS

An imprint of Elsevier

ELSEVIER

Academic Press is an imprint of Elsevier
50 Hampshire Street, 5th Floor, Cambridge, MA 02139, United States
525 B Street, Suite 1650, San Diego, CA 92101, United States
The Boulevard, Langford Lane, Kidlington, Oxford OX5 1GB, United Kingdom
125 London Wall, London, EC2Y 5AS, United Kingdom

First edition 2021

ISBN: 978-0-12-823123-4
ISSN: 0065-2458

For information on all Academic Press publications
visit our website at https://www.elsevier.com/books-and-journals

Publisher: Zoe Kruze
Developmental Editor: Tara A. Nadera
Production Project Manager: James Selvam
Cover Designer: Alan Studholme

Typeset by SPi Global, India

Working together
to grow libraries in
developing countries

www.elsevier.com • www.bookaid.org

Contents

Contributors

Oleg Alienin
National Technical University of Ukraine "Igor Sikorsky Kyiv Polytechnic Institute", Kyiv, Ukraine

Parth Bir
Electronics and Communication Department, G.L.Bajaj Institute of Technology & Management, Greater Noida, U.P., India

Amitabh Biswal
Department of Computer Science and Engineering, National Institute of Technology Silchar, Cachar, Silchar, Assam, India

Malaya Dutta Borah
Department of Computer Science and Engineering, National Institute of Technology Silchar, Cachar, Silchar, Assam, India

Francesco Daghero
Department of Control and Computer Engineering, Politecnico di Torino, Turin, Italy

Nikita Gordienko
National Technical University of Ukraine "Igor Sikorsky Kyiv Polytechnic Institute", Kyiv, Ukraine

Yuri Gordienko
National Technical University of Ukraine "Igor Sikorsky Kyiv Polytechnic Institute", Kyiv, Ukraine

Neha Gupta
Faculty of Computer Applications, Manav Rachna International Institute of Research and Studies, Faridabad, India

Dongho Ha
School of Electrical and Electronic Engineering, Yonsei University, Seoul, South Korea

Zakir Hussain
Department of Computer Science and Engineering, National Institute of Technology Silchar, Cachar, Silchar, Assam, India

Won Jeon
School of Electrical and Electronic Engineering, Yonsei University, Seoul, South Korea

Joo-Young Kim
KAIST, Daejeon, South Korea

Shiho Kim
School of Integrated Technology, Yonsei University, Seoul, South Korea

Gun Ko
School of Electrical and Electronic Engineering, Yonsei University, Seoul, South Korea

Yuriy Kochura
National Technical University of Ukraine "Igor Sikorsky Kyiv Polytechnic Institute", Kyiv, Ukraine

Hyunwuk Lee
School of Electrical and Electronic Engineering, Yonsei University, Seoul, South Korea

Jiwon Lee
School of Electrical and Electronic Engineering, Yonsei University, Seoul, South Korea

Kyuho J. Lee
The School of Electrical and Computer Engineering, The Artificial Intelligence Graduate School, Ulsan National Institute of Science and Technology, Ulsan, South Korea

Daniele Jahier Pagliari
Department of Control and Computer Engineering, Politecnico di Torino, Turin, Italy

Hyunbin Park
IT & Mobile Communications, Samsung Electronics, Seoul, South Korea

Massimo Poncino
Department of Control and Computer Engineering, Politecnico di Torino, Turin, Italy

Won Woo Ro
School of Electrical and Electronic Engineering, Yonsei University, Seoul, South Korea

Oleksandr Rokovyi
National Technical University of Ukraine "Igor Sikorsky Kyiv Polytechnic Institute", Kyiv, Ukraine

William J. Song
School of Electrical and Electronic Engineering, Yonsei University, Seoul, South Korea

Sergii Stirenko
National Technical University of Ukraine "Igor Sikorsky Kyiv Polytechnic Institute", Kyiv, Ukraine

Vlad Taran
National Technical University of Ukraine "Igor Sikorsky Kyiv Polytechnic Institute", Kyiv, Ukraine

Preface

Artificial Intelligence has witnessed tremendous growth with the advent of deep neural networks (DNNs) and Machine Learning. The algorithmic superiority of DNNs comes at extremely high computation and memory costs that pose both challenges and opportunities to the hardware platforms. GPUs (graphics processing units), NPUs (neural processing units), and specialized Hardware accelerators are the state-of-the-art in training and inference the DNNs.

This edited book consisting of 11 chapters contributed by researchers from academia and industries from different countries explores research in Hardware Accelerator Systems to improve processing efficiency and performance of GPU, NPU, approximate computing, inference accelerators, in-memory computing, machine intelligence, and quantum computing.

Chapter 1 titled *Introduction to Hardware Accelerator Systems for Artificial Intelligence and Machine Learning* discusses about the software framework for DNNs. Chapter 2 titled *Hardware Accelerator Systems for Embedded Systems* describes the considerations and constraints to deploy neural network applications in embedded systems. Chapter 3 titled *Hardware Accelerator Systems for Artificial Intelligence and Machine Learning*, discuses about the power consumption of a commercial GPU using hardware accelerators on embedded devices. Generic Quantum Hardware Accelerator (GQHA) is described via algorithms, mathematical models, and microarchitecture in Chapter 4 titled *Generic Quantum Hardware Accelerators for Conventional systems*.

Chapter 5 titled *FPGA Based Neural Network Accelerators* reviews computations in latest DNN models and their algorithmic optimizations. Device's strengths and weaknesses over other types of hardware platforms are discussed. Chapter 6 titled *Deep Learning with GPUs* presents an analysis on the evolution of GPU architectures and recent hardware and software supports for efficient acceleration of deep learning in GPUs. Chapter 7 titled *Architecture of Neural Processing Unit for Deep Neural Networks* provides a review of design of latest NPU architecture for DNN, mainly about inference engines.

Chapter 8, *Energy-Efficient Deep Learning Inference on Edge Devices*, survey the optimizations to support embedded deep learning inference with focus in hardware acceleration such as quantization and big-little architectures.

Chapter 9 titled *Last Mile Optimization of Edge Computing Ecosystem with Deep Learning Models and Specialized Tensor Processing Architectures* presents different TPA (Tensor Processing Architectures) hardware implementations including Coral Edge TPU by Google, and Movidius Neural Compute Stick by Intel.

Chapter 10 titled *Hardware Accelerator for Training with Integer Backpropagation and Probabilistic Weight Update* introduces a probabilistic weight update, also describes the hardware implementation of the probabilistic weight-update scheme. Finally, Chapter 11 titled *Music Recommender System using Restricted Boltzmann Machine with Implicit Feedback* discusses a method implementing Restricted Boltzmann machine for recommendation system. The authors have also explained how to use contrastive divergence algorithm to train Restricted Boltzmann machine and learn its parameters by collaborative filtering engine from Apache.

This book aims to offer diverse coverage of hardware accelerators of DNNs to artificial intelligent specialists as well as students who have interests in hardware accelerators. We hope the reader of book will take advantage of diverse and in-depth coverage of topics in artificial intelligence.

Prof SHIHO KIM

Seoul, S. Korea

GANESH CHANDRA DEKA

New Delhi, India

CHAPTER ONE

Introduction to hardware accelerator systems for artificial intelligence and machine learning

Neha Gupta

Faculty of Computer Applications, Manav Rachna International Institute of Research and Studies, Faridabad, India

Contents

Abstract

An AI accelerator is a category of specialized hardware accelerator or automatic data processing system designed to accelerate computer science applications, particularly artificial neural networks, machine visualization and machine learning. Typical applications embrace algorithms for AI, Internet of things and different data-intensive or sensor-driven tasks. Machine learning is widely employed in several modern artificial intelligence applications. Varied hardware platforms are enforced to support such applications. Among them, graphics process unit (GPU) is the most widely used because of its quick computation speed and compatibility with varied algorithms. Field programmable gate arrays (FPGA) show higher energy potency as compared with GPU when computing machine learning algorithm at the cost of low speed. Varied application-specific integrated circuits (ASIC) design are projected to realize the most effective energy

Advances in Computers, Volume 122
ISSN 0065-2458
https://doi.org/10.1016/bs.adcom.2020.07.001

potency at the value of less reconfigurability that makes it appropriate for special varieties of machine learning algorithms like a deep convolutional neural network. In this chapter, we will try to relate artificial intelligence and machine learning concepts to accelerate hardware resources. Chapter will discuss software framework for Deep Neural Networks and will give comparison of FPGA, CPU and GPU. At the end of the chapter future directions and conclusion will be given.

1. Introduction to artificial intelligence and machine learning in hardware acceleration

This is the age of big data. A lot of data has been created within the past 5–6 years than the whole history of the human civilization [1]. This can be primarily driven by the exponential increase within the use of sensors (10 billion each year in 2013, expected to achieve one trillion by 2020 [2]) and connected devices (6.4 billion in 2016, expected to achieve 20.8 billion by 2020 [3]). These sensors and devices generate many zetabytes (1021 bytes) of information each year—petabytes (1015 bytes) per second [3]. Machine learning is required to extract purposeful and preferably action-able information from this data. A significant amount of computation is needed to investigate this knowledge, which normally happens within the cloud. However, given the sheer volume and rate at that information is being generated, and also the high energy value of communication and sometimes restricted information measure, there is an increasing need to perform the analysis domestically close to the device instead of sending the data to the cloud. Embedding machine learning at the edge conjointly addresses necessary considerations associated with privacy, latency and security.

Machine learning (ML) is at present widely employed in several modern computer science applications [4]. Machine learning could be a thriving space within the field of computing, that involves algorithms that may learn and predict through, for instance, building models from given datasets for training. The breakthrough of the computation ability has enabled the system to calculate difficult completely different ML algorithmic rule in a com-paratively short time, providing real-time human-machine interaction similar to face detection for video surveillance, advanced driver–assistance systems (ADAS), and image recognition early cancer detection [5,6]. In recent years, several machine learning techniques, like convolution neural network (CNN) and support vector machine (SVM), have shown promise in several application domains, similar to image classification and speech recognition. Among all those applications, high detection accuracy needs difficult cubic centimeter computation that comes at the price of high machine complexity. This leads to a high demand on the hardware platform. Several hardware technologies are often employed in accelerating machine learning algorithms, like graphics process unit (GPU) and field programmable gate array (FPGA). Currently, most applications are enforced on general purpose compute engines, particularly graphics process units (GPUs).

In addition, because of its low power consumption, reconfigurability and real-time processing capacity, FPGA is a promising technology. Moreover, there are two barriers to effective implementation of the FPGA. First, architectural descriptions will capture design families with specific capacity, functionality and energy efficiency trade-offs. Second, a design flow for multiple architectures will enable the reuse and comparison of methods and techniques for optimization. In recent years, the field of artificial intelligence (AI) has seen exponential growth with the emergence of deep neural networks (DNNs), which exceed humans in a number of cognitive tasks. The algorithmic dominance of DNNs comes at exceptionally high memory and computing costs that pose significant challenges to the hardware platforms executing them [6]. GPUs and advanced wireless CMOS accelerators are actually at the cutting edge of the DNN hardware. Nevertheless, the ever-increasing complexity of DNNs and the data that they process has led to a search for the next quantum efficiency improvement in processing. The AI hardware team is exploring new tools, architectures and algorithms to improve the efficiency of processing and allow the transition from Narrow AI to Broad AI. Approximate computation, in-memory computation, artificial intelligence and quantum computing always form the basis of the computational solutions for AI workloads being explored.

Nonetheless, recently documented research from both industry and academy shows a trend in the design of integrated application-specific circuit (ASIC) for ML, especially in the field of deep neural networks (DNNs). Deep neural networks (DNNs) are now widely used in many AI applications, including computer vision, speech recognition, robotics, and so on [6]. Although DNNs provide state-of-the-art precision for many AI tasks, it comes at the expense of high computational complexity. The design of efficient hardware architectures for deep neural networks is therefore a significant step toward the widespread implementation of AI systems.

This chapter gives an overview of the hardware accelerator design, the various types of the ML acceleration, and the technique used in improving the hardware computation efficiency of ML computation. In this chapter, we will try to relate artificial intelligence and machine learning concepts to accelerate Hardware resources. Chapter will discuss software framework for deep neural networks and will give comparison of FPGA, CPU and GPU. The chapter will also give a detailed description of deep neural network acceleration beyond chips. At the end of the chapter future directions and conclusion will be given.

2. Deep learning and neural network acceleration

As a result of recent developments and exposure to reliable data in digital technology, artificial intelligence, deep learning have grown and have demonstrated its strengths and efficiency in the resolution of complex learning problems that have not previously been possible. Convolution neural networks (CNNs) have especially demonstrated their effectiveness in applications for image detection and recognition. They need intensive CPU operations and memory bandwidth, however, that makes general CPUs struggle to achieve desired performance levels. Hardware accelerators use application-specific integrated circuits (ASICs), field programmable gate arrays (FPGAs), and graphical processing units (GPUs) to improve the throughput of CNNs. More explicitly, due to their capability to optimize parallelism and energy efficiency, FPGAs have been recently adopted in order to speed up the implementation of profound learning networks.

A DNN is typically a parameterized function that takes a large number of high-dimensional entry as input to make the predictions—a classification symbol. A valid set of parameters can be obtained by training the DNN on a training dataset and optimizing parameters using methods like a stochastic gradient descent (SGD) to minimize a certain loss function. A forward pass is first performed at each training level to determine the loss, followed by a reverse pass to correct the error. The gradient is then determined and averaged for every parameter. The training cycle will take a million or more steps to completely automate a large-scale DNN.

Data dependence is twice as high in DNN training as it is inferred. Whereas the data flow of the forward pass is the same as the assumption, the reverse pass then reverses the layers. Therefore, in the backward pass, the outputs of each layer in the forward pass are reused to measure the errors (due to the chain rule of context distribution). Fig. 1 indicates the difference between the inference and the formulation data flow. A DNN may include convolution layers, completely connected (batched–matrix multiplications) layers, points like ReLU, sigmoid, max pooling and batch normalization. DNN may include convolutionary layers. The reverse pass can require

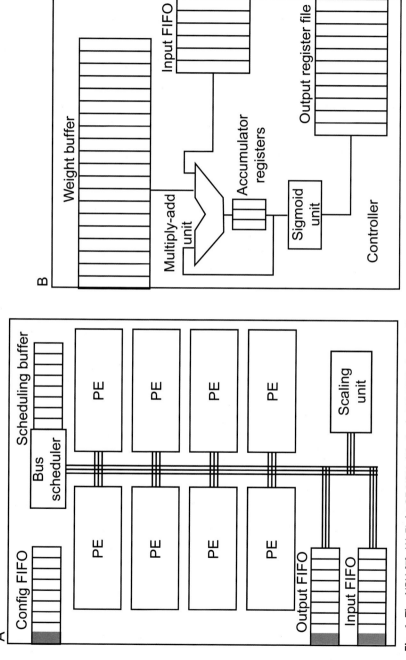

Fig. 1 The NPU [7]. (A) Eight-PE NPU; (B) single PE. FIFO: first in, first out.

point-specific operations, whose types are different from the forward pass. Matrix multiplications and convolutions keep their calculation nature in the rear pass unchanged; the only difference being that they work, respectively, on the transposed weight matrix and rotated convolutionary kernel.

Accelerators were designed for acceleration of estimated programs for general purpose [8] or small NNs in the beginning of the DNN accelerator design [9]. While on-chip accelerators have limited functionality and performance, they have revealed the basic concept of AI-specialized chips. Due to the limitations of general purpose chips, specialized chips for AI/DNN applications are often essential to design.

2.1 The neural processing unit

The neural processing unit (NPU) [7] can be used to accelerate a portion of the system instead of running on a central processing unit (CPU) using hardware-powered on-chip NNs. The NPU has a very quick hardware design. An NPU is made up of eight processing motors, as seen in Fig. 1. Each PE measures the neuron, i.e., multiplication, accumulation and sigmoid. The computation of a NN multi-layer perceptron (MLP) is therefore done by an NPU.

It was really encouraging to use MLP hardware, the NPU, to speed up some system parts. If a program segment is 1 frequently executed and 2 approximated, if 3 well defined inputs and outputs, then the NPU will accelerate that segment. Programmers need a program section that meets the requirements mentioned above must be annotated manually to run an NPU software. The compiler then compiles the section of program into the NPU instructions and transfers the calculation tasks from the CPU to the NPU at runtime. Two examples of these software segments are Sobel edge detection and fast Fourier transformation (FFT). An NPU can reduce up to 97% of the dynamic CPU instructions and achieve a speedup of up to 11.1 times.

2.2 RENO: A reconfigurable NoC accelerator

In comparison to the NPU intended to speed up programs of general interest, RENO [9] is a NNs accelerator. As seen in Fig. 2, RENO employs a similar PE definition. The PE of RENO is based on ReRAM: RENO uses a ReRAM crossbar to carry out matrix-vector multiplications as the basic computing unit. –PE is composed of four ReRAM crossbars, corresponding to positive and negative inputs and negative inputs. ROuters for coordinating

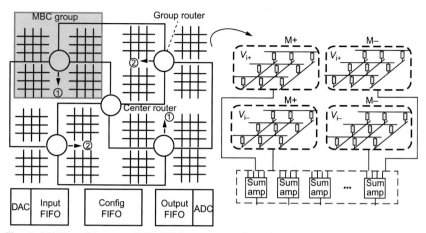

Fig. 2 RENO architecture [10]. MBC: memristor-based crossbars; $Vi+$: positive input voltage; $Vi-$: negative input voltage; M+: the MBC mapped with positive weights; M−: the MBC mapped with negative weights; Sum amp: summation amplifier.

transmission of data between PEs are used in RENO. In comparison to traditional CMOS routers, analog intermediate computing results are passed from the previous neuron to the RENO routers. At RENO, only the inputs and end outputs are digital; all the intermediate effects are analog and analog routers coordinate them. Only when transmitting data between the RENO and the CPU is data transmitted are DACs and ADCs necessary.

RENO supports auto-associated MLP and memory (AAM) processing and the corresponding instructions are intended for RENO and CPU pipeline. Because RENO is an on-chip system, the applications supported are minimal. RENO supports small datasets, for example the UCI ML repository [11] and the MNIST database (MNIST), customized for this purpose. The broad memory footprint of a DNN/CNN is typically required [7]. It is doubtful that the entire model will be projected onto the chip for large and difficult DNN/CNN models. Due to the limited bandwidth from the chip, it is necessary, in order to boost the device performance, to increase the reuse of on-chip data and the transfer of off-chip data. An analysis of the data flows is carried out during architectural design and special attention must be paid. As Fig. 3 shows [12,13], Eyeriss has been investigating the complex NN data flux in relation to spatial architecture and has suggested a row-stationary (RS) data flux to facilitate data reuse in relation to input-stationary (IS), output-stationary (OS), weight-stationary (WS), and non-locally reusing (NLR).

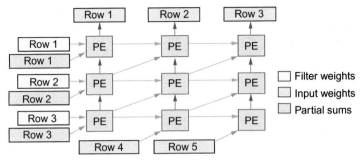

Fig. 3 Row-stationary dataflow [12,13].

3. HW accelerators for artificial neural networks and machine learning

Over many modern artificial intelligence (AI) systems, machine learning (ML) is currently commonly used [14]. The development of the computer capability has allowed the device to measure various complicated ML algorithms within a reasonably short time, providing real-time interactions between the human and the machine, for example facial recognition for video surveillance, advanced ADAS and early cancer detection [2,3]. High detection precision between all these applications requires complicated ML computation that is cost-effective. It makes the hardware design particularly challenging. Currently, the majority of applications are implemented on computer engines for general purposes, particularly GPUs. The research of both industry and academia has recently been reporting however, showing a trend in the design of an ASIC for ML application-specific, especially in the field of the deep network (DNN). During the past few years, various machine learning methods, such as CNN and SVM have proved promising in many application areas, such as image classification, and speech recognition. Some hardware technologies, for example the graphics processing unit (GPU) and field programmable gate array (FPGA), can be used to accelerate machine learning algorithms. In particular, due to its low energy consumption, reconfigurability and real-time processing capabilities, FPGA is a promising technology. Nevertheless, the successful FPGA architecture poses two challenges. First, architectural definitions will cover families of designs that are able to achieve, performance and energy efficiency with different compromises. Second, a design flow should facilitate reuse and comparison of optimization methods and tools for several architectures.

As the industry matures, programmable field gate arrays (FPGAs) are now emerging as a legitimate competition for GPUs to implement deep learning algorithms using CNN.

In this section we will present parameterized hardware architectures for two well known machine learning algorithms, convolution neural network (CNN) and support vector machine (SVM). After the discussion of traditional architectures we will be focusing on DNN based models, i.e., Eyeriss.

Hardware accelerators are uniquely based on specific architectures and blocks for various machine learning methods. The following section discusses the parameters that can be used to achieve the desired equilibrium in power, performance and energy efficiency for these architectures and building blocks. In order to improve the efficiency of FPGA architecture, two common methods exist: through parallelism and revising the bit width of the data representation. Several processing units with same functionality can be used to increase the level of parallelism. With regard to bit width, a smaller bit width leads to less resource usage of the building block and a higher degree of parallelism, but may also cause unpredictable consequences for machine learning algorithm accuracy. Hence, there is usually a tradeoff between having higher accuracy or higher processing rate. This tradeoff can also be studied by simulating the performance with different data representations.

The two architectures presented in this section take account of the above considerations. All include parallel building blocks and versatile types of data. Their definitions are parameterized to parameterize the number and bit width of the data representation.

3.1 CNN accelerator architecture

The accelerator architecture is a streaming architecture for CNN, where feature maps and weight matrices are input from off-chip memory and CNN calculations are performed by data streaming through each building block. This architecture comprises two principal types of building blocks: the CONV kernel acting as a convolution layer, and the FC kernel, performing entirely interconnected computation. In any kernel, cache input buffers and coefficients are available to boost data reuse, and computing units for different types of calculations such as chaos and dot products (Fig. 4).

Within this architecture there are two degrees of parallelism. First, arithmetic circuits can be replicated at the computation unit level in order to produce more results in each cycle. In the CONV kernel, the calculation

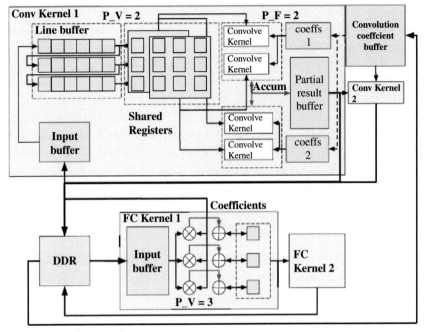

Fig. 4 CNN architectural design.

can be paralleled inside a channel or between filters. In the first example, a line buffer and a group of mutual registries can be used to measure the number of neighboring sliding windows PCV for convolution. In the second example, $P^{con}vF$ can be determined even in parallel with the number of output filters, as the heir effects are different. The dot-product operations of Pfc V can be performed simultaneously in the FC kernel. Second, if enough space exists on the same FPGA unit, there can be multiple kernels of CONv (PconvN) and FC (PfcN). In this architecture, the data representation T must not be specified. The parameters used to characterize a CNN architecture are shown in Fig. 5.

3.2 SVM accelerator architecture

There are some related features to the CNN accelerator in the SVM acceleration architecture. It is a streaming architecture, data are loaded from input interfaces and preloaded ROMs on the chip and then processed through the pipeline kernels like including the binary classifier (BC) kernels, Hamming distance computation kernels, and the Collection Kernel. Every BC kernel

Notation	Description
H^{conv}	Height of the convolution input feature map
W^{conv}	Width of the convolution input feature map
C^{conv}	Number of input channels
F^{conv}	Number of output filters
K^{conv}	Size of convolution kernels
P_V^{conv}	Number of parallel windows within one channel
P_F^{conv}	Number of parallel filters
P_N^{conv}	Number of parallel CONV kernels
P_V^{fc}	Number of parallel dot-product operations
P_N^{fc}	Number of parallel FC kernels
T	Data type

Fig. 5 Parameters used to describe a CNN architecture.

outputs information for classification of one pixel and one category pair. Each Hamming distance kernel takes vectors from the output of the KBC kernels and computes the hamming distance between K(K-1)2 components.

Second, building blocks can also be parallelly handled. The architecture of this SVM accelerator has two parameters for parallel rates, PSVM M specifies the amount of images to be processed parallel and the number of kernels of the BC to process one pixel is indicated by PSVM N (equals to K). The BC kernels are thus PSVM M as of PSVM N. Therefore The SVM model coefficients' data type T is configurable to help the balance between model precision and design parallelism. This architecture is shown in Fig. 6, and its parameters are described in Fig. 7.

3.3 DNN based hardware acceleration

A recent trend in the development of deep neural networks (DNN) is to expand deep learning applications to more resource-constrained platforms, such as mobile devices. These attempts are aimed at reducing the dimensions of the DNN model and improve the hardware efficiency which in turn will result in DNNs which are far more compact and/or highly data-sparse in their structures. These small or small models differ from the big models that are traditionally available with much more variation in the form and size of their layers and often exercise sparsity to improve performance. Many large DNN accelerators that have been designed on these models are not working well on these models. In this section, we will discuss an architecture designed

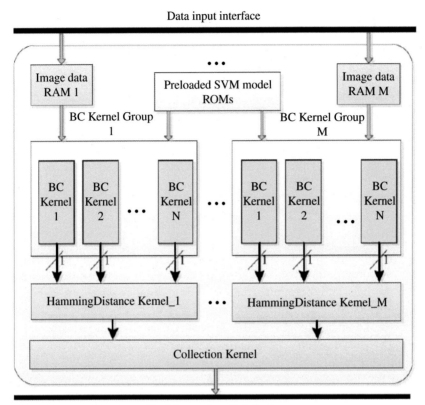

Fig. 6 SVM architectural design.

Notation	Description
K^{SVM}	Number of classes to be decided
l^{SVM}	Number of support vectors within each BC kernel
n^{SVM}	Number of dimensions in each support vector
P_M^{SVM}	Number of parallel pixel operations
P_N^{SVM}	Number of parallel BC kernels
T^{SVM}	Data type

Fig. 7 Parameters for SVM accelerated architecture.

for Eyeriss DNN accelerator which works for compact and sparse DNNs. Eyeriss introduces a highly flexible layer of on-chip network (i.e., Hierarchical mesh) that can adapt to the different amount of bandwidth & data reuse requirements. Eyeriss can also process sparse data for both weights directly inside the compact domain and activations, therefore both processing can be improved rate and efficiency of energy.

3.3.1 EyeRiss

Eyeriss is a versatile DNN processing architecture capable of adapting to different filter sizes and shapes for compact DNNs, such as MobileNet. This is done by developing a very robust on-chip network (NoC), the bottleneck for a more diverse set of DNNs. Furthermore, Eyeriss supports sparse DNNs by using the sparsity of weights and activations across various DNN layers and converts them into energy efficiency and processing speed enhancements. In the end, Eyeriss does not predict whether or not a DNN layer's total storage capacity fits into the chip; instead, it optimizes the way in which tile data from different kinds are obtained to achieve high on-chip reuse and energy efficiency. Eyeriss uses a hierarchical mesh designed to conform to the different criteria for bandwidth. If data reuse is small, a high memory hierarchy bandwidth (through unicast) can be provided to keep the PEs busy; where data reuse is high, space reuse can be optimized to achieve high energy efficiency (via multicast or broadcasting). A PE, which utilizes sparse weights and activations to improve energy efficiency and output across a range of DNN layers. For both on-chip processing as well as off-chip access data are kept in sparse compressed column (CSC) format to reduce the cost of saving and data movement. The weights are mapped to a PE by taking into account sparsity in order to increase reuse in PE, and the impact of workload imbalances can therefore reduce.

4. SW framework for deep neural networks

In the fields of computer vision and natural language processing, deep learning was effective in improving predictive ability. Modern computer vision performance is powered by the convolutional neural network model, a special feed-forward-proven type of deeper learning. The high-level concept is to learn filters from images to derive useful features and forecasts. The use of recurrent neural networks as a feedback model that is useful for organizing the learning of natural languages and sequences (As in Natural Languages) was, on the other hand, very effective in natural language processing. The complexity of network architectures and the size of the parameter space continue to increase with accuracy in both domains. A billion parameters were reached in the Google Network for unsupervised learning of image features [9], which in a separate experiment at Stanford expanded to 11 billion parameters. In NLP space, a Network of 160 billion parameters has been trained recently by digital reasoning systems. This scale of the problem includes looking beyond the single computer, which Google first proved through its distributed DistBelief [15] platform. The deep learning frameworks ecosystems with complete support for parallelism are discussed in this section. On the hardware stage, there are three levels of parallelization: within a GPU, between a single node GPU and nodes. On the application level also there are two types of parallelism: the model and the parallel data. Certain systems include release dates, main vocabulary, user-facing APIs, code models, communication models, form of profound information, paradigm programming, faulty tolerance and visualization. The following factors have not specifically been identified in the Tensorflow, CNTCK, Deeplearning4j, MXNet, H2O, Caffe, Theano and Torch deep learning areas: their open-source content, level of documentation, sophistication, and the Tensorflow are all factors. The importance of the release date, core language, and user-facing APIs are implicit. The synchronization model determines whether or not updates are synchronous or asynchronous, by executing data consistency. Synchronous execution provides more convergence in the sense of optimization kernels like the stochastic gradient descent (SGD) by preserving consistency or almost consistent with sequence. The SGD asynchronous can use more parallels and trains more quickly, but with fewer convergence speed guarantees. This compromise is laid down by frames like Tensorflow and MXNet as an choice for the user. Table 1 compares various open-source software frameworks for deep neural networks.

The communication model seeks to split the essence of machine-wide execution into popular paradigms. At a hardware level, there are three potential types of parallelism: cores inside a CPU/GPU, across multiple machines (normally deep learning GPUs), or across machines. Most of the lower-level library kernels (e.g., for linear algebra) have been configured to use several machine core by default. All frames at this point also allow multiple GPU-specific parallelism. Multimachine parallelism does not yet stand for Theano and Torch [16]. Parallel data and model are the two predominant parallel opportunities in training deep learning networks at the distributed level. Data parallels are used to train copies or parameters of the model on their own training data and to update the same on global model. The model itself is separated and trained in parallel in model parallelism.

Table 1 Open source software framework for deep neural networks.

Platform	TensorFlow	Deeplearning4j	MXNet	Caffe	Theano	Torch	CNTK	H2O
Release Date	2016	2015	2015	2014	2010	2011	2016	2014
Data Parallelism	Yes	Yes	Yes	Yes	Yes	Yes	Yes	Yes
API	C++, Python	Java, Scala	C++, Python, R, Scala	Python, Matlab	Python	Lua	NDL	Java, R, Python
Core Language	C++	C++	C++	C++	C++	C	C++	Java
Communication Model	Parameter Server	Iterative MapReduce	Parameter Server	N/A	N/A	N/A	MPI	Distributed Fork–Join
Multi-GPU	Yes	Yes	Yes	Yes	Yes	Yes	Yes	Yes
Multi-Node	Yes	Yes	Yes	No	No	No	Yes	Yes
Programming Paradigm	Imperative	Declarative	Both	Declarative	Imperative	Imperative	Imperative	Declarative
Model Parallelism	Yes	No	Yes	No	Yes	Yes	N/A	No
Deep Learning Models	DBN, CNN, RNN	DBN, CNN, RNN	DBN, CNN, RNN	DBN, CNN, RNN	DBN, CNN, RNN	DBN, CNN, RNN	DBN, CNN, RNN	DBN
Synchronization Model	Sync or Async	Sync	Sync or Async	Sync	Async	Sync	Sync	Async
Fault Tolerance	Checkpoint & Recovery	Checkpoint & Resume	Checkpoint & Resume	N/A	Checkpoint & Resume	Checkpoint & Resume	Checkpoint & Resume	N/A

5. Comparison of FPGA, CPU and GPU

To achieve high-performance embedded vision applications, runtime efficiency with power constraints must be balanced. Because of the combination of embedded computer vision hardware accelerators (e.g., CPUs, GPUs and FPGAs), and their related vendor customized visual libraries, developers have a challenge in navigating this fragmented solution domain [17]. We have performed a detailed analysis of the working time and energy efficiency of a broad variety of vision kernels in order to help assess which embedded platform is appropriate for use. In this section, we will discuss whether certain specific hardware architecture works adequately or poorly depending on the features of a number of vision kernel categories [18]. Our research explicitly refers to three widely used HW accelerators for embedded vision applications: ARM57 CPU, Jetson TX2 GPU and ZCU102 FPGA, using OpenCV, VisionWorks and xfOpenCV as vendor optimized vision libraries [19]. Based on our study, we concluded that the GPU achieves a reduction ratio of 1.1–3.2 × for energy/frame relative to other kernels. Although the FPGA performs with energy/frame reduction ratios of 1.2–22.3 × on more complicated kernels and complete visual pipelines [20]. It is also observed that the FPGA performs increasingly better as a vision application's pipeline complexity grows.

The characteristics of the evaluated hardware accelerators are discussed in this section. We have grouped the vision kernels into categories, based on their features, to understand the implications on kernel performance for their respective categories of the underlying hardware architecture.

1. *Central Processing Unit (CPU)*: The SIMD (Single Instruction, Multiple Data) instructions can be prepared by the modern CPU using several ALUs. These SIMD instruction sets are helpful in the processing of images, where they are used repeatedly on a continuous data stream. This is especially true with computer vision, where the majority of operations take place throughout the whole image. The ARM NEON SIMD and SIMD streaming extensions of Intel (SSE) are the architectural examples of SIMD.

2. *Graphic Processing Unit (GPU)*: GPUs have evolved into a specific SIMD architecture compared to general purpose CPUs that have built SIMD instruction extensions to support parallel image processing tasks. This experience has rendered GPUs with simpler processing cores than general high performance CPUs. These have simpler control logic, no

estimation or prefetching of neither branch nor a limited per-core memory. Simpler computing cores allow GPUs to fit much more cores in a chip than CPUs [21]. GPU architectures perform very well in working conditions or data dependencies that have little to no branching conditions. Furthermore, GPU architectures have a memory architecture specialized in the streaming of high-speed data to process images. For instance, the Jetson TX2 (Pascal GPU) L2, which is capable of matching the grayed image in 1080p, has 2048 KB.

3. *Field Programmable Gate Array (FPGA)*: The FPGA consists of a collection of logic blocks, DSPs, on-chap BRAMs, I/O pads and routing channels, rather than having a fixed processor-like architecture [22]. In FPGA custom data tracks can be architectured to directly stream pixels from/from external memory between computing units. Further, distributed BRAMs can be employed by holding pixels in chip (for example, Zynq UltraScale MPSoC FPGA has a 32.1 MB on-chip memory) to leverage the data locality in the vision kernels. Developers must ensure that their custom designs satisfy timing and space requirements with FPGAs.

5.1 Performance metrics

The selection of suitable metrics for energy efficiency assessment of vision kernels running on various hardware accelerators is important. These measurements should be meaningful and fair for comparison. In this section, we discuss the evaluation metrics used in our study:

(a) Run-time: Vision kernel runtime performance may be measured by calculating the time elapsed (time delay) between the start and end of a kernel code. The high resolution timer can be used to measure time spent on HW accelerators accurately. The time needed to copy images from/to external CPU and GPU memory and time to configure data moves in the FPGA shall only be taken during the execution time and exempted.

(b) Energy: Energy consumption per frame quantifies the amount of electrical energy dissipated by hardware accelerators to perform a kernel's operations on one frame. It is measured as the power consumed during the delay time to process a frame. Device power can be divided in two parts: (1) Static power: represents the amount of power consumed when no active computation is taking place (system is idle), (2) Dynamic power: represents the amount of power consumed above the static power level when the system is computing.

(c) Energy-delay product (EDP): The whole picture does not display runtime or energy per frame alone. A hardware platform can be incredibly low in power because it is too slow to use. The EDP [15] metric takes the algorithm's calculated performance into account (ms/frame) as well as the energy per frame (mJ/frame) consumed. EDP is the product of energy/frame and delay time. A fair analogy is therefore possible when selecting which hardware architecture is better suited for particular calculations. Lower EDP is better because the hardware architecture is able to complete similar computing tasks in less time using less power.

In many image processing applications, FPGAs have shown very high performance. But the new CPU and GPU have a high performance potential for these problems as well. Recent CPU supports multi-core systems, each supports improved SIMD instructions and runs up to 16 128b-data operations in one clock cycle. The recently integrated GPU supports large numbers of cores that run in parallel with CPU [23]. FPGA's high performance is due to its versatility which allows the fully optimized circuit for each application and a large number of on-chip memory banks that support high parallelism to be implemented. Thanks to these characteristics, FPGA in many applications can achieve extremely high performance despite its low operating frequency, although the designer has to work to minimize the number of operations and memory accesses in each unit so they can take advantage of more parallelism [24]. The parallelism in CPU's SIMD instructions is minimal, but the CPU operating frequency is very high and it is anticipated that CPU is highly efficient in applications with a good cache storage operation. For many image processing applications, the cache memories are large enough to store entire images and given high memory bandwidth, CPU can run the same algorithms as FPGA [25]. The comparison between CPU and FPGA thus results in an increase operating frequency and parallelism [26]. The GPU operating frequency is slightly less than CPU, but GPU supports a large number of parallel cores (240 in our target GPU) and its high-end performance is above CPU. The core is grouped and the transmission of data between groups is very slow. In addition, for each group the local memory size is very small. Due to these drawbacks, in some applications GPUs cannot execute the same algorithms as FPGA [25]. We need to carefully design the algorithm to solve these limitations in order to obtain high performance from GPU. GPU programming tools have been developed, but high performance on both GPU and CPU and FPGA remains difficult to reach.

6. Conclusion and future scope

This chapter explains how acceleration of hardware can increase the efficiency of machine learning applications. They relies on parameterized architectures and the related optimization techniques and flow for these applications. The deep learning frameworks chosen to track them among other variables for thorough study were Tensorflow, CNTK, DeepLearning4j, MXNet, H2O, Caffe, Torch, and Theano. They were contrasted by a clear collection of features from hardware and application parallelism into other details such as release date, core language, APIs, synchronization and communications models, programming paradigms, fault tolerance, and visualization. While deep learning framework provides abstraction and many are built to reach many machines, there is evidence, given the correct use of specialized hardware and attention to specific application characteristics, that a few deeper learning problems can be solved with efficiency and precision without many machines being required. The chapter compares GPU's efficiency to FPGA and CPU's (quad core). GPU has the capacity to work just as well as FPGA. FPGA output is limited by FPGA size and the bandwidth of the memory. With the latest DDR-II DRAM FPGA board and a larger FPGA, the performance can be increased by storing twice as many pixels in parallel. Our methodology can be expanded to include potential applications focused on CNN and SVM as well as other algorithms of machine learning, such as the Gaussian Mixture System [27] and Genetic Algorithms [28]. The proposed design flow will also include more FPGA-based improvements including multi-pumping [8].

References

[1] B. Marr, Big Data: 20 Mind-Boggling Facts Everyone Must Read, Forbes.com, 2015.

[2] "For a Trillion Sensor Road Map," T Sensor Summit, October 2013.

[3] Gartner Says 6.4 Billion Connected "Things" Will Be in Use in 2016, Up 30 Percent From 2015, Gartner.com, November 2015.

[4] Y. LeCun, Y. Bengio, G. Hinton, Deep learning, Nature 521 (2015) 436–444.

[5] A. Krizhevsky, I. Sutskever, G.E. Hinton, ImageNet classification with deep convolution neural networks, Adv. Neural Inf. Process. Syst. 25 (2012) 1097–1105.

[6] D. Silver, et al., Mastering the game of GO with deep neural networks and tree search, Nature 529 (7587) (2016) 484–489.

[7] H. Esmaeilzadeh, A. Sampson, L. Ceze, D. Burger, Neural acceleration for general-purpose approximate programs, Commun ACM 58 (1) (2014) 105–115.

[8] E. Fykse, Performance Comparison of GPU, DSP and FPGA Implementations of Image Processing and Computer Vision Algorithms in Embedded Systems, Master's thesis. Institute for elektronikk og telekommunikasjon, 2013.

[9] H. Niitsuma, T. Maruyama, Real-time generation of three dimensional motion fields, in: FPL, 2005, pp. 179–184.

[10] X. Liu, M. Mao, B. Liu, H. Li, Y. Chen, B. Li, et al., RENO: a high-efficient reconfigurable neuromorphic computing accelerator design, in: Proceedings of 2015 52nd ACM/EDAC/IEEE Design Automation Conference; 2015 Jun 8–12; San Francisco, CA, USA, 2015, pp. 1–6.

[11] LeCun Y, Cortes C, Burges CJC. The MNIST Database [Internet]. 2019. Available from: http://yann.lecun.com/exdb/mnist/.

[12] Y.H. Chen, T. Krishna, J.S. Emer, V. Sze, Eyeriss: an energy-efficient reconfigurable accelerator for deep convolutional neural networks, IEEE J Solid-State Circuits 52 (1) (2017) 127–138.

[13] Y.H. Chen, J. Emer, V. Sze, Eyeriss: a spatial architecture for energy-efficient dataflow for convolutional neural networks, in: Proceedings of the 2016 ACM/IEEE 43rd Annual International Symposium on Computer Architecture; 2016 Jun 18–22; Seoul, Republic of Korea, 2016, pp. 367–379.

[14] UCI Machine Learning Repository [Internet]. Irvine: University of California 2019. Available from: http://archive.ics.uci.edu/ml/.

[15] A. Darabiha, et al., Video-rate stereo depth measurement on programmable hardware, Comput. Vis. Pattern Recognit. 1 (2003) 203–210.

[16] R. Collobert, K. Kavukcuoglu, C. Farabet, Torch7: a matlab-like environment for machine learning, in: BigLearn, NIPS Workshop, Number EPFL-CONF192376, 2011.

[17] T. Saegusa, T. Maruyama, An FPGA implementation of realtime K-means clustering for color images, J. Real Time Image Proc. 2 (4) (2007) 309–318.

[18] M. Estlick, M. Leeser, J. Theiler, J.J. Szymanski, Algorithmic transformations in the implementation of K-means clustering on reconfigurable hardware, FPGA 1 (2001) 103–110.

[19] J. Qiu, J. Wang, S. Yao, K. Guo, B. Li, E. Zhou, J. Yu, T. Tang, N. Xu, S. Song, Y. Wang, H. Yang, Going deeper with embedded FPGA platform for convolutional neural network, FPGA 1 (2016) 26–35.

[20] J. Diaz, et al., FPGA-based real-time optical-flow system, IEEE Trans. Circuits Syst. Video Technol. 16 (2) (2006) 274–279.

[21] S. Che, J. Li, J.W. Sheaffer, K. Skadron, J. Lach, Accelerating compute-intensive applications with GPUS and FPGAS, in: Symposium on Application Specific Processors, 2008. SASP 2008, IEEE, 2008, pp. 101–107.

[22] B. Maliatski, O. Yadid-Pecht, Hardware-driven adaptive k-means clustering for real-time video imaging, IEEE TCSVT 15 (1) (2005) 164–166.

[23] Y. Jia, E. Shelhamer, J. Donahue, S. Karayev, J. Long, R. Girshick, S. Guadarrama, and T. Darrell. Caffe: convolutional architecture for fast feature embedding. In Proceedings of the 22nd ACM International Conference on Multimedia, pages 675–678. ACM, 2014.

[24] F.N. Iandola, K. Ashraf, M.W. Moskewicz, K. Keutzer, Firecaffe: near-linear acceleration of deep neural network training on compute clusters, arXiv 1 (2015) 2592–2600. preprint arXiv:1511.00175.

[25] T. Saegusa, T. Maruyama, Y. Yamaguchi, How fast is an FPGA in image processing? in: FPL, 2008, pp. 77–82.

[26] J. Donahue, Y. Jia, O. Vinyals, J. Hoffman, N. Zhang, E. Tzeng, T. Darrell, Decaf: a deep convolutional activation feature for generic visual recognition, ICML 32 (2014) 647–655.

[27] C. Guo, H. Fu, W. Luk, A fully-pipelined expectation maximization engine for Gaussian mixture models, in: FPT, 2012.

[28] L. Guo, C. Guo, D.B. Thomas, W. Luk, Pipelined genetic propagation, in: FCCM, 2015.

About the author

Dr. Neha Gupta is currently working as an Associate professor, Faculty of Computer Applications at Manav Rachna International Institute of Research and Studies, Faridabad campus. She has done her PhD from Manav Rachna International University, Faridabad. She has total of 13+ years of experience in teaching and research. She is a Life Member of ACM CSTA, Tech Republic and Professional Member of IEEE. She has authored and coauthored 30 research papers in SCI/SCOPUS/Peer Reviewed Journals and IEEE/IET Conference proceedings in areas of Web Content Mining, Mobile Computing, and Web Content Adaptation. She has also authored books with international publisher like IGI Global and Pacific International and has also authors various book chapters with publishers like Elsevier, IGI Global, CRC Press, etc. Her research interests include ICT in Rural Development, Web Content Mining, Cloud Computing, Data Mining and NoSQL Databases. She is a technical programme committee (TPC) member in various conferences across globe. She is an active reviewer for International Journal of Computer and Information Technology and in various IEEE Conferences around the world. She is one of the Editorial and review board members in International Journal of Research in Engineering and Technology. Recently she has completed her certification as blockchain professional from CIALFORE, Delhi.

CHAPTER TWO

Hardware accelerator systems for embedded systems

William J. Song
School of Electrical and Electronic Engineering, Yonsei University, Seoul, South Korea

Contents

Abstract

This chapter describes various engineering considerations and constraints to deploy neural network applications in embedded systems and presents a variety of processing solutions to accelerate neural network computations in the embedded hardware. Deep learning on embedded systems has potentially many advantages for security, privacy, latency, energy, power, etc. However, deploying the deep neural networks in embedded systems imposes numerous hardware challenges on the resource-limited embedded edge devices. Embedded systems for deep learning typically target on providing rapid inferences, and thus latency rather than throughput in general becomes the primary objective for the executions of embedded hardware. The central point of hardware acceleration in embedded systems is to place neural network computations closer to I/Os and sensors to provide fast inferences. With continued advances in processor technologies, embedded edge devices evolve to become capable of handling compute-intensive workloads at low power. Such a trend propels integrating the hardware acceleration of deep neural networks into the embedded systems. There is not a universal solution for all different kinds of embedded systems. Different embedded processing solutions can be employed to accelerate the neural network applications depending on their performance requirements, operating conditions (e.g., network connectivity, power and thermal constraints), costs, etc. These considerations leave a wide range of hardware options

Advances in Computers, Volume 122
ISSN 0065-2458
https://doi.org/10.1016/bs.adcom.2020.11.004

23

for the embedded systems. The embedded hardware to accelerate neural network applications ranges from single-board devices such as Google Edge TPU to high-performance processors such as Intel Xeon and AMD EPYC CPUs, NVIDIA GPUs with Tensor Cores. All of these hardware choices for the neural network acceleration provide distinct features and computational capabilities in the embedded systems.

1. Introduction

Deep neural networks (DNNs) have become important applications in a wide range of domains encompassing natural language processing [1], image classification [2,3], video recognition [4], surveillance cameras [5], autonomous driving [6], and diverse Internet of Things (IoT) devices [7]. The great success of neural networks and deep learning is attributed to unparalleled accuracy in numerous application domains. Deep learning on embedded systems has potentially many advantages for security, privacy, latency, energy, power, etc. However, deploying neural network applications in embedded systems imposes great hardware challenges on the resource-limited embedded edge devices. This chapter discusses engineering issues and technological trends related to the hardware acceleration of deep neural networks in embedded systems.

Steps to deploy a neural network in an embedded system is abstracted in Fig. 1. The neural network first needs to be sufficiently *trained* with a large amount of data. The training data set has a decisive influence on the neural network, and numerous methodologies have been proposed to improve the quality of training process [8–12]. Training the neural network is essentially the process of finding right quantities for individual weights (i.e., weighted connections between neurons). The performance (i.e., accuracy) of neural

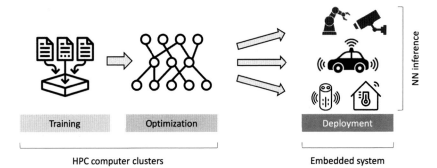

Fig. 1 Deployment of a neural network in an embedded system. Hardware acceleration in the embedded system primarily targets on inferences (i.e., predictions) [18].

network is consequently contingent on how well the network is trained. To amortize the computational and memory costs of neural network, it can be compacted by leveraging various optimization techniques such as pruning [13], compression [14], quantization [15], low and mixed precisions [16,17], etc. The variety of optimization techniques attempt to reduce redundancy in neural network computations based on observations that the neural network computing is *approximate* rather than *deterministic*; the output of neural network is a predicted result with probability. Training and optimization are often processed in a tightly coupled manner such that the neural network is retrained after applying optimizations to fine-tune the weights and minimize the loss of information [13,14]. Neural network training typically relies on high-performance computing (HPC) systems for the batch processing of sizable training data set. Latency and power are generally less of a concern during the training phase, but *throughput* becomes the primary objective or constraint. Graphics processing units (GPUs) excel in this aspect, and thus the neural network training is almost always conducted by using high-performance GPUs at large scales.

After the neural network is trained and optimized, it is then deployed in the embedded system for real uses. It performs a specific algorithm known as *inference* in the embedded system, such as inspecting products in an assembly line, tracking cars on the road, interpreting voice commands, etc. Once the neural network is deployed in the embedded system, it executes fixed operations for different inputs unless the network is updated or replaced with new algorithms. Implementing fast inferences (i.e., predictions) of the neural network imposes a distinct set of challenges on the embedded hardware. Operations in the embedded system are typically realtime-constrained in that inference executions must be completed within specified time frames. Thus, *latency* rather than throughput matters when it comes to the inferences. Neural network computing is a data-intensive task that needs to execute a large number of operations to process sizable data (i.e., weights and neurons). Handling rapid inferences requires low-latency compute capabilities. However, elevating the operating point (i.e., voltage and clock frequency) of processing units may result in conflicting decisions in an embedded edge device that is often battery-operated and possibly be equipped only with a passive cooling method (i.e., cooling through natural convection). Such operating conditions consequently limit the power density and thermal envelopes that the edge device can sustain. Thus, the physical limitations are important factors to consider since they constrain the operations and performances of embedded edge devices [19].

2. Neural network computing in embedded systems

Deploying neural networks in embedded systems has its own set of requirements that are distinguished from what are needed for deployment in HPC systems (e.g., data centers, servers). The central point of integrating neural network acceleration into the embedded systems is to place neural network computing closer to I/Os and sensors for faster inferences. Since neural network operations are computationally intensive, the traditional solution to this problem is to offload the heavy computations to remote HPC clusters such as cloud computing systems as depicted in Fig. 2. Such an execution model alleviates the hardware requirements of embedded edge devices, but they are traded with increased concerns for network stability, latency, security, etc. With continued advances in processor technologies and innovations in hardware management techniques such as dynamic power management (DPM), embedded edge devices become increasingly capable of handling more compute-intensive workloads at low power. Such trends motivate and drive modern embedded systems to incorporate neural network computing capabilities into their local processors.

2.1 Driving neural network computing into embedded systems

Reasons for pushing neural network computing directly into embedded systems vary depending on target applications, markets (e.g., industry, consumers), etc. There is not a universal solution for all different kinds of

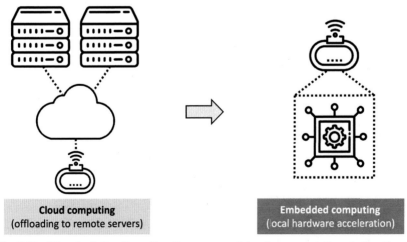

Cloud computing
(offloading to remote servers)

Embedded computing
(local hardware acceleration)

Fig. 2 Traditional solution (i.e., offloading resource-intensive computations to the cloud) versus modern embedded systems with built-in neural network acceleration capability.

embedded systems. Different embedded processing solutions can be employed to execute neural network algorithms depending on performance requirements, operating conditions, and costs, which leave a wide range of hardware options for developers to choose from [20]. The following explains the reasons behind pushing the neural network acceleration into the embedded systems.

- *Reliability*: Relying on a traditional solution that offloads neural network computations to remote computing clusters (e.g., cloud computing) requires a constant and stable network connection to transfer a large volume of data. In such an execution environment, losing the network connection translates to service unavailability. The disconnected network and thus disrupted services may simply mean unpleasant user experiences to typical end-users. However, if the neural network is deployed on a mission-critical system such as an autonomous vehicle, unmanned aerial vehicle (UAV), smart factory, power plant, and satellite, the network disruption (possibly caused by security attacks, blind spots, network congestions, unexpected failures, etc.) will be detrimental to the operability and safety of the embedded system. Thus, bringing parts or all of neural network computations into local processors in the embedded system can alleviate such concerns since the system becomes less dependent on external factors such as the network connections and remote computer systems.
- *Security*: Security is another biggest concern when an embedded system relies on the network connections and remote computing clusters for neural network computations. Since the data packets of neural network algorithm have to be transferred through an open network (e.g., Internet), they become vulnerable to security attacks such as eavesdropping, alteration, etc. Such security attacks can distort the results of neural network computations [21] and induce faulty operations in the embedded system. Integrating neural network computing directly into embedded processors can dismiss such concerns by isolating neural network computations within the embedded hardware. Therefore, accelerating neural networks in the embedded system helps eliminate the possibilities of security attacks in the network and thus can build a more robust system.
- *Privacy*: Even with secure network connections, privacy can be another concern in the traditional approach since private or confidential information has to be handed over to remote computing clusters to process the neural network data and provide related services. Deep learning in local edge devices can enable embedded systems to provide users with more secure and private services without transferring a large volume

of personal, confidential, or sensitive information over open networks. While the embedded systems typically target on offering fast inferences, neural network training is generally done by remote, high-performance computing clusters. Thus, requesting the neural network training to the remote computer systems still leaves aforementioned security and privacy concerns. However, bringing the training features of neural network computing into the embedded systems is a highly challenging problem because of scarce hardware resources and inherent physical limitations (e.g., power limits) of the embedded systems. Tackling this challenge will require substantial innovations in various aspects of neural network computing from software to hardware to lighten neural network algorithms and strengthen hardware computing capabilities to build complete autonomous systems.

- *Latency*: Latency is the primary constraint in realtime-constrained embedded edge devices. The latency is directly related to the quality of service (QoS) since long-latency inferences degrade user experiences and thus the QoS. With unstable network connections, guaranteeing the minimum QoS becomes problematic for varying network latencies. The latency becomes especially critical to embedded systems that need to process streaming data such as monitoring devices or surveillance cameras. These types of systems typically have minimum performance requirements (e.g., frames per second), and thus the latency determines the speed of embedded systems and fulfillment of the performance requirements [19]. Since embedded processors in the past were not capable enough to execute data-intensive neural network algorithms, it was faster to transfer a bulk of data over the network to remote computer systems and then receive only final results from them than directly calculating the neural network algorithms on the local embedded processors of limited compute capabilities. In other words, network latencies were smaller than the delays of embedded processing. With continued advances in processor technologies, embedded systems have become capable of handling data-intensive neural network inferences at similar or even faster speed than the traditional approaches.
- *Bandwidth*: Neural network computing is a data-intensive task, and thus executing the neural network via cloud computing requires a large amount of data transfers to offload its computations to the remote computing clusters. Hence, network bandwidth in addition to the latency becomes a critical factor that affects the performance in such an execution environment. Network congestions and thus insufficient network

bandwidth add extra latencies to the baseline round-trip time, thereby limiting the overall performance of an embedded system and degrading the QoS. Placing neural network computations in or near the embedded system can effectively eliminate the network bandwidth concerns and unleash the network from a burden to transfer the large volume of inference data. In contrast, when the neural network needs to be trained in remote computing clusters, the network bandwidth is generally less of a concern because the training is not a realtime-constrained task and thus can be executed when the remote clusters and connecting networks are idle such as in night times.

- *Energy and power.* Energy and power are important elements of an embedded system. In the traditional execution model where neural network computations are offloaded to remote computing clusters, the embedded system spends energy on transferring data over the network rather than computations. On the other hand, pushing neural network computations into the embedded system shifts energy spending from the data transfers to direct computations of the data. Since neural networks are computationally intensive workloads, finding the optimal operating point for the embedded system is an essential part of system operations to minimize the energy and power consumption as shown in Fig. 3. Determining the optimal operating point (i.e., operating voltage and clock frequency) of the embedded system varies depending on application requirements and working conditions. As depicted in the figure, lowering the operating point helps the embedded system reduce the power dissipation, but in turn prolonged execution time may result in larger energy consumption. If the embedded edge device operates on

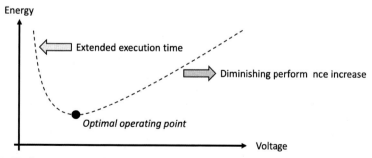

Fig. 3 Finding an optimal operating point depends the requirement of underlying embedded system. Lowering the operating point reduces the power dissipation of embedded system, but prolonged execution time may result in larger energy consumption.

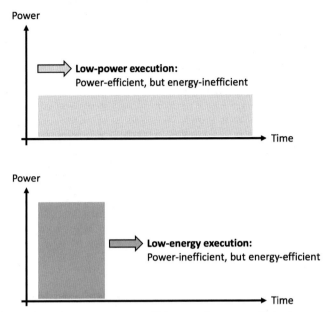

Fig. 4 Comparison between power-efficient and energy-efficient executions. An embedded system prefers a different mode of operation depending on its working conditions and application requirements.

a battery, increased energy consumption leads to shorter battery time. Although energy and power are often interchangeably used in many literatures, it is worth noting that they have very different implications as illustrated in Fig. 4. The energy represents the total amount of work done, whereas the power indicates the intensity of work executions. Thus, an energy-efficient operation is not necessarily translated to a power-efficient execution, and vice versa. In a system that operates 24/7 and is plugged into a stable power source (i.e., wall power), low-power execution is generally a preferred mode of operation to minimize thermal dissipation and resulting impacts on lifetime reliability [22,23]. However, a battery-operated system may prefer the opposite mode of execution if such an operation ends up saving more energy [24,25].

2.2 Considerations for choosing embedded processing solutions

Deploying neural networks in embedded systems needs to consider a variety of factors to satisfy the requirements of target applications and also to minimize the costs of deployment in the embedded hardware. In general, many of the requirements for executing the neural networks locally within the embedded processors may overlap with what are already available in the embedded hardware. The following lists the factors that need to be accounted for when choosing embedded processing solutions for accelerating neural networks.

- *Understanding end-to-end application executions*: Neural network computing is a data-intensive task that generates a large number of operations to process sizable data. However, an application to deploy in an embedded system may also demand other acceleration capabilities besides the neural network. Notably, an accelerator is not a standalone processing unit in the embedded system, and the neural network is not solely executed by the accelerator. The accelerator is dependent on generic host central processing units (CPUs) to initiate the application, handle host operations, and process irregular or serialized executions [19]. In addition, the target application may not necessarily exploit only a single type of accelerator but multiple different types of processing units to speed up different parts of the application. Modern embedded system-on-chip (SoC) devices increasingly incorporate various types of accelerators and processing units to provide applications with diverse functionalities and hardware acceleration choices as shown in Fig. 5. In addition, Fig. 6 plots the execution phases of a few representative nerual networks. The figure shows that the workloads are comprised of multiple phases alternating between serialized and parallelizable executions. The serial executions can be heavy-lifted by generic CPU cores, and the parallelizable phases are boosted by specialized accelerators [26]. This kind of execution method

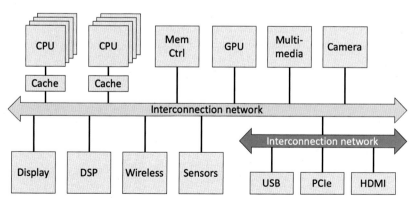

Fig. 5 A system-on-chip (SoC) design comprised of various accelerators and processing units.

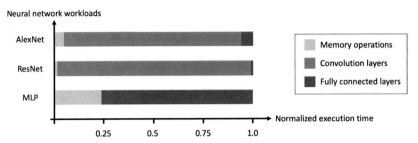

Fig. 6 The breakdown of execution progresses along the normalized timeline for a few representative neural networks [27]. Different neural networks spend different amount of time on distinct phases.

represents a typical heterogeneous computing model. In the presence of heterogeneous processing units in the embedded system, it is important to understand the entire execution flow of the target application to identify (i) which parts of the application need to be accelerated and (ii) what are the best processing options to speed up the identified parts. These considerations will help select the most appropriate execution methods in the embedded system.

- *Characterizing target workloads and their performance requirements*: The domain-specific nature of embedded processing makes it distinguished from other computer systems for general-purpose computing. When choosing an embedded processing solution for neural network computing, it is essential to understand the characteristics of application that the embedded system targets on. The target application needs to be thoroughly analyzed to identify required specifications and performance needs. The result of workload characterization is then used to guide selecting the right embedded hardware for the target application. Fig. 7 plots the performance and operational intensity of various representative neural networks on a roofline graph [19,27]. Plotting the roofline graph helps us understand the performance limitations of neural networks on the executed hardware [28]. Workloads closer to the slope (e.g., MLP, RNN) are known to show more memory-bound characteristics, and thus improving the memory speed (i.e., bandwidth) helps enhance the overall performance of corresponding workloads. On the other hand, applications near the roof (i.e., flat line in the graph) are known to be bounded by the computational capabilities of executing hardware. The performance of these workloads can benefit from employing more powerful processing options or escalating the operating points. Since different neural network

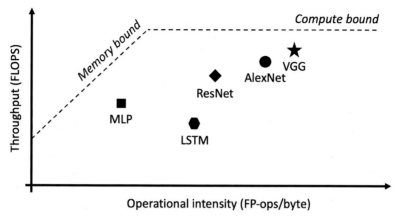

Fig. 7 Performance and operational intensity of various neural networks on a roofline graph [19,27,28]. The roofline graph reveals the hardware bounds of measured workloads. Workloads closer to the slope tend to show memory-bound characteristics, whereas the ones closer to the roof (i.e., flat line) are bounded by the computational capabilities of executing hardware.

applications exhibit distinct hardware performances and traits as shown in Fig. 7, the characterization of target workloads is important to understand and specify the required features of embedded system.

- *Determining optimal operating points*: In an embedded computing environment, performance is not the only determinant of system operations. The operations of embedded system are also governed by many physical factors including energy, power, thermal, and reliability. As illustrated in Fig. 3, pushing the operating point toward the maximum performance increases the energy consumption, and the elevated power dissipation results in thermal increases. The power dissipation of computing system can be decomposed into two parts; dynamic and leakage powers. The dynamic power is related to the computational activities of devices (i.e., switching of transistors), whereas the leakage power has exponential dependency on the temperature increase irrespective of the switching activities. The increased temperature due to intensified power density thus raises the leakage component of power dissipation, forming a positive feedback loop between the power and temperature. The thermal increase also exacerbates aging phenomena such as electromigration and bias temperature instability (BTI) and consequently threatens the lifetime reliability of embedded processor. As shown in Fig. 8, there exist complex interactions between various physical phenomena. With continued miniaturization of device technologies, the operations of computing systems

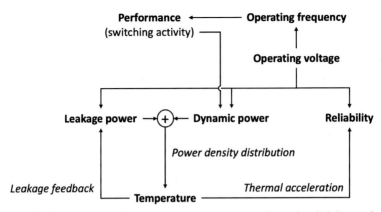

Fig. 8 Multi-physics interactions among energy, power, thermal, reliability and their compound impacts on the performance of executing hardware [22,23].

are increasingly governed by such physical phenomena, and their performance impacts become more prominent in the embedded systems.

Importantly, the performance change of computing system is not proportional to the increase of operating clock frequency, since the growth rate of performance diminishes as the clock frequency increases. Thus, the increased power dissipation by pushing the operating point to a higher voltage and clock frequency deteriorates the power efficiency of embedded hardware. Since embedded edge devices are often powered by batteries, the power efficiency also has a decisive impact on determining the operating point of embedded system.

- *Minimizing the total costs of ownership*: Determining an embedded processing solution for a neural network needs to find a proper balance between the hardware requirements (e.g., performance, power, reliability) and total costs of ownership (TCO) [29,30]. The costs encompass not only initial investments into the hardware deployment (affecting memory and storage capacities, processor options, computing power, etc.) but also various other aspects such as costs involved in the development, maintenance, replacement, and enhancement of deployed hardware over the lifetime.

3. Hardware acceleration in embedded systems

3.1 Hardware acceleration options

There exist numerous hardware options to accelerate neural network computing in embedded systems. As discussed in Section 2, processing solutions should differ depending on the desired requirements of target applications and embedded systems such as performance, energy, power, thermal, reliability, cost, form factor, etc. The hardware options encompass conventional processing units such as CPUs and GPUs, field-programmable gate arrays (FPGAs), specialized accelerators built on application-specific integrated circuit (ASIC) implementations. Table 1 summarizes the embedded processing options by hardware types. Numbers and features in the table represent general cases and do not reflect those of particular products of certain vendors.

CPUs for high-performance computing can achieve fast inference executions of deep neural networks with small batch inputs [31]. HPC CPUs include up to several tens of cores in a single chip, where each core is often equipped with simultaneous multithreading (SMT) capabilities and single-instruction, multiple-data (SIMD) units for vector operations. For instance, AMD EPYC 7002-series processors [32] include 64 cores in a chip and support up to 128 hardware threads. The CPU as a general-purpose processing unit facilitates the programming and development of neural network applications, and the use of SMT and SIMD features in the multicore processor helps enhance the power efficiency for providing greater throughput per power [26]. However, CPUs are known to have inferior power efficiency

Table 1 Hardware options for neural network acceleration.

Hardware types	Power efficiency (TFLOPS/W)	# of processing units	Hardware features
CPU (HPC)	∼0.01	Several tens of cores	General-purpose multicores with SMT and SIMD features
CPU (mobile/edge)	∼0.02	A few cores	Asymmetric general purpose
GPU (HPC)	∼0.05	Several thousands of cores	Massively parallel SIMT executions for general-purpose computing
GPU	∼0.05	Several hundreds of cores	SIMT executions for general-purpose computing
FPGA	0.3 or higher	Customized compute logic	Customized designs on programmable hardware
ASIC	1.5 or higher	Customized compute logic	Customized designs

than other hardware options and thus in general have limited uses for neural network acceleration in the embedded systems.

CPUs for mobile or edge computing often incorporate a mixture of asymmetric cores such as ARM big.LITTLE architecture [33] into a single chipset to meet the varying demands of compute-intensive or latency-sensitive workloads with high-performance big cores and also to provide power-efficiency executions via little cores. For example, Qualcomm Snapdragon 845 Mobile Platform [34] is built on the ARM big.LITTLE architecture that houses four high-performance cores and four high-efficiency cores. In resource-limited mobile or edge devices, little performance difference is observed between the CPUs and GPUs [35,36]. Therefore, such devices without dedicated hardware accelerators for neural network computing typically exploit the CPUs to execute the inference operations.

GPUs for high-performance computing are superior to other computing platforms in terms of achieving high-throughput computations. Since the neural network computing is a highly data-intensive task that needs to execute a large number of operations repeating over multi-dimensional data, the computations are well fitted for the single-instruction, multiple-threads (SIMT) executions of GPUs. The HPC GPUs incorporate several thousands of cores into a single processor package with large on-chip memory components including registers, caches, and shared memory. For instance, NVIDIA A100 GPU [37] contains 6,912 CUDA cores and provides the peak throughput of 19.5 TFLOPS for 32-bit single-precision (SP) calculations. Such abundant compute and memory resources enable the GPU to execute deep neural networks in massively parallel manner, thereby delivering greater throughput than other computing platforms (e.g., CPUs). Therefore, the GPUs have been serving as de facto hardware for neural network computing, especially for training the deep neural networks. The GPUs are known to offer better power efficiency than the conventional CPUs with general-purpose programmability, but they still fall short of desirable power efficiency for neural network computing. The GPUs are generally used for training the deep neural networks at large scales, and the processing solutions to implement fast inferences in embedded systems are sought after by leveraging other hardware acceleration options.

Embedded GPUs in mobile SoC typically include only a handful number of cores in a processor package, which are insufficient to deliver enough computing power to neural network computations. For example, Samsung Exynos 9810 Mobile Processor [38] integrates Mali-G72 MP18 GPU containing 18 cores. Hence, the mobile devices, if not equipped with dedicated accelerators, usually leverage CPU cores to execute the deep neural networks, where the CPUs show comparable performance as embedded GPUs [35,36]. When it comes to higher-end embedded devices such as NVIDIA Jetson-series processors [39], a GPU accommodates up to several hundreds of cores to offer high-throughput computing capability in the embedded processor. For example, NVIDIA Jetson AGX Xavier includes a Volta-architecture GPU with 512 CUDA cores and 8-core ARM v8.2 64-bit CPU. By incorporating comparably large GPU in the embedded processor, it can provide much higher throughput than other embedded GPUs found in typical mobile devices.

Recent trends for both high-performance and embedded GPUs (e.g., NVIDIA A100 and Jetson AGX Xavier) supplement the conventional GPU microarchitecture with specialized matrix/vector solvers such as the Tensor Cores found in recent NVIDIA GPUs. A Tensor Core is combinational logic specialized for processing matrix-multiply-and-accumulate (MMA) operations in a processing block of streaming multiprocessor (SM) [40]. Fig. 9 illustrates the block diagram of processing block, where a pair of Tensor Cores share the scheduling resources of processing block such as the register file and warp scheduler with other conventional GPU pipeline elements including floating-point and inter ALUs. A Tensor Core consists of four-element dot products (FEDPs) that collectively

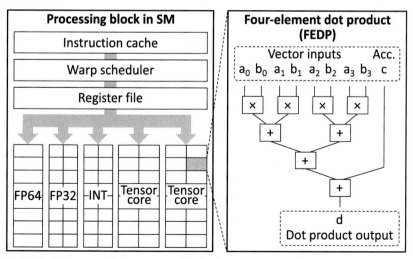

Fig. 9 Block diagram of a processing block of a streaming multiprocessor in GPU. The tensor cores executing four-element dot product (FEDP) operations provide greater throughput than the conventional CUDA cores [40].

execute 4×4 MMA operations. A group of four threads in a warp is called a *threadgroup*, and a threadgroup works on a 4×8 data points using the tensor cores. A couple of threadgroups form an *octet* and collective produce a 8×8 tile of data. Four octets in a 32-thread warp consequently generate an MMA output for a 16×16 matrix. The NVIDIA Volta architecture that first introduced Tensor Cores in its architecture had strict restrictions on the size of matrix inputs to the Tensor Cores [41], but the successive Turing and Ampere architectures have loosen the restrictions to allow more diverse forms (i.e., dimensions) of data inputs. The Tensor Cores in the processing block provides greater computational intensity than the conventional CUDA cores by densely packing multipliers and adders into the FEDPs. In particular, NVIDIA Tesla V100 [41] has 16 single-precision floating-point units in a processing block, whereas the Tensor Cores can yield 8x greater throughput for the same precision [42]. Thus, the recent NVIDIA GPUs incorporating Tensor Cores offer computational advantages for neural networks in both high-performance and embedded computing domains [43].

FPGAs have rapidly emerged as an hardware alternative to accelerate neural network computations. The FPGA is inherently programmable to accelerate the inference of a particular neural network [44]. It can implement the optimized datapath of neural network algorithm, and the synthesized design mapped to reconfigurable logic offers greater power efficiency and lower latency than instruction-based pipeline executions found in the conventional CPUs and GPUs. Importantly, the reconfigurable logic and routing in the FPGA enables various different forms of neural network accelerators to be efficiently implemented on the FPGA fabric [45–50]. Despite the programmability support in hardware, exploiting the FPGA for neural network acceleration incurs more extensive development time and steeper learning curve compared to using the conventional general-purpose processing options. A single FPGA chip in general does not have enough logic counts and memory spaces (e.g., registers and SRAMs) to accommodate the sizable volume of neural network data. Hence, the FPGA is generally used for implementing rapid inference executions in an embedded system rather than devising a high-throughput computing environment for neural network training.

Accelerators based on ASIC implementations provide highly efficient processing solutions for neural network computing. The ASIC implementations aim at optimizing the dataflow of a neural network algorithm to compose a tailored design for the target workload. Neural network accelerators on ASIC implementations are similar in this aspect to what drives using FPGAs for hardware acceleration (i.e., fitted accelerator implementations for different workloads). The ASIC implementations provide much superior performance and power efficiency [51–57] to other processing solutions including FPGAs, GPUs, and CPUs. However, the unparallel efficiency is traded with significantly greater development time and costs, and the customized design makes it relatively harder to program and use. Thus, the ASIC implementations may not be the best choice for products in competitive low-margin markets where minimizing the time to market (TTM) is critical, or in environments that require frequent software and hardware updates and maintenances.

3.2 Commercial options for neural network acceleration

Numerous hardware solutions have been proposed to accelerate neural network computations at various levels of computing domains from mobile to HPC systems. Besides the research prototypes introduced in academic papers, Table 2 summarizes off-the-shelf commercial devices readily available for neural network acceleration in embedded systems.

- *Raspberry Pi 4*: Raspberry Pi 4 Model B [58] boasts a small form factor and is an affordable, single-board computing device. It uses a Broadcom processor that has quad-core Cortex-A72 (ARM v8) 64-bit CPU, Videocore VI GPU, 2–8GB LPDDR4-3200 SDRAM, and a variety of I/Os on the board including Wireless LAN, Bluetooth, Gigabit Ethernet, HDMI, USB-C, etc. The GPU in Raspberry Pi 4 has no general-purpose graphics processing unit (GPGPU) supports or specialized accelerators in the SoC.

- *Arduino Portenta H7*: Similar to Raspberry Pi 4, Arduino Portenta H7 [59] offers an affordable hardware solution for embedded processing. It includes dual-core ARM Cortex-M7 CPU and a 32-bit M4 MCU with Chrom-ART GPU. Arduino Portenta H7 also has no GPGPU supports or dedicated accelerators for neural network computing in the SoC. Both Raspberry Pi and Arduino processors provide low-end embedded or IoT devices with handy processing solutions with similar hardware features and specifications.

- *NVIDIA Jetson TX2*: NVIDIA Jetson TX2 [60] is a high-performance embedded processing platform with quad-core ARM Cortex-A57 and dual-core Denver D15 CPUs, Pascal-architecture GPU of 256 CUDA cores integrated with 8GB 128-bit LPDDR4 memory. The memory is hard-wired to a memory controller in the processor and shared with the host CPU (i.e., unified memory) [61]. NVIDIA Jetson TX2 does not have dedicated hardware accelerators such as Tensor Cores found in its successors.

Table 2 Configuration of commercial hardware options for neural network computing.

Commercial device	CPU	Accelerator	Max power
Raspberry Pi 4 Model B [58]	4 ARM Cortex-A72 cores	None	10 W
Arduino Portenta H7 [59]	2 ARM Cortex-M7 cores 1 Cortex-M4 MCU core	None	N/A
NVIDIA Jetson TX2 [60]	4 ARM Cortex-A57 cores 2 Denver D15 cores	256-core Pascal GPU	15 W
NVIDIA AGX Xavier [39]	8 ARM v8.2 cores	512-core Volta GPU 64 Tensor Cores	30 W
Google Edge TPU [62]	4 ARM Cortex-A53 cores 1 Cortex-M4F core	TPU coprocessor with systolic array MXU	2 W
Intel Neural Compute Stick 2 [64]	2 SPARC v8 RISC cores	16 SHAVE cores with VLIW SIMD units	2.5 W
Xilinx PYNQ-Z1 FPGA [65]	2 ARM Cortex-A9 cores	Programmable fabric	5 W
Tesla FSD [66]	12 ARM Cortex-A72 cores	Neural processing unit with 96×96 MACs	100 W
AMD EPYC 7742 CPU [32]	64 Infinity architecture CPU cores	None	225 W
NVIDIA A100 GPU [37]	None	6912-core Ampere GPU 432 Tensor Cores	250 W

- *NVIDIA AGX Xavier*: NVIDIA Jetson AGX Xavier [39] is a higher-end embedded processing platform than its predecessor, NVIDIA Jetson TX2. It has octa-core ARM v8.2 CPU, Volta-architecture GPU with 512 CUDA cores and 64 Tensor Cores integrated with 32GB 256-bit LPDDR4 memory. The Tensor Cores introduced in the Volta architecture delivers greater throughput for neural network computations. Similar to NVIDIA Jetson TX2, it also supports the unified memory shared between the host CPU and GPU.
- *Edge TPU*: Edge tensor processing unit (TPU) [62] is a downsized ASIC implementation of Google TPU [55] to provide a handy hardware solution for neural network acceleration in low-power edge devices. Similar to Raspberry Pi and Arduino processors, edge TPU is implemented as a

single-board computing device. The edge TPU co-processor is integrated with quad-core ARM Cortex-A53 and Cortex-M4F CPU, GC7000 Lite Graphics, 1GB LPDDR4 memory along with a variety of I/O peripherals including USB-C, Gigabit Ethernet, HDMI, etc. The matrix unit (MXU) of Cloud TPU [63] used in data centers implements a 128×128 systolic array with INT8 data types. It is conjectured that the edge TPU has inherited the similar design from Cloud TPU, but the details of TPU co-processor (i.e., accelerator) have not been disclosed yet.

- *Intel Neural Compute Stick 2*: Intel neural compute stick (NCS) 2 [64] is a small-factor computing device enclosed in a thumb-drive chassis. It works as a plug-and-play device that technically can be connected to any platforms via the USB-3.0 interface. Intel NCS 2 is built on the Intel Movidius Myriad X vision processing unit (VPU) that encloses 16 cores of streaming hybrid architecture for vector engine (SHAVE) integrated with dual-core v8 RISC CPU. Each SHAVE core is a very-long instruction word (VLIW) programmable engine with SIMD functional units, and it natively supports mixed precisions of 32, 16, and 8-bit data types [61].

- *Xilinx Zynq-family FPGA*: Xilinx Zynq-family FPGA [65] implements a programmable SoC that integrates reconfigurable fabric and dual-core ARM Cortex-A9 or quad-core ARM Cortex-A53 CPU into a single package. Python productivity for Zynq (a.k.a. PYNQ) aims at facilitating the use of Xilinx Zynq FPGA by enabling the programmability using Python on the board without having to use ASIC-style design tools to devise programmable logic circuits. In particular, PYNQ-Z1 FPGA board has a dual-core ARM Cortex-A9 CPU and supports 13,330 logic slices (or 85K logic cells), 220 digital signal processing (DSP) slices, and 630KB fast block RAM (BRAM).

- *Tesla FSD*: Tesla full self-driving (FSD) [66] is an embedded processing solution developed by Tesla for its autonomous driving system. It is not a commercially available off-the-shelf product but represents a good example of embedded processing for deep neural networks. A Tesla FSD chip encloses 12 Cortex-A72 CPU cores and 96×96 multiply-and-accumulate (MAC) units for neural processing.

- *HPC processor*: In addition to the listed embedded processing options of small form-factor, single-board computing devices such as Raspberry Pi 4, Arduino Portenta H7, and edge TPU as well as powerful embedded processors such as Tesla FSD, conventional CPUs and GPUs can also be employed in embedded systems to provide processing solutions for neural network computing. For instance, AMD EPYC 7742 CPU [32] has 64 cores in a processor package, and it supports up to 128 hardware threads. The processor runs at 2.25 GHz base clock frequency and allows boosted executions at the 3.4 GHz clock rate. In contrast, NVIDIA A100 GPU [37] contains 6912 CUDA cores in the processor along with 432 Tensor Cores. The processor is coupled with 40 GB on-board memory that delivers 1.6 TB/s of memory bandwidth.

4. Software frameworks for neural networks

Software frameworks play an essential role in executing neural networks on hardware accelerators. They facilitate the composition of diverse neural network structures, use of acceleration libraries and algorithms, and deployment on the hardware accelerators. This section summarize popularly used software frameworks to deploy deep neural networks on the embedded hardware.

- *TensorFlow*: TensorFlow [67] is one of the most popular machine learning frameworks, developed by Google. TensorFlow uses dataflow graphs to represent both computations of an algorithm and states that the algorithm operates on. The computational engine of TensorFlow is written in C/C++, and the framework can execute on multicore CPUs, general-purpose GPUs, and Google's custom-designed ASIC (i.e., TPU). TensorFlow supports a variety of applications with focuses on facilitating the training and inference of deep neural networks. Its Python-based scripting interface provides users with an easy-to-program environment, and thus it has been widely adopted since it was released as an open-source project.

- *TensorFlow Lite*: TensorFlow Lite [68] is an open-source framework for deep neural networks to execute TensorFlow models on mobile, IoT, and edge devices. To support on-device executions (i.e., low-latency inferences), TensorFlow Lite optimizes computational graphs and the weights of neural networks to reduce the computational and memory requirements of TensorFlow models. In particular, TensorFlow Lite provides the models with various optimization techniques including pruning and quantization. TensorFlow Lite consists of two main components: (i) interpreter and (ii) converter. The interpreter of TensorFlow Lite enables executing optimized models on many different hardware platforms such as mobile, IoT, and edge devices. The converter of TensorFlow Lite optimizes the TensorFlow models for use by the interpreter.

- *TensorRT*: NVIDIA TensorRT [69] is a deep learning inference framework to provide neural network applications with low-latency, high-performance operations on NVIDIA GPUs. TensorRT implements parsers to import models trained on other platforms and plugins to optimize the imported neural networks. The neural networks can be optimized to support mixed-precision computations (e.g., INT8, INT16 data types), and TensorRT helps minimize memory footprints on the GPUs.
- *Caffe*: Caffe2 [70] is supported by Facebook, and it extends its preceding academic effort, Caffe [71]. The first version of Caffe (or Caffe 1.0) mainly targeted on large-scale usecases of typical convolutional neural network (CNN) applications. With rapid evolutions in deep learning technologies, more diverse forms of neural networks have emerged, and computing demands on various hardware platforms besides conventional high-performance GPUs have become necessary. Caffe2 improves the shortcomings of Caffe 1.0 by adding the supports for large-scale distributed training, mobile deployment, model optimizations (e.g., quantization), etc. Caffe2 has become officially deprecated and merged to PyTorch [72].
- *PyTorch*: PyTorch [72] is an open-source framework for the training and inference of deep neural networks. Supported by Facebook, PyTorch is one of the most popularly used neural network frameworks in the machine learning community. PyTorch provides a Python package for high-level programming features such as tensor communications (e.g., NumPy, SciPy, Cython) with GPU acceleration. In contrast to other deep learning frameworks, PyTorch features dynamic construction of computation graphs instead of using pre-defined graphs with fixed functionalities. This feature is valuable for cases when the required memory space is unknown for creating a neural network.
- *Intel Movidius Neural Compute SDK*: The Intel Movidius Neural Compute SDK (NCSDK) [73] enables the construction and deployment of deep neural networks on Intel Movidius Neural Compute Stick devices. The NCSDK provides Neural Compute API (NCAPI) for developing applications in C/C++ or Python and various software tools to compile, profile, validate the neural network applications.
- *Darknet*: Darknet [74] is an open-source neural network framework written in C and CUDA for CPU and GPU supports. The overall implementation of Darknet is analogous to that of Caffe [71], but its lighter codebase makes it easier to understand and modify the framework for various hardware and software testings.

• *FPGA frameworks*: Utilizing vendor-provided automation frameworks facilitates the deployment of deep neural networks on FPGA boards unless low-level customization of dataflow is necessary. FINN [75] is one of the exemplary software frameworks for Xilinx PYNQ FPGA. It is an open-source framework from Xilinx to facilitate neural network inferences on Xilinx PYNQ FPGAs. It specifically targets on quantized neural networks (QNNs) and generates dataflow-oriented architectures for the target QNNs. FINN provides a PyTorch library named Brevitas for training the QNNs. A recent Xilinx FPGA such as Zynq 7000-series MPSoC incorporates a dedicated acceleration engine named deep learning processor unit (DPU) [76] for deep-learning inferences. The DPU can be exercised via deep neural network development kit (DNNDK) provided by Xilinx. DNNDK supports both Caffe [71] and TensorFlow [67] within a unified framework.

Acknowledgments

This work was supported by the Ministry of Science and ICT, South Korea, through the Information Technology Research Center (ITRC)-supported program supervised by the Institute of Information and Communications Technology Planning and Evaluation (IITP) under Grant #2020-0-01847.

References

[1] Y. Bengio, R. Ducharme, P. Vincent, C. Janvin, A neural probabilistic language model, J. Mach. Learn. Res. 3 (2003) 1137–1155.

[2] K. He, X. Zhang, S. Ren, J. Sun, Deep residual learning for image recognition, in: International Conference on Computer Vision and Pattern Recognition, June, 2016, pp. 770–778.

[3] A. Krizhevsky, I. Sutskever, G. Hinton, ImageNet classification with deep convolutional neural networks, in: International Conference on Neural Information Processing Systems, December, 2012, pp. 1097–1105.

[4] J. Redmon, A. Farhadi, YOLO9000: better, faster, stronger, in: IEEE Conference on Computer Vision and Pattern Recognition, June, 2016, pp. 6517–6525.

[5] W. Sultani, C. Chen, M. Shah, Real-world anomaly detection in surveillance videos, in: IEEE/CVF Conference on Computer Vision and Pattern Recognition, June, 2018, pp. 6479–6488.

[6] M. Yang, S. Wang, J. Bakita, T. Vu, F. Smith, J. Anderson, J. Frahm, Re-thinking CNN frameworks for time-sensitive autonomous-driving applications: addressing an industrial challenge, in: IEEE Real-Time and Embedded Technology and Application Symposium, April, 2019, pp. 305–371.

[7] M. Mohammadi, A. Al-Fuqaha, S. Sorour, M. Guizani, Deep learning for IoT big data and streaming analytics: a survey, IEEE Commun. Surv. Tutorials 20 (4) (2018) 2923–2960.

[8] S. Ioffe, S. Christian, Batch normalization: accelerating deep network training by reducing internal covariate shift, in: Proceedings of the 32nd International Conference on International Conference on Machine Learning, 2015, pp. 448–456.

[9] M. Olson, A. Wyner, R. Berk, Modern neural networks generalize on small data sets, in: Advances in Neural Information Processing Systems, December, 2018, pp. 3619–3628.

[10] F. Pineda, Generalization of back-propagation to recurrent neural networks, Phys. Revi. Lett. 59 (19) (1987) 2229–2232.

[11] J. Schmidhuber, Deep learning in neural networks, Neural Netw. 61 (C) (2015) 85–117.

[12] D. Wilson, T. Martinez, The general inefficiency of batch training for gradient descent learning, Neural Networks 16 (10) (2003) 1429–1451.

[13] S. Han, H. Mao, W.J. Dally, Deep compression: compressing deep neural networks with pruning, trained quantization and Huffman coding, in: International Conference on Learning Representations, May, 2016.

[14] S. Han, J. Pool, J. Tran, W.J. Dally, Learning both weights and connections for efficient neural networks, in: International Conference on Neural Information Processing Systems, December, 2015, pp. 1135–1143.

[15] B. Jacob, S. Kligys, B. Chen, M. Zhu, M. Tang, A. Howard, H. Adam, D. Kalenichenko, Quantization and training of neural networks for efficient integer-arithmetic-only inference, in: IEEE/CVF Conference on Computer Vision and Pattern Recognition, June, 2018, pp. 2704–2713.

[16] S. Jung, S. Moon, Y. Lee, J. Kung, MixNet: an energy-scalable and computationally lightweight deep learning, in: IEEE/ACM International Symposium on Low Power Electronics and Design, July, 2019, pp. 1–6.

[17] H. Sharma, J. Park, N. Suda, L. Lai, B. Chau, V. Chandra, H. Esmaeilzadeh, Bit fusion: bit-level dynamically composable architecture for accelerating deep neural networks, in: International Symposium on Computer Architecture, June, 2018, pp. 764–775.

[18] M. Nadeski, Bringing machine learning to embedded systems, March, 2019. Available: https://www.ti.com/lit/wp/sway020a/sway020a.pdf (Online).

[19] K. Bogil, L. Sungjae, T. Amit, S. William, Energy-efficient acceleration of deep neural networks on realtime-constrained embedded edge devices, IEEE Access 8 (2020) 216259–216270.

[20] L. Gwennap, M. Demler, L. Case, A Guide to Processors for Deep Learning, first ed., The Linley Group, 2017.

[21] M. Ozdag, Adversarial attacks and defenses against deep neural networks: a survey, Proc. Comput. Sci. 140 (2018) 152–161.

[22] W. Song, S. Mukhopadhyay, S. Yalamanchili, KitFox: multi-physics libraries for integrated power, thermal, and reliability simulations of multicore microarchitecture, IEEE Trans. Compon. Packag. Manuf. Technol. 5 (11) (2015) 1590–1601.

[23] S. William, M. Saibal, Y. Sudhakar, Managing performance-reliability tradeoffs in multicore processors, IEEE International Reliability Physics Symposium, 2015, pp. 3C.1.1–3C.1.7.

[24] R. Efraim, R. Ginosar, C. Weiser, A. Mendelson, Energy-aware race to halt: a down to EARth approach for platform energy management, IEEE Comput. Archit. Lett. 13 (1) (2014) 25–28.

[25] A. Raghavan, Y. Luo, A. Chandawalla, M. Papaefthymiou, K. Pipe, T. Wenisch, M. Martin, Computational sprinting, in: IEEE International Symposium on High-Performance Computer Architecture, February, 2012, pp. 1–12.

[26] W. Song, C. Cher, A. Buyuktosunoglu, P. Bose, Measurement-driven methodology for evaluating processor heterogeneity options for power-performance efficiency, in: IEEE International Symposium on Low Power Electronics and Design, August, 2016, pp. 284–289.

[27] B. Kim, S. Lee, C. Park, H. Kim, W. Song, The Nebula benchmark suite: implications of lightweight neural networks, IEEE Trans. Comput. (2020) 1–15, https://doi.org/10.1109/TC.2020.3029327.

[28] S. Williams, A. Waterman, D. Patterson, Roofline: an insightful visual performance model for multicore architectures, Commun. ACM 52 (4) (2009) 65–76.

[29] J. Chiang, T. Schulte, S. Zammattio, Lowering the total cost of ownership in industrial applications, 2017. White Paper WP-01230-2.1, Intel Programmable Solutions Group.

[30] B. Grot, D. Hardy, P. Lotfi-Kamran, B. Falsafi, C. Nicopoulos, Y. Sazeides, Optimizing data-center TCO with scale-out processors, IEEE Micro 32 (5) (2012) 52–63.

[31] K. Hazelwood, S. Bird, D. Brooks, S. Chintala, U. Diril, D. Dzhugakov, M. Fawzy, B. Jia, Y. Jia, A. Kalro, J. Law, K. Lee, J. Lu, P. Noordhuis, M. Smelyanskiy, L. Xiong, X. Wang, Applied machine learning at Facebook: a datacenter infrastructure perspective, in: IEEE International Symposium on High Performance Computer Architecture, February, 2018, pp. 620–629.

[32] K. Lepak, G. Talbot, S. White, N. Beck, S. Naffziger, The next generation and enterprise server product architecture, in: IEEE Hot Chips Symposium, August, 2017.

[33] I. Lin, B. Jeff, I. Rickard, ARM platform for performance and power efficiency—hardware and software perspectives, in: International Symposium on VLSI Design, Automation and Test, April, 2016, pp. 1–5.

[34] K. Hawkins, The impact of big core and little core architecture on application development, 2018. Available: https://developer.qualcomm.com/blog/impact-big-core-little-core-architecture-application-development (Online) Qualcomm Developer Network, February.

[35] C. Wu, D. Brooks, K. Chen, D. Chen, S. Choudhury, M. Dukhan, K. Hazelwood, E. Isaac, Y. Jia, B. Jia, T. Leyvand, H. Lu, Y. Lu, L. Qiao, B. Reagen, J. Spisak, F. Sun, A. Tulloch, P. Vajda, X. Wang, Y. Wang, B. Wasti, Y. Wu, R. Xian, S. Yoo, P. Zhang, Machine learning at Facebook: understanding inference at the edge, in: IEEE International Symposium on High Performance Computer Architecture, February, 2019, pp. 331–344.

[36] W. Siqi, A. Gayathri, Z. Yifan, G. Neeraj, P. Anuj, M. Tulika, High-throughput CNN inference on embedded ARM Big.LITTLE multicore processors, IEEE Trans. Comput. Aided Des. Integr. Circuits Syst. 39 (10) (2019) 2254–2267.

[37] R. Krashinsky, O. Giroux, S. Jones, N. Stam, S. Ramaswamy, NVIDIA Ampere Architecture In-Depth, May, 2020. Available: https://developer.nvidia.com/blog/nvidia-ampere-architecture-in-depth (Online) NVIDIA Develop Blog.

[38] Samsung, The Samsung Exynos 9 Series: Spec Comparison, 2018. Available: https://www.samsung.com/semiconductor/minisite/exynos/newsroom/blog/the-samsung-exynos-9-series-spec-comparison (Online) Samsung Newsroom Blog.

[39] M. Ditty, A. Karandikar, D. Reed, NVIDIA's Xavier SoC, in: IEEE Hot Chips Symposium, August, 2018.

[40] M. Raihan, N. Goli, T. Aamodt, Modeling Deep learning accelerator enabled GPUs, in: IEEE International Symposium on Performance Analysis of Systems and Software, March, 2019, pp. 79–92.

[41] J. Choquette, Volta: programmability and performance, in: IEEE Hot Chips Symposium, August, 2017.

[42] H. Kim, S. Ahn, B. Kim, Y. Oh, W. Ro, W. Song, Duplo: lifting redundant memory accesses of neural networks for GPU Tensor Cores, in: IEEE/ACM International Symposium on Microarchitecture, October, 2020, pp. 725–737.

[43] K. Hyeonjin, A. Sungwoo, O. Yunho, K. Bogil, R. Won Woo, S. William, Duplo: lifting redundant memory accesses of deep neural networks for GPU tensor cores, IEEE/ACM International Symposium on Microarchitecture, 2020, pp. 725–737.

[44] A. Ling, M. Abdelfattah, A. Bitar, D. Capalija, G. Chiu, Flexibility: FPGAs and CAD in deep learning acceleration, 2016. White Paper WP-01283-1.0, Intel Programmable Solutions Group.

[45] L. Lu, J. Xie, R. Huang, J. Zhang, W. Lin, Y. Liang, An efficient hardware accelerator for sparse convolutional neural networks on FPGAs, in: IEEE International Symposium on Field-Programmable Custom Computing Machines, April, 2019, pp. 17–25.

[46] Y. Ma, Y. Cao, S. Vrudhula, J. Seo, Optimizing loop operation and dataflow in FPGA acceleration of deep convolutional neural networks, in: ACM/SIGDA International Symposium on Field-Programmable Gate Arrays, February, 2017, pp. 45–54.

[47] H. Sharma, J. Park, D. Mahajan, E. Amaro, J. Kim, C. Shao, A. Mishra, H. Esmaeilzadeh, From high-level deep neural models to FPGAs, in: IEEE/ACM International Symposium on Microarchitecture, October, 2016, pp. 1–12.

[48] N. Suda, V. Chandra, G. Dasika, A. Mohanty, Y. Ma, S. Vrudhula, J. Seo, Y. Cao, Throughput-optimized OpenCL-based FPGA accelerator for large-scale convolutional neural networks, in: ACM/SIGDA International Symposium on Field-Programmable Gate Arrays, February, 2016, pp. 16–25.

[49] C. Zhang, P. Li, G. Sun, Y. Guan, B. Xiao, Optimizing FPGA-based accelerator design for deep convolutional neural networks, in: ACM/SIGDA International Symposium on Field-Programmable Gate Arrays, February, 2015, pp. 161–170.

[50] C. Zhang, D. Wu, J. Sun, G. Sun, G. Luo, J. Cong, Energy-efficient CNN implementation on A deeply pipelined FPGA cluster, in: International Symposium on Low Power Electronics and Design, August, 2016, pp. 326–331.

[51] Y. Chen, J. Emer, V. Sze, Eyeriss: a spatial architecture for energy-efficient dataflow for convolutional neural networks, in: ACM/IEEE International Symposium on Computer Architecture, June, 2016, pp. 367–379.

[52] Y. Chen, T. Luo, S. Liu, S. Zhang, L. He, J. Wang, L. Li, T. Chen, Z. Xu, N. Sun, O. Temam, DaDianNao: a machine-learning supercomputer, in: IEEE/ACM International Symposium on Microarchitecture, December, 2014, pp. 609–622.

[53] Y. Chen, T. Yang, J. Emer, V. Sze, Eyeriss v2: a flexible accelerator for emerging deep neural networks on mobile devices, IEEE J. Emerg. Selec. Top. Circuits Syst. 9 (2) (2019) 292–308.

[54] S. Han, X. Liu, H. Mao, J. Pu, A. Pedram, M. Horowitz, W. Dally, EIE: efficient inference engine on compressed deep neural network, in: ACM/IEEE International Symposium on Computer Architecture, June, 2016, pp. 243–254.

[55] N. Jouppi, C. Young, N. Patil, D. Patterson, G. Agrawal, R. Bajwa, S. Bates, S. Bhatia, N. Boden, A. Borchers, R. Boyle, P. Cantin, C. Chao, C. Clark, J. Coriell, M. Daley, M. Dau, J. Dean, B. Gelb, T. Ghaemmaghami, R. Gottipati, W. Gulland, R. Hagmann, C. Ho, D. Hogberg, J. Hu, R. Hundt, D. Hurt, J. Ibarz, A. Jaffey, A. Jaworski, A. Kaplan, H. Khaitan, D. Killebrew, A. Koch, N. Kumar, S. Lacy, J. Laudon, J. Law, D. Le, C. Leary, Z. Liu, K. Lucke, A. Lundin, G. MacKean, A. Maggiore, M. Mahony, K. Miller, R. Nagarajan, R. Narayanaswami, R. Ni, K. Nix, T. Norrie, M. Omernick, N. Penukonda, A. Phelps, J. Ross, M. Ross, A. Salek, E. Samadiani, C. Severn, G. Sizikov, M. Snelham, J. Souter, D. Steinberg, A. Swing, M. Tan, G. Thorson, B. Tian, H. Toma, E. Tuttle, V. Vasudevan, R. Walter, W. Wang, E. Wilcox, D. Yoon, In-datacenter performance analysis of a tensor processing unit, in: ACM/IEEE International Symposium on Computer Architecture, June, 2017, pp. 1–12.

[56] S. Liu, Z. Du, J. Tao, D. Han, T. Luo, Y. Xie, Y. Chen, T. Chen, Cambricon: an instruction set architecture for neural networks, in: ACM/IEEE International Symposium on Computer Architecture, June, 2016, pp. 393–405.

[57] S. Zhang, Z. Du, L. Zhang, H. Lan, S. Liu, L. Li, Q. Guo, T. Chen, Y. Chen, Cambricon-X: an accelerator for sparse neural networks, in: IEEE/ACM International Symposium on Microarchitecture, October, 2016, pp. 1–12.

[58] R.P. Foundation, Raspberry Pi 4, 2019. Available: https://www.raspberrypi.org/products/raspberry-pi-4-model-b (Online).

[59] Arduino, Arduino Portenta H7, 2020. Available: https://www.arduino.cc/pro/hardware/product/portenta-h7 (Online).

[60] A. Skende, Introducing parker: next-generation tegra system-on-chip, in: IEEE Hot Chips Symposium, August, 2016.

[61] R. Hadidi, J. Cao, Y. Xie, B. Asgari, T. Krishna, H. Kim, Characterizing the deployment of deep neural networks on commercial edge devices, in: IEEE International Symposium on Workload Characterization, November, 2019, pp. 35–48.

[62] Google, Edge TPU, in: Google Cloud Products–Internet of Things (IoT), January, 2019. Available: https://cloud.google.com/edge-tpu (Online).

[63] C. Chao, Cloud TPU: Codesigning Architecture and Infrastructure, in: IEEE Hot Chips Symposium, 2019.

[64] Intel, Intel Neural Compute Stick 2 (Intel NCS2), 2019. Available: https://software.intel.com/content/www/us/en/develop/hardware/neural-compute-stick.html (Online).

[65] Xilinx, Python Productivity for Zynq (Pynq), 2018. Available: https://pynq.readthedocs.io/en/v2.5.1 (Online).

[66] E. Talpes, D. Sarma, G. Venkataramanan, P. Bannon, B. McGee, B. Floering, A. Jalote, C. Hsiong, S. Arora, A. Gorti, G. Sachdev, Compute solution for tesla's full self-driving computer, IEEE Micro 40 (2) (2020) 25–35.

[67] M. Abadi, P. Barham, J. Chen, Z. Chen, A. Davis, J. Dean, M. Devin, S. Ghemawat, G. Irving, M. Isard, M. Kudlur, J. Levenberg, R. Monga, S. Moore, D. Murray, B. Steiner, P. Tucker, V. Vasudevan, P. Warden, M. Wicke, Y. Yu, X. Zheng, TensorFlow: a system for large-scale machine learning, in: USENIX Symposium on Operating Systems Design and Implementation, November, 2016, pp. 265–283.

[68] Google, ML for mobile and edge devices–TensorFlow lite, 2020. Available: https://www.tensorflow.org/lite (Online).

[69] H. Abbasian, J. Park, S. Sharma, S. Rella, Speeding Up Deep Learning Inference Using TensorRT, April, 2020. Available: https://developer.nvidia.com/blog/speeding-up-deep-learning-inference-using-tensorrt/ (Online).

[70] Facebook Research, Caffe2: a new lightweight, modular, and scalable deep learning framework, April, 2017. Available: https://caffe2.ai (Online).

[71] Y. Jia, E. Shelhamer, J. Donahue, S. Karayev, J. Long, R. Girshick, S. Guadarrama, T. Darrell, Caffe: convolutional architecture for fast feature embedding, in: ACM International Conference on Multimedia, November, 2014, pp. 675–678.

[72] A. Paszke, S. Gross, F. Massa, A. Lerer, J. Bradbury, G. Chanan, T. Killeen, Z. Lin, N. Gimelshein, L. Antiga, A. Desmaison, A. Kopf, E. Yang, Z. DeVito, M. Raison, A. Tejani, S. Chilamkurthy, B. Steiner, L. Fang, J. Bai, S. Chintala, PyTorch: an imperative style, high-performance deep learning library, in: Advances in Neural Information Processing Systems, November, 1987, pp. 2229–2232.

[73] Intel, Intel Movidius Neural Compute SDK, 2019. Available: https://movidius.github.io/ncsdk (Online).

[74] J. Redmon, Darknet: Open Source Neural Networks in C, 2013. Available: http://pjreddie.com/darknet (Online).

[75] Y. Umuroglu, N. Fraser, G. Gambardella, M. Blott, P. Leong, M. Jahre, K. Vissers, FINN: a framework for fast, scalable binarized neural network inference, in: ACM/SIGDA International Symposium on Field-Programmable Gate Arrays, February, 2017, pp. 65–74.

[76] Xilinx, DNNDK User Guide, 2019. Available: https://www.xilinx.com/support/documentation/user_guides/ug1327-dnndk-user-guide.pdf (Online) UG13327 (v1.5).

About the author

William J. Song is currently an Assistant Professor with the School of Electrical Engineering, Yonsei University in Seoul, South Korea. He earned his PhD degree in Electrical and Computer Engineering from Georgia Institute of Technology, Atlanta, GA, and BS degree in Electrical Engineering from the Yonsei University, Seoul, South Korea. His research focus lies in the challenges of heterogeneous architectures and processing near data for deep neural networks. His interests also include solutions to power, thermal, and reliability issues in many-core microarchitectures and 3D-integrated packages. Prior to joining the faculty of Yonsei University, he was a senior engineer at Intel in Santa Clara, CA. He was a graduate research intern at Qualcomm, San Diego, CA (2015 summer), IBM T.J. Watson Research Center, Yorktown Heights, NY (2014 summer and fall), AMD Research, Bellevue, WA (2013 summer), and Sandia National Labs, Albuquerque, NM (2012, 2011, and 2010 summers). He received a Distinguished Faculty Award from Yonsei University in 2018 and Teaching Excellence Awards in 2018 and 2020. He was a recipient of IBM/SRC graduate fellowship from 2012 to 2015. He received the Best Student Paper Award at IEEE International Reliability Physics Symposium (IRPS) in 2015 and Best in Session Award at SRC TECHCON in 2014.

CHAPTER THREE

Hardware accelerator systems for artificial intelligence and machine learning

Hyunbin Park[a] and Shiho Kim[b]
[a]IT & Mobile Communications, Samsung Electronics, Seoul, South Korea
[b]School of Integrated Technology, Yonsei University, Seoul, South Korea

Contents

Abstract

Recent progress in parallel computing machines, deep neural networks, and training techniques have contributed to the significant advances in artificial intelligence (AI) with respect to tasks such as object classification, speech recognition, and natural language processing. The development of such deep learning-based techniques has enabled AI-based networks to outperform humans in the recognition of objects in images. The graphics processing unit (GPU) has been the primary component used for parallel computing during the inference and training phases of deep neural networks. In this

Advances in Computers, Volume 122
ISSN 0065-2458
https://doi.org/10.1016/bs.adcom.2020.11.005

study, we perform training using a desktop or a server with one or more GPUs and inference using hardware accelerators on embedded devices. Performance, power consumption, and requirements of embedded system present major hindrances to the application of deep neural network-based systems using embedded controllers such as drones, AI speakers, and autonomous vehicles. In particular, power consumption of a commercial GPU commonly surpasses the power budget of a stand-alone embedded system.

To reduce the power consumption of hardware accelerators, reductions in the precision of input data and hardware weight have become popular topics of research in this field. However, precision and accuracy share a trade-off relationship. Therefore, it is essential to optimize precision in a manner that does not degrade the accuracy of the inference process. In this context, the primary issues faced by hardware accelerators are loss of accuracy and high power consumption.

1. Introduction

Recent progress in deep learning and convolutional neural networks (CNNs) has contributed to the advances in artificial intelligence with respect to tasks such as object classification [1] and detection [2], speech recognition [3], and natural language processing [4]. Deep learning-based methods have proven to be capable of even restoring specific regions deleted from images [5] or creating compelling fake videos of any person [6,7].

Deep neural networks (DNNs) often require more than millions of data points for training, which often require a day or more of computation time [8]. In addition, the amount of power consumed by such architectures to perform GPU-based computations for inference and training often exceeds the power budget of the embedded systems. To address these difficulties, a number of studies have proposed training the network using a desktop or a server and performing the inference phase using hardware accelerators on embedded devices [8–10]. In particular, artificial neurons that comprise DNNs are required to compute dot products between pairs of vectors. Therefore, to reduce the power consumption of the hardware of the artificial neuron to satisfy the power budget of the embedded systems, it is essential to reduce the complexities of the adder and multiplier circuits within the neuron. Certain studies [11–13] have managed to decrease the power consumption of artificial neurons by reducing the precision of input vectors and weight vectors compared to the standard floating point 32-bit (FP32) precision. Any reduction in the precision of hardware decreases the hardware resources required by the computational circuits and the memory. However, excessive reduction in precision is liable to degrade the inference accuracy of the parent deep convolutional neural network (DCNN). Several DCNNs have been reported to have reduced the precision of the weight vectors of their artificial neurons to 1 bit [14,15]. However, quantization of weights to 1 bit degrades the inference accuracy of the DCNN by approximately 20% compared to DCNNs utilizing FP32 [14]. Therefore, it is important to strike a balance between hardware precision and inference accuracy. The utilization of gradient descent, which is a widely used training technique that accumulates the gradients of the loss function with respect to all the weight variables, has considerably accelerated the development of deep learning-based methods [16]. In networks employing the gradient descent technique, more precise gradients are expected to induce better performances during training and inference. Therefore, such networks commonly employ GPUs with FP32 during training, instead of using training hardware accelerators with reduced precision.

The various architectures and power saving strategies employed by hardware inference accelerators are discussed in this chapter. Further, digital neurons and hardware inference and training accelerators for embedded systems in DCNNs are also discussed. The digital neuron is observed to exhibit a structural advantage in terms of power consumption compared to artificial neurons, while exhibiting identical precision and negligibly degraded accuracy. In addition, it allows users to designate custom weights and inputs in exchange of particular degrees of loss in accuracy.

The remainder of the chapter is divided into the following sections. In Section 2, we present a brief background of CNNs, quantization of weights and activations, and computational elements for hardware accelerators. In Section 3, we discuss hardware inference accelerators for DNNs. In Section 4, we present hardware inference accelerator with digital neurons in detail. Finally, in Section 5, we summarize the contents of the chapter.

2. Background

In this section, we present an overview of CNNs and hardware accelerators developed in prior works. There are four primary considerations during the design of hardware accelerators—throughput, power consumption, precision of weights and activations, and inference accuracy. These four factors share trade-off relationships. For example, an increase in the performance of computational engines raises the throughput, but also increases power consumption. In contrast, a decrease in the precision of weights and activations reduces power consumption at the cost of degrading inference accuracy.

2.1 Overview of convolutional neural networks

Frank Rosenblatt proposed a mathematical model for a single layer neuron, named the perceptron, shown in Fig. 1 [17]. The perceptron computed the dot product of two vectors—an input vector, **X**, and a weight vector, **w**— and then adds the bias, b, to it. The activation function outputs 1 if the result of the dot product is greater than a predefined threshold, θ, and outputs 0 otherwise. The operation can be expressed by the following equation.

$$\text{Output} = \begin{cases} 0 & \left(b + \sum_{i=1}^{N} X_i \cdot w_i \le \theta \right) \\ 1 & \left(b + \sum_{i=1}^{N} X_i \cdot w_i > \theta \right) \end{cases} \tag{1}$$

The weight, summation function, and activation function imitate the integrate-and-fire operation of a biological neuron. The perceptron was capable of manipulating logical classifications of AND, NAND, and OR functions. However, in 1969, Marvin Minsky et al. reported that it was incapable of executing the XOR function [18]. DNNs with multiple hidden layers have overcome this limitation of single–layer neural networks.

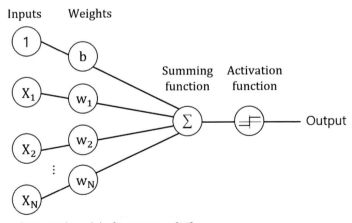

Fig. 1 Mathematical model of perceptron [17].

A CNN employs a 2-D or 3-D convolution to extract features from input data. The equation of 3-D convolution can be expressed as follows.

$$F_{\text{Out}}[x][y][z] = \sum_{k=0}^{M-1} \sum_{j=0}^{W-1} \sum_{i=0}^{H-1} \mathbf{X}[x+i][y+j][z+k]$$
$$\cdot \mathbf{w}[x+i][y+j][z+k] + b$$
$$= \vec{\mathbf{X}} \cdot \vec{\mathbf{w}} + b \qquad (2)$$

3-D convolution is simplified by adding a constant and the dot product between two vectors, and it extracts one output pixel of the feature map, as shown in Fig. 2. Thus, a stack of deep convolutional layers extracts a stack of feature maps corresponding to the input data. As the layers become deeper, greater amounts of abstracted information are extracted [19]. For example, feature maps of shallow layers respond only to edges or blobs, whereas feature maps of deeper layers respond to the overall shape of an object.

An overview of the classification of CNNs is presented in this section based on the exposition of a representative CNN, LeNet-5 [16]. As depicted in Fig. 3, LeNet-5 consists of the following sequence of components—a convolutional layer (C1), pooling layer (S1), convolutional layer (C2), pooling layer (S2), fully connected layer (FC1), fully connected layer (FC2), fully connected layer (FC3), and softmax layer. Pooling is utilized to reduce computational requirements and prevent over-fitting of the model to the

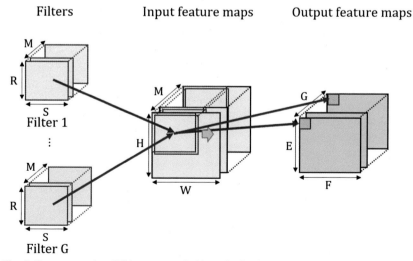

Fig. 2 Representative CNN structure, LeNet-5 [16].

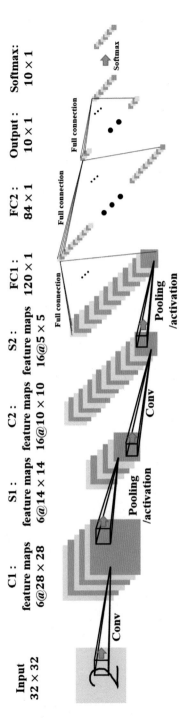

Fig. 3 3D convolution sweeping 3D filters in CNN.

training data. Pooling reduces the size of the extracted feature map using certain pooling techniques. The maxpooling technique is a representative pooling technique. It selects the maximum-valued neuron out of each neuron cluster and connects the selected neuron to the output. In each fully connected layer, all artificial neurons between the layers are connected across a layer. Fully connected layers act as linear classifiers, while reducing the dimensions of the inputs. The softmax layer also functions as a linear classifier that converts each output of the final fully connected layer to its corresponding inference probability, which is expressed by the following.

$$\sigma\left(y_j\right) = \frac{e^{y_j}}{\sum_{i=1}^{K} e^{y_i}} \text{ for } j = 1, ..., K. \tag{3}$$

where the inference probability corresponding to each output, y_j, has been normalized using the sum of all outputs.

2.2 Quantization of weights and activations

Since the 2010s, GPUs have become the most widely used computing devices employed during the inference and training phases of deep learning-based architectures. Whereas the 32-bit floating point format (FP32), based on the IEEE standard for floating-point arithmetic (IEEE 754) [20], is generally used in such systems, many artificial neural networks do not require precise arithmetic calculations using 32-bit floating point. Rather, they employ pooling and dropout techniques to reduce the precision of computations to avoid overfitting, which may optimize the model excessively with respect to training data [21–28]. Therefore, a number of studies [12,13,29,30] have proposed methods to reduce the precision of weights, inputs, and activations during forward propagation.

Lai et al. reduced weights to 11 bits and the activations to 16 bits, and the resulting degradations in the inference accuracies of AlexNet [31], SqueezNet [32], GoogLeNet [33], and VGG-16 [34] were observed to be within 1% [11]. Gysel limited weights to 4 bits and activations to 8 bits, inducing only 0.3% and 2% additional error in the recognition accuracies on the MNIST and CIFAR 10 datasets, respectively [30].

Courbariaux et al. reduced weights to 1 bit [14]—taking values of −1 or 1. The network based on binary weights is called BinaryConnect, and it is known to exhibit 98.82% and 91.73% inference accuracy on MNIST and CIFAR-10, respectively. Binarized neural networks (BNNs) [15] represented both weights and activations by 1-bit variables. However, the use of binary weights induces excessive loss in weight precision. BinaryConnect and BNN reduce inference accuracy by 19% and 29.8% (in Top-5 inference) with respect to object recognition on the ImageNet database compared to their counterparts using normal representations [35].

This section proposes forward propagation with quantized activations and weights using partial sub-integers. This technique allows designers and users to choose appropriate bit-widths for weights, as well as bit-costs of representations, in exchange for acceptable losses in inference accuracy. The quantization of weights using the sum of a limited number of partial sub-integers allows multipliers to be composed of a limited number of barrel shifters. This could also may reduce computational power consumption.

The inference accuracy of MNIST handwritten number is simulated according to precision of weights and activation, where the neural network used is LeNet-5 [16] and pre-trained 32-bit weights are quantized by round-to-nearest [29] scheme, as shown in Table 1. Quantization of pre-trained FP32 weights to 8-bit integer weights does not degrade the inference accuracy compared to the single precision. In addition, inference degradations caused by the adoption of 5-bit and 8-bit weights are also almost negligible.

If the multiplication of each component in the inner product is implemented using the arithmetic shift and addition operations, a simpler multiplier circuit than the Booth multiplier can be designed. Booth multiplier based on a 5-bit multiplier produces four partial products, including the addition of 1 for the two's complement [36]. However, neural networks do not require exact algebraic calculations. Therefore, multiplication via a 5-bit multiplier can be replaced by the addition of two partial sub-integers ($w_n 2^k$), as expressed by Eq. (4). For example, $7 \cdot X = 8 \cdot X - X$.

$$w \cdot X = \left(w_1 2^i + w_2 2^j \right) \cdot X \; (w_{1,2} \in -1, 0, 1) \tag{4}$$

Eq. (4) allows the multiplier to be implemented solely based on two-barrel shift circuits, which considerably reduces circuit complexity and power consumption. The proposed alternative supports the multiplication of signed

Table 1 Simulated inference accuracy of handwritten MNIST number according to weight precision, where simulated neural network is LeNet-5.

Weight precision	Activation precision	Inference accuracy
FP32	FP32	99.10%
8-bit integer	8-bit integer	99.10%
5-bit integer	8-bit integer	98.95%

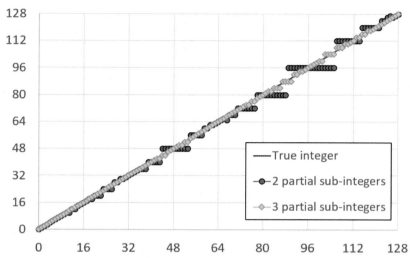

Fig. 4 Representation of integers ranging from 0 to 128 partitioned into two or three partial sub-integers.

5-bit integers ranging between −16 and 15, with the exceptions of ±13 and ±11. In these cases, the weight value 10.9 should be quantized to 10, and the weight value 11.1 should be quantized to 12.

Fig 4 depicts the comparison between true 8-bit integers and their representations by partitioning them into two or three partial sub-integers. The maximum error between a true 8-bit integer and two corresponding partial sub-integers is approximately 9%, and that between a true 8-bit integer and three corresponding partial sub-integers is approximately 2%. A maximum error of 9% in the representation precision induces insignificant error in the inference accuracy [21–28].

2.2.1 Performance of neural networks using quantized weights and activations based on arithmetic binary shift operations

• *Simulation of the inference accuracy of LeNet-5 on the MNIST database*

Fig. 5 illustrates a comparison between the simulated inference accuracy of LeNet-5 using FP32 and that of the proposed quantized LeNet-5 using 5-bit integer weights partitioned into two partial sub-integers each and 8-bit integer activation. Between epoch 1 to epoch 6, the accuracy of LeNet using FP32 is observed to be higher than that of the proposed variant. This can be attributed to the wider variation exhibited by weights initialized using values from the normal distribution compared to those initiated using FP32. However, after epoch 19, the inference accuracy of the proposed

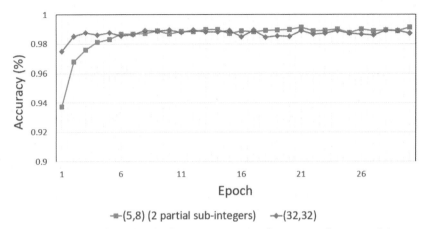

─■─(5,8) (2 partial sub-integers)   ─◆─(32,32)

Fig. 5 Comparison of simulated inference accuracies of LeNet-5 with FP32 and the proposed quantized LeNet-5 with 5-bit integer weights except ±11 and ±13 and 8-bit integer activation.

quantized neural network is observed to be higher than that of the FP32 variant, despite the lower resolution. The quantization simulates the effect of noise, which prevents overfitting. The maximum inference accuracy attained by the proposed scheme is also observed to be higher than that of the FP32 variant—99.16% compared to 98.96%, respectively.

• *Simulation of the inference accuracy of AlexNet on the IMAGENET database* We also trained AlexNet [31] using quantized 5-bit integer weights partitioned into two partial sub-integers each and 8-bit integer activations on the IMAGENET database. It exhibited a Top-1 inference accuracy of 52.7% and Top-5 inference accuracy of 76.3% using 5-bit integer weights (two partial sub-integers) and 8-bit integer activations, and a Top-1 inference accuracy of 56.4% and a Top-5 inference accuracy of 80.1% using 8-bit integer weights (three partial sub-integers) and 8-bit integer activations.

Fig. 6 shows a comparison of the inference accuracy of the proposed scheme and those of existing methods on MNIST. BNN and BC, which employed MLP, are observed to exhibit inference accuracies of 98.6% and 98.82%. The MLPs of BNN and BC consisted of three hidden layers each, comprising 4096 binary units and 1024 binary units, respectively. The accuracy degradations for BNN and BC are observed to be within 1% compared to the accuracy achieved by the FP32 variant.

The proposed schemes using two partial sub-integers (5, 8) and three partial sub-integers (8, 8) are observed to exhibit inference accuracies of

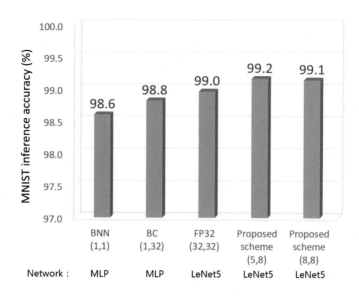

precision of (weight, activation)

Fig. 6 Comparison on MNIST inference accuracy of the proposed scheme and the prior arts (FP32, BinaryConnect (BC) [14], and BNN [15]).

99.16% and 99.14%, respectively, both of which surpass that of the FP32 variant. As in the previous case, quantization simulates the effect of noise, thereby preventing overfitting. Another study [15] also reported that quantization improves inference accuracy.

Based on the simulation results presented in Fig. 6, even 1-bit resolution of weights can be considered to be sufficient for inference on the MNIST database. However, excessive precision reduction in weights and activations would induce greater losses in accuracy in the case of more complex databases (e.g., comprising objects and terrains, rather than numbers, with a greater number of categories, including colors rather than grayscale images).

Fig. 7 shows a comparison between the inference accuracy of the proposed scheme and those of existing methods (FP32, BinaryConnect (BC) [14], and BNN [15]) on the IMAGENET dataset. BNN exhibits an additional error of 28.7%/29.8% in Top-1/Top-5 accuracy, and BC exhibits an additional error of 21.2%/19.2% in Top-1/Top-5 accuracy compared to the FP32 variant. In comparison, the proposed quantized version (5, 8) of AlexNet exhibits an additional error of 3.9%/3.9% in Top-1/Top-5 accuracy, whereas the quantized version (8, 8) version of AlexNet yields an

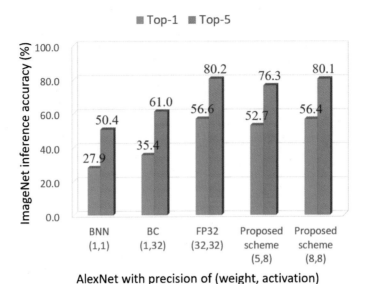

Fig. 7 Comparison on IMAGENET inference accuracies of the proposed scheme and existing ones (FP32, BinaryConnect (BC) [14], and BNN [15]).

additional error of only 0.2%/0.1% in Top-1/Top-5 accuracy compared to the FP32 variant. The additional error in each case can be reduced by increasing the hardware precision.

In summary, the proposed architectures using two partial sub-integers (5, 8) and three partial sub-integers (8, 8) exhibits inference accuracies of 99.16% and 99.14% on MNIST, and 3.9%/3.9% and 0.2%/0.1% in Top-1/ Top-5 accuracy on IMAGENET.

2.3 Computational elements of hardware accelerators in deep neural networks

Computational elements of hardware accelerators for DNNs are responsible for the computation of dot product of pairs of vectors. Therefore, they are required to execute arithmetic operations such as multiplication and addition. Two representative computational elements of hardware accelerators are presented in this subsection. The multiply-and-accumulation (MAC) structure shown in Fig. 8A is adopted in multiple widely used methods. The MAC structure contains a binary multiplier-adder unit and an accumulator (ACC). It multiplies an input vector with a weight vector, and stores the product in a register to produce the partial sum (Psum) for the dot

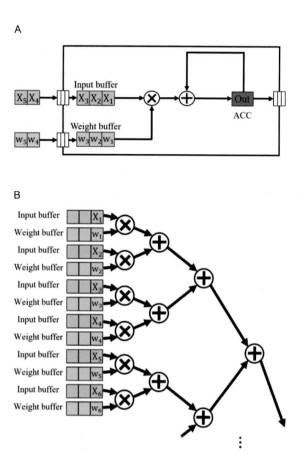

Fig. 8 Two representative computational elements for computing dot product in hardware accelerators. (A) MAC structure, and (B) adder-tree structure.

product computation. Subsequently, the Psum is accumulated recursively. Hardware accelerators presented in [37–43] employed a parallel MAC-architecture to increase throughput. Some hardware accelerators employ a modified version of the MAC structure, which receives Psums from an adjacent MAC structure and accumulates it [44–50].

The second representative computational element is an adder-tree structure, which is shown in Fig. 8B. Each multiplier in the adder-tree structure receives an input and a weight in parallel, and each binary adder in the tree structure adds one group comprising two multiplications. Some hardware accelerators with temporal architectures employ adder-tree -based computational elements [10,51,52].

The booth multiplier is now reviewed as a representative multiplier. We also describe the Wallace Tree adder, which aggregates partial products efficiently.

• *Booth multiplier*

An arithmetic binary multiplication of an M-bit multiplicand by an R-bit multiplier yields R partial products. These R partial products are then summed to calculate the actual product. Therefore, multiplication requires greater gate delay, larger circuit volume, and higher power consumption than addition, which makes the calculation of the inner product in neural processing expensive.

Booth recoding reduces the number of partial products to less than or equal to half of the original number [53–55]. It transforms arithmetic multiplication by an R-bit multiplier into the sum of $(R - 1)/2$ partial products in base-4, which can be expressed as follows.

$$A = \sum_{j=0}^{\frac{R-1}{2}} A_j 4^j \text{ with } \left(A_j \in -2, -1, 0, 1, 2\right) \tag{5}$$

For example, 5-bit arithmetic multiplication produces five partial products. On the other hand, the Booth multiplier theoretically produces only three partial products. In practice, however, the partial products can be negative, in which case the two's complement of the partial products need to be performed. The signed 2's complement addition requires sign-extended partial products, as depicted in Fig. 9. The reduction in the number of partial products by the Booth multiplier reduces the hardware resources required for the computation of partial products. Fig. 9A illustrates the extension of the sign bits of the partial products to the MSB bit of output obtained from the Booth multiplier. The partial product appearing in the first line is extended to 16 bits. Extension by a greater number of bits corresponds to a greater number of gates used in the implementation of the circuit. To reduce circuit complexity arising from sign extension, a simplified sign extension method was introduced [36,56], as depicted in Fig. 9B.

• *Wallace tree adder*

Wallace tree adder [57–62] is a circuit that sums partial products arranged in a trapezoidal format using a multi-operand adder, which generates partial products during the multiplication of pairs of binary numbers. The Wallace tree sums the partial products using $\log_{2/3}(2N)$ addition stages and reduces the number of partial products to two-thirds of the original

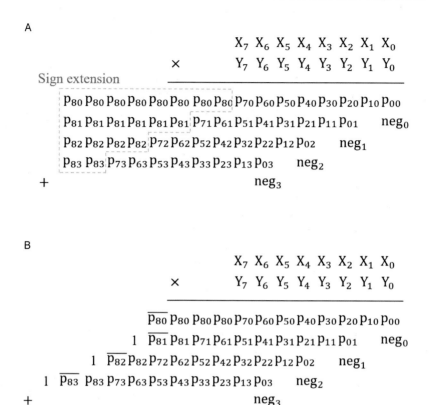

Fig. 9 Example of partial products generated in the process of multiplication in an 8 × 8 Booth multiplier via the (A) sign extension method (B) simplified sign extension method.

number per stage. Fig. 10 shows an example of a Wallace tree that aggregates eight partial products in parallel generated by multiplying two 8-bit binary integers. During each stage, full adders or half adders arranged in parallel receive three or two bits in the same column of the partial products, as depicted in Fig. 10B. The aggregated outputs obtained by the full adders and the half adders are transmitted to the same column of the following stage, and the carry outputs are transmitted to the upper part of the following stage, as shown in Fig. 9B. During the output of the final stage, the number of partial products is reduced to two, and the CLA adder finally sums these two vectors. The Wallace tree exhibits an $O(\log_{2/3}(2N))$ propagation delay. The Wallace tree multiplier exhibits advantages in terms of both power consumption and gate delay compared to the carry-save multiplier or the array multiplier [63].

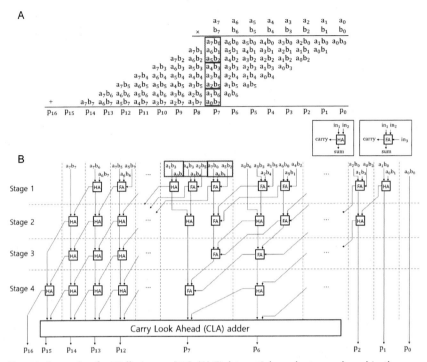

Fig. 10 Example of a Wallace tree [57]. (A) Eight partial products produced in the process of multiplying two 8-bit binary numbers, and (B) schematic view showing summation of the partial products of Wallace tree.

3. Hardware inference accelerators for deep neural networks

Representative existing hardware inference accelerators for DNNs with spatial architectures are reviewed in this section. In addition, hardware accelerators that adopt a shift-based multiplier instead of a Booth multiplier to reduce power consumption are also discussed. Since the 2010s, the primary hindrance to the application and commercialization of artificial neurons in embedded systems has been the high power consumption of constituent deep learning-based hardware accelerators that perform the computations required by artificial neurons. For example, thermal design power (TDP) of the commercially sold Graphics Processing Units (GPU),

Tesla P100, Intel Knights Landing, and GeForce GTX 1070, have been reported to be 300, 200, and 150 W, respectively [38,39,64]. To reduce the power consumption, researchers have attempted to reduce the resolutions of inputs, activations, and weights [12,37,44,65–73]. Lai et al. proposed an inference accelerator using 8-bit weights, which exhibited 36% storage reduction for weights and 50% power reduction in multipliers [11] compared to architectures based on single precision. Renzo et al. proposed and implemented an artificial neuron architecture called YodaNN [68] that computes the dot product between 1-bit weight vectors and 12-bit input vectors. In YodaNN, multiplication using 1-bit weight vectors allows the multiplier hardware to be implemented using only two's complement circuits and 2–1 Muxes, which significantly simplifies the structure of the multiplier. The power consumption of the manufactured chip of hardware inference accelerator in YodaNN was reported to be 895 μW with UMC 65-nm technology at 0.6 V. However, as mentioned in the previous chapter, CNNs with binary weights perform well in terms of inference on MNIST/CIFAR-10, but exhibit accuracies degraded by approximately 20% on IMAGENET.

3.1 Architectures of hardware accelerators

Architectures of hardware inference accelerators can be classified into two categories—temporal architectures, as depicted in Fig. 11A, and spatial architectures, as shown in Fig. 11B. GPUs [66–79] and hardware accelerators [9–10,51,68] typically comprise temporal architectures. In temporal architectures, all CEs receive data solely from registers and employ centralized control. Further, the external memory (e.g., DRAM) transmits data for inputs of CEs to the memory hierarchy considering filter size, sweep, and strides. Libraries such as cuBLAS [80] and cuDNN [74] are used for data transmission. These libraries transform 3-D convolutions of input feature maps and a filter into dot products of two 2-D matrices. This transformation increases the throughput. However, it simultaneously unfolds the elements overlapped during filter sweeping, and therefore, increases the memory access of external memories and the storage requirements of hardware accelerators, resulting in an overall increase in power consumption. On the other hand, CEs of the hardware accelerators proposed in [29,37,44–49,69] adopt spatial architectures to enhance data reuse. In a spatial architecture, a CE stores data such as Psums, weights, or activations

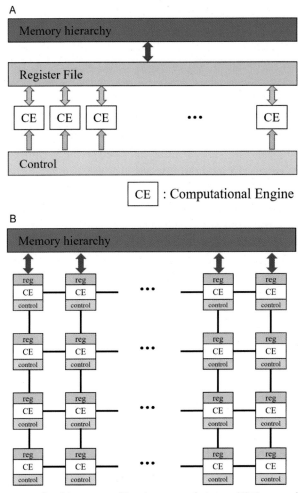

Fig. 11 Two types of architectures of hardware accelerators. (A) Temporal architecture, and (B) Spatial architecture.

and transmits them directly to adjacent CEs. For example, a spatial architecture with a stationary weight allows each CE to store the elements of weights and deliver the Psum to an adjacent CE. In addition, a spatial architecture with a stationary output allows each CE to hold Psum in a stationary fashion and deliver an element of the input feature map to an adjacent CE. Direct data transmission between CEs helps to reduce external memory access.

3.2 Eyeriss: hardware accelerator using a spatial architecture

The normalized energy cost required by computational elements to access the DRAM is approximately 33 times greater than that required to access buffers within an accelerator. In turn, the latter is approximately 3 times greater than that required to access an adjacent computational element. Therefore, it is important to reuse data that has been transferred from the DRAM to reduce power consumption. The Eyeriss Hardware Accelerator [37] is based on this principle and its operation has been depicted in Fig. 12, alongside the processing steps of 1–D convolution. Buffers arranged in a MAC structure store the weights of a row of filters and several inputs of the input feature map for data reuse. On each cycle, one input in the buffer is updated using the input to be computed in the following cycle. The MAC structure in a MAC array transmits the Psum to an adjacent MAC. However, the MAC structure in Eyeriss requires buffer registers to store the input data for data communication with other MACs. Eyeriss transmits inputs diagonally; weights, horizontally; and partial sums, vertically, as illustrated in Fig. 13. Data transmission between adjacent MACs allows data reuse.

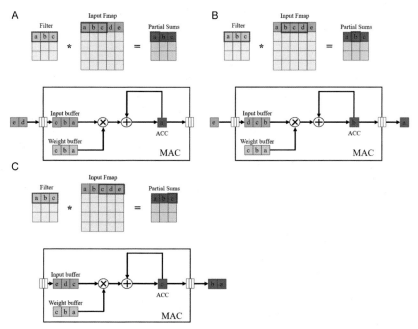

Fig. 12 Operation of Eyeriss [37]. (A) 1-D convolution step 1, (B) 1-D convolution step 2, (C) 1-D convolution step 3.

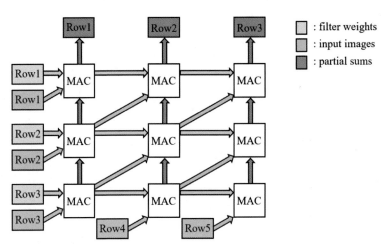

Fig. 13 Example sequence of Eyeriss for computing 3×3 convolution with 3×3 MAC array [37].

3.3 UNPU and BIT FUSION: hardware accelerators using shift-based multiplier

To reduce circuit areas of multipliers, UNPU [81] and BIT FUSION [82] adopted shift circuits instead of traditional multipliers to perform arithmetic multiplication. UNPU produces 16 partial products with 16 sets comprising AND gates, binary adders, and 1-bit shift circuits, as depicted in Fig. 14A. It can support multiplication corresponding to weights of bit-widths varying from 1 bit to 16 bits. The BIT FUSION scheme [82] decomposes multiplication operations into smaller ones, e.g., it decomposes 4-bit multiplication into four 2-bit multiplications, as shown in Fig. 14B. By adjusting the combinations of the component multiplications, it can also support weights of varying bit-widths.

3.4 Digital neuron: a multiplier-less massive parallel processor

An artificial neuron in the Integrate-and-Fire model [83,84] performs a convolution by calculating the inner product between pairs of input and weight vectors.

$$
\begin{aligned}
\mathrm{Out}[x][y] &= \mathbf{F}\left(b + \sum_{i=0}^{R-1} \sum_{j=0}^{S-1} \mathbf{X}[x+i][y+j] \cdot \mathbf{w}[x+i][y+j] \right) \\
&= \mathbf{F}\left(\vec{\mathbf{X}} \cdot \vec{\mathbf{w}}^{\mathrm{T}} + b \right)
\end{aligned}
\tag{6}
$$

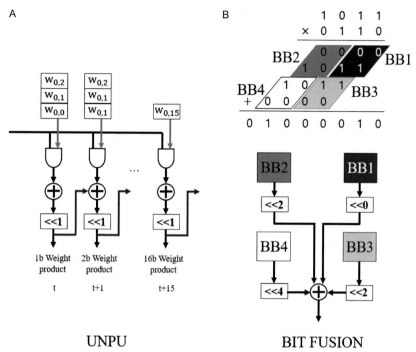

Fig. 14 Computation engine with shift-based multiplier in prior works in (A) UNPU [81], and (B) BIT FUSION [82].

where $\mathbf{X}$, $\mathbf{w}$, b, $\mathbf{F}$ and Out denote an input vector, a weight vector, a bias, an activation function, and an output, respectively.

The architecture of a digital neuron is described in this section, which operates by calculating the dot products between 8-bit integer inputs and two-sub-integer-based 5-bit integer weights [85]. It consists of multipliers comprising two-barrel shift circuits and multi-operand adders, and employs 5-bit weights that are produced by adding two 2-base partial sub-integers, as indicated by Eq. (4). The proposed digital neuron is an artificial neuron that digitally computes the dot products between pairs of vectors. Fig. 15 depicts a block diagram of the proposed digital neuron. It performs convolutions using a three-dimensional filter, in which the number of the input channels is denoted by N_ch and the number of the filtered inputs in each channel is denoted by N.

The digital neuron includes a neural element, which calculates convolutions using a two-dimensional filter with N inputs and N weights

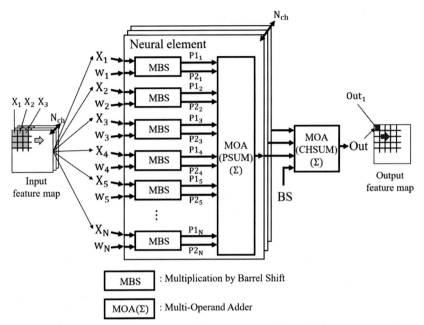

Fig. 15 Block diagram of the proposed digital neuron. The digital neuron calculates convolution using a three-dimensional filter.

corresponding to each channel. Therefore, the number of the neural elements in the digital neuron is equal to the number of input channels. The neural element consists of a multiplication by barrel shift (MBS) block and a multi-operand adder (MOA) block. Each MBS block contains two-barrel shift circuits, which produce two partial sub-integers, P1 and P2, denoted by $w_1 2^i$ and $w_2 2^j$ in Eq. (4), respectively.

The MOA (PSUM) block collects a total of 2N outputs from the MBS blocks, and aggregates them. Finally, the outputs obtained from all neural elements and the bias, BS, are collected and summed by the MOA (CHSUM) block to yield the final dot product of the two vectors.

Fig. 16 shows a block diagram of the proposed digital neuron with 8-bit integer inputs, X[7:0], and 5-bit integer weights, w[4:0], including blocks to calculate two's complements and zeroes to produce negative and zero partial sub-integers, respectively. Each MBS block includes two-barrel shifter blocks for arithmetic shift, a one's complement block for required two's complement inversions, and a Zero block to convert the bits of a partial product to zero. Control signals SH1/SH2, ALZ1/ALZ2, and INV1/INV2 are produced based on look-up table circuits.

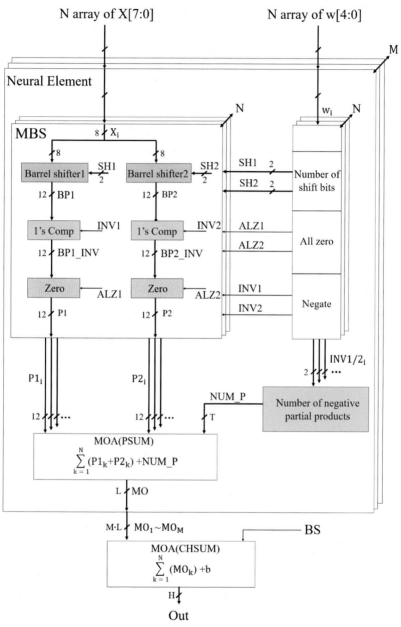

Fig. 16 Block diagram of the proposed digital neuron including blocks for two's complements and zeroing for producing negative or zero partial products.

To represent negative partial sub-integers, two's complement circuits are required. To this end, 1 is added to BP1_INV or BP2_INV if P1 or P2 is required to be negative, respectively. This addition can be realized using a binary adder such as a ripple carry adder or a Carry Look Ahead (CLA) adder. However, each neural element then requires 2N additional binary adders, which drastically increases circuit area and power consumption. The number of 1s added, and therefore the sum of all the 1's, is equal to the number of negative partial products appearing in M neural elements. Therefore, the MOA (SUM1s) block may be appended to the structure to calculate the number of negative partial products by summing INV1 and INV2 signals in parallel.

The proposed MOA (PSUM) block arithmetically sums $N + 1$ inputs, which comprise N 12-bit P1 signals, N 12-bit P2 signals, and the NUM_P signal. Fig. 17 depicts a schematic diagram of the proposed MOA (PSUM)

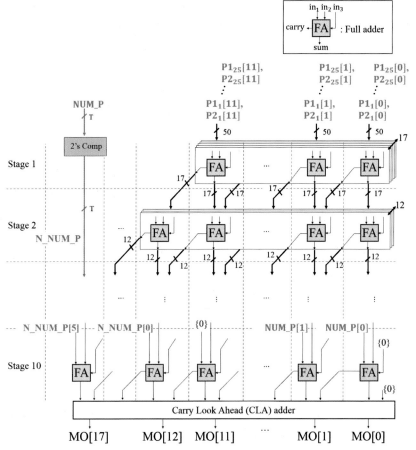

Fig. 17 Schematic view of the proposed MOA (PSUM) block, where N is 25.

block, where $N = 25$. The operational principle of the MOA block is similar to that of the Wallace tree adder [57]. It groups together three bits belonging to the same column of P1 and P2 vectors along the vertical direction. Following that, the number of input vectors at each stage is reduced by two-thirds using full adder arrays and their bit-widths are simultaneously expanded by one bit. By the output phase of stage 9, the number of P1 and P2 vectors is reduced to two. The NUM_P vector is delayed compared to P1 and P2. Therefore, it is added to the two output vectors of stage 9 during the final stage.

To enable the summation of numbers with different signs, a sign-extension of the MOA inputs to 18 bits is required, to correspond to the bit-width of MOA outputs. However, sign-extension increases circuit area by approximately 21%. As an alternative, the proposed MOA adds the negative of the NUM_P vector to convert the 12-bit value to an 18-bit value during Stage 10.

Fig. 18 illustrates the principle of the proposed sign extension procedure. Summation of bits with extended signs is equivalent to performing the summation by replacing the bits with negative operands by their two's complements. Therefore, sign extension can be replaced by addition of altered binary number. This procedure only requires the implementation of a two's complement circuit, and enables the summation of numbers with positive or negative signs. The proposed MOA block reduces the numbers of critical paths and total gates by 42% and 36%, respectively, compared to the conventional adder-tree structures with Booth multipliers [10].

The proposed digital neuron operates solely based on combinational logic. Therefore, it requires no internal storage, and, thus, does not consume any power for memory access of intermediate data.

To compare the performances of the aforementioned multipliers, we simulate three types of dot product-computing artificial neurons—adder

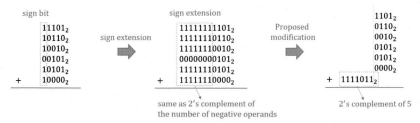

Fig. 18 Example of summation of six 5-bit binary numbers. This example describes the equivalence of the summation of extended sign bits with the two's complement of the number of negative operands.

Table 2 Comparison of performances of artificial neurons in existing systems during the computation of dot product of two vectors with 50 components.

Weight precision	Adder tree with booth multiplier ($N = 50$) [10]	YodaNN ($N = 50$) [79]	Proposed digital neuron ($N = 25$, $N_{ch} = 2$)	
			5-bit (two partial sub-integers)	8-bit (three partial sub-integers)
Weight precision	5-bit (exact)	1-bit	5-bit (two partial sub-integers)	8-bit (three partial sub-integers)
Activation precision	8 bit	8 bit	8 bit	8 bit
Critical path	57	40	41	46
Simulated power consumption	19.758 mW	5.808 mW	10.305 mW	15.810 mW

tree with Booth multiplier [10], YodaNN (BinaryConnect) [68], and the digital neuron proposed in this work. Table 2 summarizes the comparison. The difference between the number of critical paths in the 5-bit variant of the proposed digital neuron and that in YodaNN is only 1, which is insignificant. The difference between the numbers of critical paths in the 5-bit and 8-bit versions of the proposed digital neuron is only 5, even though the precision increases by 3 bits from the first to the second. The 5-bit variant of the proposed digital neuron has 16 more critical paths than the adder tree with Booth multiplier.

To simulate the power consumption of the multipliers, a 65 nm CMOS process is selected on HSPICE at an operational voltage of 1 V. The simulated power consumption of the 5-bit variant of the digital neuron is observed to be approximately 1.77 times greater than that of YodaNN, and its precision is five times higher. Further, the simulated power consumption of the 8-bit variant of the Digital Neuron is approximately 2.72 times greater than that of YodaNN, and its precision is eight times higher. The power consumption of the adder tree is observed to be approximately 191% of that of the Digital Neuron.

The number of critical paths and simulated power consumption of the 5-bit variant of the Digital Neuron are 41 gates and 10.305 mW, respectively, whereas those of the 8-bit variant are 46 gates and 15.810 mW. The proposed scheme is confirmed to improve the number of critical paths and power consumption compared to artificial neurons based on adder trees with Booth multipliers.

3.5 Power saving strategies for hardware accelerators

The power consumed to transfer data from a DRAM to an artificial neuron and that consumed to transfer data from a buffer to an artificial neuron have been reported to be approximately 200 times and 6 times that required to transfer data from a resistor to an artificial neuron [37]. This exemplifies the importance of reducing data transfer from external memories to registers to reduce power consumption.

Therefore, hardware inference accelerators [9,10,37,51,72], GPUs [78,79], and TPUs [38,39] store data in intermediate buffers from external memory, and then transmit them to artificial neurons. In the aforementioned architectures, data containing the same information is transferred from external memories to buffers, or between buffers, multiple times. Evidently, this increases the power consumption.

Three power-saving strategies for hardware accelerators are described in this section to enable accelerators to adhere to the power budgets of the embedded systems.

The first strategy is to reduce the precisions of constituent weights and activations. Most CPUs and GPUs employ FP32 precision for close-to-ideal accuracy. Certain studies have demonstrated that even if weights and activations are quantized in neural networks using a lower number of bits than FP32, the inference accuracy is not degraded [12,13,29,30]. In addition, [86] proposed a flexpoint format, which adopts a 16-bit integer format with a shared 5-bit exponent. Although the storage requirements of the shared bits are negligible, this format increases dynamic range. The SiMul scheme [87] controls the number of accumulated partial products required during multiplication. However, it increases the number of cycles required during multiplication by increasing the number of partial products. Reducing the precision appropriately helps to reduce circuit complexity of computational engines and storage requirements compared to that utilizing FP32. However, excessive precision reduction causes degradation of inference accuracy. Thus, optimal precision is required to maintain an inference accuracy within the designed range.

The second strategy is to minimize external memory accesses. DRAM access by a computational engine has been reported to consume 200 times more power than register access by a computational engine [37]. Certain hardware accelerator architectures do not reuse data in the output feature map [9,10,37,52,88,89], as shown in Fig. 19A, which necessitates external memory access to load the input feature map and store output feature maps corresponding to every layer. On the other hand, TPUs reuse data of the

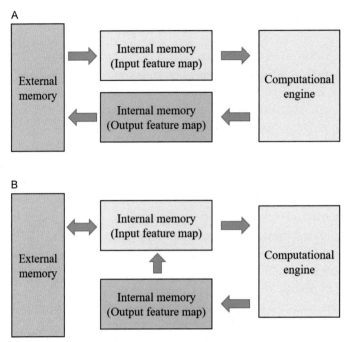

Fig. 19 (A) Output feature map data not reused, and (B) output feature map data reused.

output feature map to minimize external memory access [38,39], as depicted in Fig. 19B. The proposed inference accelerator also employs an architecture for data reuse of the output feature map.

The third strategy is the application of zero-aware architecture [65,90,91] that does not compute MAC operations when the weight or an activation is zero. In AlexNet, the ratio of zero weights and zero activations between layer 2 to layer 5 is 62%–65% and 50%–76% [31], respectively. The computational savings using zero-skipping computations can be further accelerated via the pruning technique [92,93], in which a neural network is trained to prune any connection with a near-zero weight.

4. Hardware inference accelerators using digital neurons

In this chapter, a system architecture using digital neurons [85] is proposed, which transfers data from external memories to registers only once and reuses the data afterward. Further, a configurable architecture to adjust the filter size of digital neurons is also proposed.

The power consumed to transfer data from a DRAM and from a buffer to an artificial neuron have been reported to be approximately 200 times and 6 times, respectively, that required to transfer data from a resistor to an artificial neuron [37]. Thus, to reduce power consumption, it is important to reduce data transfer from external memories to registers. Therefore, hardware inference accelerators [9,10,37,51,72], GPUs [78,79], and TPUs [38,39] store data in intermediate buffers from external memories, and then transmit them to artificial neurons. In the aforementioned architectures, data containing the same information is transferred from external memories to buffers, or between buffers, multiple times. This radically increases power consumption.

4.1 System architecture

Fig. 20 shows a system architecture based on digital neurons. NT1, NT2, NT3, and NT4 compute the 3-D convolutions for the convolutional layer, and NT_FCs compute the dot products between pairs of vectors for the fully connected layer. Further, the neural tile (NT) comprises eight neural elements. Fig. 21 depicts a block diagram of the NT. The 8 constituent neural elements compute the dot products of 25 components in parallel, and their outputs are subsequently aggregated by MOA block. The outputs of MOA block are stored within registers. The outputs of registers are then transmitted back to the MOA block for accumulation. A mux in the NT determines whether the outputs are to be accumulated.

Fig. 22 depicts a timing diagram of the system presented in Fig. 20. As data loaded from an external memory is generally 32-bit, four input pixels, four biases, or six weights are loaded into one clock. All image pixels are initially loaded onto the **X** bank from the external memory. Then, the weights and biases of all filters belonging to layer 1 are loaded onto the **w** bank (conv). Following the completion of the transfer of biases corresponding to layer 1, the weights of all filters belonging to layer 2 are stored within the **w** bank (conv). One clock cycle after the completion of the transfer of biases corresponding to layer 1, outputs of registers in NTs are stored within the output feature map. Once all the weights have been successfully stored within the w bank (conv) and **w** bank (FC), they no longer need to be loaded from the external memory for the inference of other images. Further, once all weights and biases corresponding to all layers are loaded within the **w** bank, during the inference of each image, the following one is loaded onto the X bank.

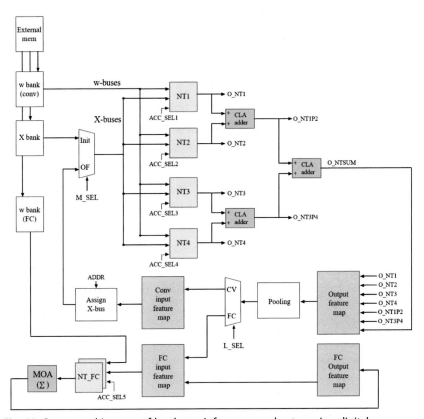

Fig. 20 System architecture of hardware inference accelerator using digital neurons.

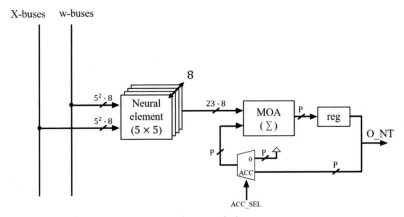

Fig. 21 Neural tile (NT) comprising eight neural elements.

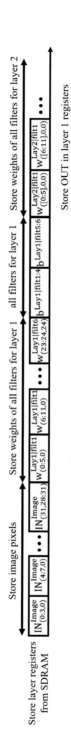

Fig. 22 Timing diagram of the system architecture.

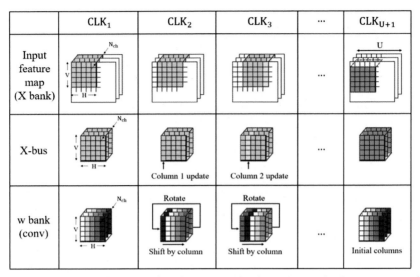

Fig. 23 Data transfer from input feature map or **X** bank to X-bus, and column rotation of **w** bank.

Elements of filtered region that require the computation of dot products is assigned to X-bus from the input feature map or the **X** bank, as shown in Fig. 23. Corresponding to each unit increase in a clock, the filtered region within the input feature map is translated to the right by one column. This is implemented by updating a column in the X-bus, deleting the leftmost column in the input feature map in the previous clock, and replacing it by loading the rightmost column of the following clock. This final step can be performed by increasing the address of the column by 5. Columns that are not updated do not consume power. Therefore, this process helps to reduce the overall power consumption. The **w** bank sequentially rotates to the right following the sweep of the filter. This is implemented using shift registers, by connecting them horizontally and connecting the rightmost columns to the leftmost columns. At the beginning of the next row of the filter sweep, the entire filtered area of the input feature map is assigned to the X-bus, and the w-bus is updated using the initial columns of CLK1.

Four outputs obtained from the NTs—O_NT1, O_NT2, O_NT3, and O_NT4—are added by the two CLA adders. The outputs of the two CLA adders—O_NT1P2 and O_NT3P4—are added by the other CLA adder, yielding O_NTSUM. Following activation, O_NT1, O_NT2, O_NT3, O_NT4, O_NT1P2, O_NT3P4, and O_NTSUM are all truncated to 8 bits and stored in the output feature map. The hardware implementation of the activation ReLU and truncation is simple, and, therefore, they are omitted from Fig. 20.

Examples of the utilization of these signals have been illustrated in Fig. 24. Case 1 depicts an application of a $5 \times 5 \times 32$ filter, in which each of the four NTs receives a quarter of the depth-wise divided filter and calculates the dot product of that quarter. Then, the outputs of the four NTs are summed together by the three CLA adders. In this case, the output feature map collects four O_NTSUMs and transmits them to the Pooling block.

Case 2 depicts an application of a $7 \times 7 \times 16$ filter, in which NT1 and NT2 receive the half and the remainder of the 7×7 filter, which are $[0:24] \times [0:7]$ and $[0:48] \times [0:7]$. Similarly, NT3 and NT4 receive the half and the remainder of the 7×7 filter, which are $[0:24] \times [8:15]$ and $[0:48] \times [8:15]$. The four outputs obtained from the four NTs are summed by the three CLA adders. As in the previous case, the output feature map collects four O_NTSUMs and transmits them to the Pooling block.

Case 4 depicts an application of a $7 \times 7 \times 8$ filter, in which two groups of two NTs separately receive two $7 \times 7 \times 8$ regions in the input feature map. The outputs obtained from NT1 and NT2 and those obtained from NT3 and NT4 are separately added by CLA adders. These two outputs— O_NT1P2 and O_NT3P4—are stored in the output feature map. After collecting four O_NT1P2/O_NT3P4s, the output feature map transmits them to the Pooling layer.

Case 4 depicts an application of a $7 \times 7 \times 8$ filter, in which two groups of two NTs separately receive two $7 \times 7 \times 8$ regions in the input feature map. The outputs obtained from NT1 and NT2 and those obtained from NT3 and NT4 are separately added by CLA adders. These two outputs— O_NT1P2 and O_NT3P4—are stored in the output feature map. After collecting four O_NT1P2/O_NT3P4s, the output feature map transmits them to the Pooling layer.

Case 6 depicts an application of a $5 \times 5 \times 128$ filter, in which the depth of the filter exceeds the combined depth of the four NTs. In this case, the ACC_SEL signal is used to accumulate the outputs obtained from the NTs. The depth of the filter is 128. Therefore, the outputs are accumulated four times in total. However, in this case, all signals of the X-bus and the w-bus need to be appropriately updated. Thus, although the calculation is performed, this case consumes additional power.

The output obtained from the Pooling block is transmitted to the demux, as depicted in Fig. 20. If the following layer is a convolutional layer,

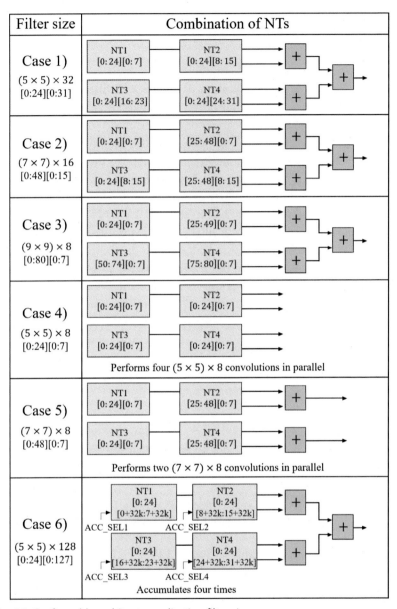

Fig. 24 Configurable architecture adjusting filter size.

the demux transmits the output to the Conv input feature map. On the other hand, if the following layer is a fully connected layer, the demux transmits the output to the FC input feature map. Finally, the Assign X-bus block transforms the filtered area of the input feature map into the X-bus, as shown in Fig. 23.

The dot product computation required in a fully connected layer is performed by the NT_FC. If the number of neurons of a fully connected layer exceeds the capacity of the NT_FC, the ACC_SEL5 signal is used to accumulate the outputs obtained from the NT_FC. The outputs are then aggregated by the MOA block, and transmitted to the FC output feature map.

4.2 Implementation and experimental results

Fig. 25 illustrates a block diagram of a hardware inference accelerator implementing LeNet-5 on a DE-115 FPGA board. The pre-trained weights, biases, and input images are transmitted to the FPGA board via a USB connection by the Eclipse software using the Universal Asynchronous Receiver/ Transmitter (UART) protocol. 5-bit weights, partitioned to two sub-integers, are taken. This allows the consideration of 5-bit numbers between -16 and 15 except ± 11 and ± 13, as described in Section 3. The Advanced Reduced Instruction Set Computer (RISC) Machine (ARM) processor is used to transmit the training data to the on-chip memory.

A convolutional layer of the implemented system architecture has been illustrated in Fig. 26. The scale-down node is implemented by truncating the rightmost bits. For instance, a scale-down by $\frac{1}{16}$ is implemented by truncating the rightmost 4-bits. The Activation node is implemented using ReLU equipped with an upper limit. As the precision of the activation is taken to be 8-bit, the upper limit is set to 255 (11111111_2).

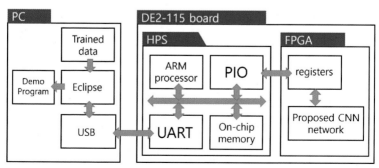

Fig. 25 Implementation of digital neuron hardware accelerator on DE-115 FPGA board.

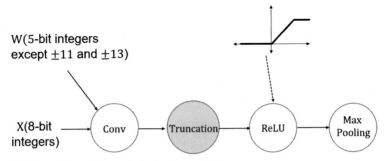

Fig. 26 Convolutional layer of the implementation.

Table 3 Performance of system architecture on handwritten MNIST number recognition by the proposed implementation, operating LeNet-5 [16].

Performance of the system architecture implemented in DE2-115 board	
Inference accuracy of MNIST database	98.95%
Number of clocks required in computing dot product of NTs (25 MHz)	1 clock
Total gates	201,796
Number of clocks/computation time (25 MHz) for inference of one image	2384 clocks/95 µs
Operations per Second (OPS)	40 GOPS
Simulated Max power consumption	121 mW
OPS/W	330 OPS/W

Table 3 summarizes the performance of the proposed hardware inference accelerator. It is observed to exhibit an inference accuracy of 98.95% on the MNIST database. A total of 201,796 gates is used to implement the LeNet-5 network on the DE2 FPGA Board. The inference of one image is observed to require 2384 clocks cycles, and the corresponding temporal delay is 95 µs on a 25-MHz clock.

The proposed hardware inference accelerator avoids accessing the external memory for storing intermediate feature maps in registers. This helps to reduce its power consumption.

Fig. 27 shows that an increase in the dimension of the neural elements increases the throughput of the entire system. If one neural element is assumed to be capable of computing two-dimensional convolutions corresponding to a

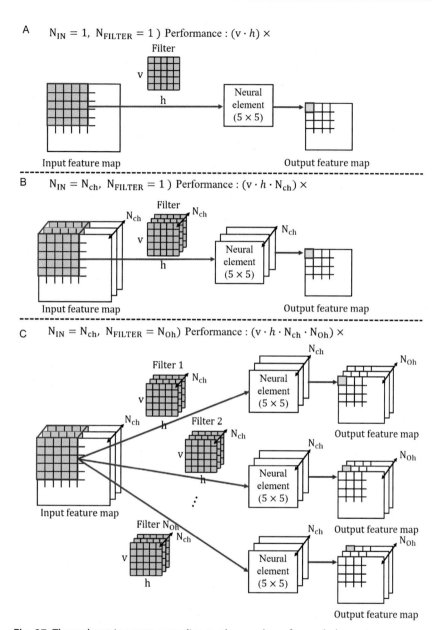

Fig. 27 Throughput increase according to the number of neural elements.

filter area of $\mathbf{v} \cdot \mathbf{h}$ filter area, its performance is given by $(\mathbf{v} \cdot \mathbf{h})\times$. $\mathbf{N_{ch}}$ neural elements are capable of computing three-dimensional convolutions of $\mathbf{v} \cdot \mathbf{h} \cdot \mathbf{N_{ch}}$ in parallel. Thus, their performance is given by $(\mathbf{v} \cdot \mathbf{h} \cdot \mathbf{N_{ch}})\times$. If the number of a filter is $\mathbf{N_{oh}}$, $\mathbf{N_{ch}} \cdot \mathbf{N_{oh}}$ neural element bundles are capable of computing convolutions of $\mathbf{v} \cdot \mathbf{h} \cdot \mathbf{N_{ch}} \cdot \mathbf{N_{oh}}$ elements, and their performance, in this case, is given by $(\mathbf{v} \cdot \mathbf{h} \cdot \mathbf{N_{ch}} \cdot \mathbf{N_{oh}})\times$. In the final case, it is not necessary to update the filter during the computation of a convolutional layer. This helps to reduce the overall power consumption.

5. Summary

The implementation of hardware accelerators for the purpose of inference has been widely studied [8–10,37,69,72,73] and a commercial hardware accelerator for inference has been developed for embedded systems [114], which exhibited lower power consumption compared to conventional GPUs. However, till date, frame rates and resolutions of input images during real-time inference have remained limited by the unsatisfactory performance of inference accelerators. The biggest constraint on their performance is the limited bandwidth of the memory. For example, although Dynamic Random Access Memory (DRAM) is capable of transmitting only 32-bit data to the hardware accelerator every clock, realization of fully random access should be possible. Therefore, this limited nature of the memory bandwidth should be investigated in future works. However, in deep learning, the order data import from memory is pseudo-determined. Therefore, radical development in deep learning is necessary to produce advanced hardware accelerators that do are not constrained by the limited bandwidth of memory.

All layers of the proposed inference accelerator share a single digital neuron due to the limited logic capacity of the DE2-115 FPGA board used for the implementation. In the future, layers should be structured to use independent digital neurons and pipelined architecture should be implemented to improve the frame rates and resolutions of inputs, using advanced FPGA with high logic capacity.

A digital neuron was used in the proposed accelerator to partition weights into limited numbers of 2-base partial sub-integers. The limitations imposed by the use of partial sub-integers reduced the circuit complexity and, in turn, the overall power consumption. As evidenced by a simulation, the power consumed by the digital neuron was lower than that consumed by a conventional artificial neuron.

We executed the inference phase using the digital neuron in a python simulation with 8-bit activations and 5-bit integer weights with two partial sub-integers, and 8-bit activations and 8-bit integer weights with three partial sub-integers. Training was performed via gradient descent with FP32 precision. The 5-bit-weight variant exhibited an inference accuracy of 99.15% on the MNIST database, surpassing that exhibited by the FP32 variant. The 5-bit- and 8-bit-weight variants exhibited additional degradations of 3.9%/3.9% and only 0.2%/0.1% in Top-1/Top-5 accuracy, respectively, on the IMAGENET database, compared to those exhibited by the FP32 variant.

The inference accuracy of the implemented accelerator was observed to be 98.95%, and each inference required 2384 clock cycles. The simulated power consumption of the implementation was 121 mW, which allowed it to be utilized in embedded systems with power budgets lower than 1 W.

Acknowledgments

This work was supported by the Institute of Information & Communications Technology Planning & Evaluation (IITP) grant funded by the Korea government (MSIT) (No. 2020-0-00056, To create AI systems that act appropriately and effectively in novel situations that occur in open worlds), and partially by (No. 2017-0-00244, HMD Facial Expression Recognition Sensor and Cyber-interaction Interface Technology).

Key terminology and definitions

Hardware accelerator A hardware accelerator is a hardware system specially constructed to perform certain functions more efficiently than possible on software running on a general-purpose processing unit, such as CPU or GPU. An Artificial Intelligence accelerator is a type of specialized hardware accelerator designed to accelerate Artificial Intelligence-based applications, especially deep neural networks and machine learning.

Quantization Quantization is the process of constraining data representation from that based on real numbers to that based on a discrete set such as the integers. In the context of deep learning, quantization denotes the process of approximating the variables or parameters of a neural network represented by floating-point numbers using low bit-width binary numbers or integers.

Digital neurons A digital neuron is a hardware inference accelerator for convolutional deep neural networks with integer inputs and integer weights for embedded systems proposed by H. Park and S. Kim. The fundamental idea is the reduction of circuit area and power consumption by manipulating the dot products between input features and weight vectors using barrel shifters and parallel adders. This reduction allows a greater number of computational engines to be mounted on the inference accelerator, resulting in higher throughput compared to existing HW accelerators.

Inference accelerator An inference accelerator is an optimized hardware accelerator for inferencing via deep neural networks and other performance-critical AI functions.

References

[1] C.R. Qi, et al., Volumetric and multi-view CNNs for object classification on 3D data, in: Proc. CVPR 2016, Las Vegas, NV, USA, 2016, pp. 5648–5656.

[2] W. Ouyang, et al., DeepID-Net: deformable deep convolutional neural networks for object detection, in: Proc. CVPR 2015, Boston, MA, USA, 2015, pp. 2403–2412.

[3] D. Amodei, et al., End-to-end speech recognition in english and mandarin, in: Proc. 33rd International Conference on Machine Learning 2016, New York, NY, USA, 2016, pp. 173–182.

[4] J. Hirschberg, C.D. Manning, Advances in natural language processing, Science 349 (6245) (2015) 261–266.

[5] Y. Li, et al., Generative face completion, in: Proc. CVPR 2017, Honolulu, Hawaii, USA, 2017, pp. 1–9.

[6] R. Kumar, et al., ObamaNet: photo-realistic lip-sync from text, in: Proc. NIPS 2017, Long Beach, CA, USA, 2017, pp. 1–4.

[7] Tractica, Deep Learning Intelligence for Enterprise Applications Report, 2018.

[8] J. Kim, B. Grady, R. Lian, J. Brothers, J.H. Anderson, FPGA-based CNN inference accelerator synthesized from multi-threaded C software, in: Proc. 30th IEEE International System-on-Chip Conference (SOCC) 2017, Munich, Germany, 2017, pp. 268–273.

[9] M. Peemen, A.A. Setio, B. Mesman, H. Corporaal, Memory-centric accelerator design for convolutional neural networks, in: Proc. 2013 IEEE 31th International Conference on Computer Design (ICCD), Asheville, NC, USA, 2013, pp. 13–19.

[10] C. Zhang, et al., Optimizing FPGA-based accelerator design for deep convolutional neural networks, in: Proc. 2015 ACM/SIGDA International Symposium on Field-Programmable Gate Arrays, Monterey, CA, USA, 2015, pp. 161–170.

[11] L. Lai, N. Suda, V. Chandra, Deep convolutional neural network inference with floating-point weights and fixed-point activations, in: Proc. 34rd International Conference on Machine Learning (ICML 2017), Sydney, Australia, 2017.

[12] P. Judd, et al., Reduced-precision strategies for bounded memory in deep neural nets, in: Proc. 6th International Conference on Learning Representations (ICLR 2016), San Juan, Puerto Rico, 2016.

[13] D.D. Lin, S.S. Talathi, V.S. Annapureddy, Fixed point quantization of deep convolutional networks, in: Proc. 33rd International Conference on Machine Learning (ICML 2016), New York, NY, USA, 2016, pp. 2849–2858.

[14] M. Courbariaux, Y. Bengio, J.-P. David, BinaryConnect: training deep neural networks with binary weights during propagations, in: Proc. Advances in Neural Information Processing Systems (NIPS 2015), Montreal, Montreal, Canada, 2015.

[15] M. Courbariaux, I. Hubara, D. Soudry, R. Yaniv, and Y. Bengio, (2016), Binarized Neural Networks: Training Deep Neural Networks With Weights and Activations constrained to +1 and −1, arXiv: 1602.02830.

[16] Y. LeCun, L. Bottou, Y. Bengio, P. Haffner, Gradient-based learning applied to document recognition, Proc. IEEE 86 (11) (1998) 2278–2324.

[17] F. Rosenblatt, The perceptron: a probabilistic model for information storage and organization in the brain, Psychol. Rev. 65 (6) (1958) 386–408.

[18] M. Minsky, S. Papert, Perceptrons: An Introduction to Computational Geometry, MIT Press, Cambridge, 1969.

[19] A. L. Katole, et al., (2015), Hierarchical Deep Learning Architecture for 10K Objects Classifications, arXiv: 1509.01951.

[20] IEEE Standard for Floating-Point Arithmetic (IEEE 754-2008), 2008.

[21] N. Srivastava, et al., Dropout: a simple way to prevent neural networks from overfitting, J. Mach. Learn. Res. 15 (1) (2014) 1929–1958.

[22] J. Nagi, et al., Max-pooling convolutional neural networks for vision-based hand gesture recognition, in: Proc. Signal and Image Processing Applications (ICSIPA 2011), Kuala Lumpur, Malaysia, 2011, pp. 342–347.

[23] Y. Kim, Convolutional Neural Networks for Sentence Classification, 2014. srXiv: 1408.5882v2.

[24] T. Sainath, A. Mohamed, B. Kingsbury, B. Ramabhadran, Deep convolutional neural networks for LVCSR, in: Proc. Acoustics, Speech, and Signal Processing (ICASSP 2013), Vancouver, Canada, 2013, pp. 8614–8618.

[25] P. Sermanet, S. Chintala, Y. LeCun, Convolutional neural networks applied to house numbers digit classification, in: Proc. Pattern Recognition (ICPR 2012), Tsukuba, Japan, 2012, pp. 3288–3291.

[26] G. Dahl, T. Sainath, G. Hinton, Improving deep neural networks for LVCSR using rectified linear units and dropout, in: Proc. Acoustics, Speech, and Signal Processing (ICASSP 2013), Vancouver, Canada, 2013, pp. 8609–8613.

[27] G. Levi, T. Hassner, Age and gender classification using convolutional neural networks, in: Proc. Computer Vision and Pattern Recognition (CVPR 2015), Boston, USA, 2015, pp. 34–42.

[28] M. Zeng, et al., Convolutional neural networks for human activity recognition using mobile sensors, in: Proc. Mobile Computing, Applications and Services, Austin, TX, USA, 2014, pp. 197–205.

[29] S. Gupta, A. Agrawal, K. Gopalakrishnan, Deep learning with limited numerical precision, in: Proc. 32nd International Conference on Machine Learning (ICML 2015), Lille, France, 2015.

[30] P.M. Gysel, Ristretto: Hardware-Oriented Approximation of Convolutional Neural Networks, M.S. thesis, Dept. Electrical and Computer Engineering, University of California, CA, USA, 2016.

[31] A. Krizhevsky, I. Sutskever, G.E. Hinton, ImageNet classification with deep convolutional neural networks, in: Proc. 2012 Advances in Neural Information Processing Systems (NIPS 2012), Lake Tahoe, NV, USA, 2012, pp. 1097–1105.

[32] F. Iandola et al., February 2016. SqueezeNet: AlexNet-Level Accuracy With 50x Fewer Parameters and <0.5 MB Model Size, arXiv: 1602.07360 (accessed October 13, 2020).

[33] C. Szegedy, et al., Going deeper with convolutions, in: Proc. Computer Vision and Pattern Recognition (CVPR 2015), Boston, MA, USA, 2015, pp. 1–9.

[34] K. Simonyan, A. Zisserman, September 2014, Very Deep Convolutional Networks for Large-Scale Image Recognition, arXiv: 1409.1556 (accessed October 13, 2020).

[35] V. Sze, Y. Chen, T. Yang, and J. Emer, March 2017. Efficient Processing of Deep Neural Networks: A Tutorial and Survey, arXiv: 1845160 (accessed October 13, 2020).

[36] R. Hussin, et al., An efficient modified booth multiplier architecture, in: Proc. Electronic Design, Penang, Malaysia, 2008.

[37] Y. Chen, T. Krishna, J. Emer, V. Sze, Eyeriss: an energy efficient reconfigurable accelerator for deep convolutional neural networks, in: Proc. IEEE International Solid-State Circuits Conference (ISSCC 2016), San Francisco, USA, 2016, pp. 262–263.

[38] N. Jouppi, et al., In-datacenter performance analysis of a tensor processing unit, in: Proc. International Symposium on computer Architecture (ISCA 2017), Toronto, ON, Canada, 2017.

[39] E. Olsen, June 2017, Proposal for a High Precision Tensor Processing Unit, arXiv: 1706.03251 (accessed October 13, 2020).

[40] T. Nowatzki, et al., Pushing the limits of accelerator efficiency while retaining programmability, in: Proc. IEEE International Symposium on High Performance Computer Architecture (HPCA 2016), Barcelona, Spain, 2016.

[41] Y. Shen, M. Ferdman, P. Milder, Maximizing CNN accelerator efficiency through resource partitioning, in: Proc. 2017 International Symposium on Computer Architecture (ISCA 2017), Toronto, ON, Canada, 2017.

[42] J. Albericio, et al., Bit-pragmatic deep neural network computing, in: Proc. International Symposium on Microarchitecture (MICRO 2017), Massachusetts, USA, 2017.

[43] M. Alwani, H. Chen, M. Fredman, P. Milder, Fused-layer CNN accelerator, in: Proc. International Symposium on Microarchitecture (MICRO 2016), Taipei, Taiwan, 2016.

[44] M. Sankaradas, et al., A massively parallel coprocessor for convolutional neural networks, in: Proc. International Conference on Application-specific Systems, Architectures and Processors, Boston, MA, USA, 2009.

[45] V. Sriram, D. Cox, K.H. Tsoi, W. Luk, Towards an embedded biologically-inspired machine vision processor, in: Proc. 2010 International Conference on Field-Programmable Technology (FPT 2010), 2010, pp. 273–278.

[46] S. Chakradhar, M. Sankaradas, V. Jakkula, S. Cadambi, A dynamically configurable coprocessor for convolutional neural networks, in: Proc. 2010 International Symposium on Computer Architecture (ISCA 2010), Saint-Malo, France, 2010.

[47] Z. Du, et al., ShiDianNao: shifting vision processing closer to the sensor, in: Proc. 2015 International Symposium on Computer Architecture (ISCA 2015), Portland, Oregon, 2015.

[48] T. Chen, et al., DianNao: a small-footprint high-throughput accelerator for ubiquitous machine-learning, in: Proc. International Conference on Architectural Support for Programming Languages and Operating Systems (ASPLOS 2014), Salt Lake City, Utah, USA, 2014.

[49] Y. Chen, et al., DaDianNao: a machine-learning supercomputer, in: Proc. International Symposium on Microarchitecture (MICRO 2014), Cambridge, UK, 2014.

[50] W. Lu, et al., Flexible dataflow accelerator architecture for convolutional neural networks, in: Proc. IEEE International Symposium on High Performance Computer Architecture (HPCA), Austin, TX, USA, 2016.

[51] M. Motamedi, P. Gysel, V. Akella, S. Ghiasi, Design space exploration of FPGA-based deep convolutional neural networks, in: Proc. 2016 21st Asia and South Pacific Design Automation Conference, Macau, China, 2016, pp. 575–580.

[52] D. Mahajan, TABLA: a unified template-based architecture for accelerating statistical machine learning, in: Proc. IEEE International Symposium on High Performance Computer Architecture (HPCA 2016), Barcelona, Spain, 2016.

[53] H. Lee, A power-aware scalable pipelined booth multiplier, in: Proc. International SOC Conference. Santa Clara, CA, USA, 2004.

[54] S. Kuang, J. Wang, Design of power-efficient configurable booth multiplier, J. Trans. Circuits Syst. 1 57 (3) (2009) 568–580.

[55] M.O. Lakshmanan, M. Ali, High performance parallel multiplier using Wallace-booth algorithm, in: Proc. International Conference on Neural Information Processing. Computational Intelligence for the E-Age. Penang, Malaysia, 2003.

[56] T.L.M.D. Ercegovac, Digital Arithmetic, Morgan Kaufmann Publishers, California, USA, 2003.

[57] A. Dandapat, S. Ghosal, P. Sarkar, D. Mukhopadhyay, A 1.2-ns 16x16-bit binary multiplier using high speed compressors, Int. J. Electr. Electron. Eng. 4 (3) (2010) 234–239.

[58] C. Vinoth, et al., A novel low power and high speed Wallace tree multiplier for RISC processor, in: Proc Electronics Computer Technology. Kanyakumari, India, 2011.

[59] K. Prasad, K. Parhi, Low-power 4-2 and 5-2 compressors, in: Proc Asilomar conference on Signals, Systems and Computers. Pacific Grove, CA, USA, 2001.

[60] N. Itoh, et al., A 600-MHz 54/spl times/54-bit multiplier with rectangular-styled Wallace tree, J. Solid-State Circuits 36 (2) (2001) 249–257.

[61] M. Rao, S. Dubey, A high speed and area efficient Booth recoded Wallace tree multiplier for fast arithmetic circuits, in: Proc Asia Pacific Conference on Postgraduate Research in Microelectronics and Electronics. Hyderabad, India, 2012.

[62] J. Fadavi-Ardekani, M*N Booth encoded multiplier generator using optimized Wallace trees, IEEE Trans. Very Large Scale Integr. VLSI Syst. 1 (2) (1993) 120–125.

[63] A. Raj, T. Latha, VLSI Design, PHI Learning, New Delhi, 2008.

[64] W. Lu, et al., FlexFlow: a flexible dataflow accelerator architecture for convolutional neural networks, in: Proc. IEEE International Symposium on High Performance Computer Architecture (HPCA 2017), Austin, TX, USA, 2017.

[65] D. Kim, J. Ahn, S. Yoo, A novel zero weight/activation-aware hardware architecture of convolutional neural network, in: Proc. Design, Automation & Test in Europe Conference & Exhibition (DATE 2017), Lausanne, Switzerland, 2017.

[66] J. Qiu, J. Wang, S. Yao, K. Guo, B. Li, E. Zhou, J. Yu, T. Tang, N. Xu, S. Song, Y. Wang, H. Yang, Going deeper with embedded FPGA platform for convolutional neural network, in: Proc. of the 2016 ACM/SIGDA International Symposium on Field-Programmable Gate Arrays, New York, NY, USA, 2016, pp. 26–35.

[67] C. Zhang, Z. Fang, P. Zhou, P. Pan, J. Cong, Caffeine: towards uniformed representation and acceleration for deep convolutional neural networks, in: Proc. the 35th International Conference on Computer-Aided Design (ICCAD'16). New York, NY, USA, 2016, pp. 12:1–12:8.

[68] R. Andri, L. Cavigelli, D. Rossi, L. Benini, YodaNN: an architecture for ultralow power binary-weight CNN acceleration, IEEE Trans. Comput. Aided Des. Integr. Circuits Syst. 37 (1) (2018) 48–60.

[69] L. Cavigelli, D. Gschwend, C. Mayer, S. Willi, B. Muheim, L. Benini, Origami: a convolutional network accelerator, in: Proc. of the 25th Edition on Great Lakes Symposium on VLSI (GLSVLSI '15), New York, NY, USA, 2015, pp. 199–204.

[70] S. Han, et al., EIE: efficient inference engine on compressed deep neural network, in: Proc. Symposium on Computer Architecture (ISCA 2016), Seoul, Korea, 2016.

[71] R. Zhao, et al., Accelerating binarized convolutional neural networks with software-programmable, in: Proc. International Symposium on Field-Programmable Gate Arrays, Monterey, California, USA, 2017.

[72] H. Li, et al., A high performance FPGA-based accelerator for large-scale convolutional neural networks, in: Proc. International Conference on Field Programmable Logic and Applications (FPL 2016), Lausanne, Switzerland, 2016.

[73] V. Gokhale, et al., Snowflake: an efficient hardware accelerator for convolutional neural networks, in: Proc. International Symposium on Circuits and Systems (ISCAS 2017), Baltimore, MD, USA, 2017.

[74] S. Chetlur, et al., October 2014, cuDNN: Efficient Primitives for Deep Learning, arXiv: 1410.0759 (accessed October 13, 2020).

[75] J. Nickolls, I. Buck, M. Garland, K. Skadron, Scalable parallel programming with CUDA, in: Proc. ACM SIGGRAPH 2008, Los Angeles, CA, USA, 2008, pp. 40–53.

[76] D. Leubke, CUDA: scalable parallel programming for high-performance scientific computing, in: Proc. International Symposium on Biomedical Imaging: From Nano to Macro. Paris, France, 2008, pp. 836–838.

[77] J. Nickolls, W. Dally, The GPU computing era, IEEE Micro 30 (2) (2010) 56–69.

[78] V. Narasiman, et al., Improving GPU performance via large warps and two-level warp scheduling, in: Proc. IEEE/ACM International Symposium on Macroarchitecture, Porto Alegre, Brazil, 2011, pp. 308–317.

[79] L. Luo, M. Wong, W. Hwu, An effective GPU implementation of breadth-first search, in: Proc. Design Automation Conference (DAC '10). Anaheim, CA, USA, 2010, pp. 52–55.

[80] B. Zhang, et al., The CUBLAS and CULA based GPU acceleration of adaptive finite element framework for bioluminescence tomography, Opt. Express 18 (19) (2010) 20201–20214.

[81] J. Lee, et al., UNPU: an energy-efficient deep neural network accelerator with fully variable weight bit precision, IEEE J. Solid State Circuits (2018) 1–13.

[82] H. Sharma et al., May 2018, Bit Fusion: Bit-Level Dynamically Composable Architecture for Accelerating Deep Neural Networks, arXiv: 1712.01507v2 (accessed October 13, 2020).

[83] R. Jolivet, T.J. Lewis, W. Gerstner, Generalized integrate-and-fire models of neuronal activity approximate spike trains of a detailed model to a high degree of accuracy, J. Neurophysiol. 92 (2) (2004) 959–976.

[84] A. Krenker, J. Bester, A. Kos, Introduction to the artificial neural networks, in: Artificial Neural Networks, INTECH Open Access Publisher, Rijeka, Croatia, 2011, pp. 3–18.

[85] H. Park, D. Kim, S. Kim, December 2018, Digital Neuron: A Hardware Inference Accelerator for Convolutional Deep Neural Networks, arXiv: 1812.07517v2 (accessed October 13, 2020).

[86] U. Koster, et al., Flexpoint: an adaptive numerical format for efficient training of deep neural networks, in: Proc. NIPS 2017, Long Beach, CA, USA, 2017.

[87] Z. Liu, et al., SiMul: an algorithm-driven approximate multiplier design for machine learning, IEEE Micro 38 (2018) 50–59.

[88] J. Park, et al., Scale-out acceleration for machine learning, in: Proc. International Symposium on Microarchitecture (MICRO 2017), Massachusetts, USA, 2017.

[89] S. Venkataramani, et al., SCALEDEEP: a scalable compute architecture for learning and evaluating deep networks, in: Proc. 2017 International Symposium on Computer Architecture (ISCA 2017), Toronto, ON, Canada, 2017.

[90] H. Kang, April 2018, Accelerator-Aware Pruning for Convolutional Neural Networks, arXiv: 1804.09862 (accessed October 13, 2020).

[91] S. Mittal, A survey of FPGA-based accelerators for convolutional neural networks, Neural Comput. Applic. 32 (2018) 1–31.

[92] S. Han, J. Pool, J. Tran, W. J. Dally, June 2015, Learning Both Weights and Connections for Efficient Neural Networks, arXiv: 1506.02626 (accessed October 13, 2020).

[93] J. Yu, et al., Scalpel: customizing DNN pruning to the underlying hardware parallelism, in: Proc. International Symposium on computer Architecture (ISCA 2017), Toronto, ON, Canada, 2017.

About the authors

Hyunbin Park received his B.S. degree in Electrical and Electronic engineering from Yonsei University, Seoul, South Korea, in 2013. He received his Ph.D. degree from an M.S.-Ph.D. joint course at the School of Integrated Technology from Yonsei University, in 2019. When he was a graduate student, his primary topics of research included system design and system architecture of Neural Processing Units (NPUs) with integral weights and activations, and training techniques for deep learning-based methods for integer gradients. Between March, 2019, and June, 2019, he was employed as a Postdoctoral Researcher at Yonsei Institute of Convergence Technology (YICT) in Yonsei University. He is currently involved in the analysis and benchmarking of NPUs for mobile devices at Mobile Communications Business, Samsung Electronics Co., Ltd.

Shiho Kim is a professor at the School of Integrated Technology at Yonsei University, Seoul, Korea. His previous titles include a System-on-chip design engineer at LG Semicon Ltd. (currently SK Hynix), Korea, Seoul [1995–1996], Director of RAVERS (Research center for Advanced hybrid electric Vehicle Energy Recovery System), a government-supported IT research center, Associate Director of the Yonsei Institute of Convergence Technology (YICT), where he conducted the Korean National ICT consilience program, which is a Korean national program for cultivating talented engineers in the field of information and communication technology [2011—2012], and Director of the Seamless Transportation Lab, at Yonsei University, Korea [since 2011 to present].

His primary research interests include the development of software and hardware technologies for intelligent vehicles, blockchain technology for intelligent transportation systems, and reinforcement learning for autonomous vehicles. He is a member of the editorial boards and a reviewer for various Journals and International Conferences. Till date, he has organized two International Conferences as the Technical Chair/General Chair. He is a member of IEIE (Institute of Electronics and Information Engineers of Korea), KSAE (Korean Society of Automotive Engineers), vice president of KINGC (Korean Institute of Next Generation Computing), and a senior member of IEEE. He is the coauthor for over 100 papers and holds more than 50 patents in the field of information and communication technology.

The authors would like to thank for IITP (Institute for Information & communications Technology Planning & Evaluation). This work was support by MSIT (Ministry of Science and ICT), Korea, under the "ICT Consilience Creative Program" (IITP-2019-2017-0-01015) supervised by IITP.

CHAPTER FOUR

Generic quantum hardware accelerators for conventional systems

Parth Bir*

Electronics and Communication Department, G.L.Bajaj Institute of Technology & Management, Greater Noida, U.P., India

Contents

Abstract

Quantum mechanics proposes, universe is a sum of a generic building block. Different orientation (i.e., angle, phase, amplitude, etc.) and summation of blocks forms entities. When differentiated, building blocks used for formation of entity are termed as basis. Following computational theory, these basis are termed as computational basis. Classical computers possess binary basis. Quantum system possess exponential

---

*Current address: Research & Development Engineer, Oneirix Labs, Karve Nagar, Pune, India

Advances in Computers, Volume 122
ISSN 0065-2458
https://doi.org/10.1016/bs.adcom.2021.01.007

97

computational power because of infinite computational basis. When computing solution to a problem, it's found in state space. Deterministic model (Conventional) requires both correct and incorrect solution set. For problems of probabilistic nature with plenty of variables (NP and P problems), computing solution requires exponential time, as entire state space is scanned. Furthermore, if solution is incomputable, the computation will never complete as solution is missing from both sets. Probabilistic model (Quantum) conducts a guided state space search and possess greater information carrying capacity per bit. Therefore, Quantum Accelerators (QA) are ideal for solving such problems. Resulting implementation of a Generic Quantum Hardware Accelerator (GQHA) is described via algorithms, mathematical models and microarchitecture. Next, a competitive industrial analysis and virtual implementation in a cloud environment is defined. Finally, it's proven that GQHA can replace conventional accelerators to produce faster and reliable results.

1. Introduction

In this chapter, we explore hardware accelerations theory, data handling and implementation principles with respect to quantum mechanical model of universe. Initially, we begin with a basic introduction to theory of computation. Next, we determine the need, design and constraints of Hardware Accelerators (HA) in common. Finally, we discuss the theory and implementation details for Quantum Hardware Accelerators (QHA) [1–5]. Firstly, only the impact of quantum natured principles (i.e., software based quantum accelerator) is conferred. Next, impact of solemnly hardware based systems is considered. Finally, a hybrid (i.e., software and hardware) quantum accelerator system is explained to demonstrate the next generic solution to hardware acceleration.

2. Principles of computation

The process of computation refers to defining a set of definite procedures on some data to achieve desired or correct output. The set of rules governing the allowed procedures are termed as principles of computation. These set of rules are in turn defined by laws of nature. Comprehending nature in different manners lead to unique models of computation. Therefore, currently two models of computation exist, namely classical and quantum model of computation. Furthermore, if comprehended mathematically, prior is of deterministic nature [6,7], whereas the latter is of probabilistic nature [8,9] (Their mathematical nature determines how they calculate solution).

Next, the two different models adopt unique data handling and visualization pattern as follow:
- **Deterministic model of computation:** Systems based on a deterministic model of computation are designed such that both input and output

of the system is predetermined (i.e., for a given input, output is set). Since the transfer function (mathematical function defining the relationship between the input and the output) is defined, required system only replicates its possible outcomes.

Present hardware accelerators such as Graphical Processing Units (GPU's), Field Programmable Gate Arrays (FPGA), Digital Signal Processors (DSP) and Tensor Processing Unit (TPU) [10–14], etc., all come under the domain of deterministic models. A general purpose Central Processing Unit (CPU) is able to perform all computations by differentiating the assigned operation into basic assembly language logic instructions and logic processing units (namely, ALU, FPU, etc.). However, due to differential process and additional load management, net throughput or output speed is very slow. Additionally, this requires a constant high Cycles per Instruction (CPI) value, thus higher frequency and power consumption. Therefore, performing specialized or heterogeneous data intensive operations on CPU is a costly. Therefore, hardware accelerators are necessary for improving performance and imperative in time or power-sensitive systems. Though based on deterministic model, HA are not implemented in a generic manner as CPU, but rather they are built on the principle with respect to their respective data type (For example, graphics data or video or vectors, etc.) and specific algorithms. Therefore, HA are data specific accelerators. Besides, data nature and algorithm set, major reason for exceptional performance of HA [15] is their microarchitecture. The internal microarchitecture defines the actual data path, pipelines, threads, memory architecture and thus determine overall efficiency of HA. Therefore, different accelerators implement specific architectures and algorithms to achieve better and unique data handling and processing, thus accelerating application specific operations. This enables higher order speedup, reduced power consumption, lower latency and increased parallelism, It is to be noted, while using HA, manufacturer must ensure that its particular data type requirements peaks or else it simply ends with wasted wafer space and zero throughput.

- **Non-Deterministic or Probabilistic model of computation:** Systems based on non-deterministic model are not determinable beforehand. Therefore, output for the given input is determined based on probabilistic model [16] (i.e., for a given input, output is unknown). Consequently, along with transfer function, monitoring the frequency or rather determining the most probable solution from the array of solutions depict the outcome or solution to given problem.

Presently, quantum computers [17–19] are still in research phase. Major companies namely IBM, Google and Honeywell Quantum Solutions and academic institutions like TU Delft, etc., are some of the players at the forefront of quantum research. The general idea behind quantum computers or rather accelerating a process using a quantum hardware accelerator is to find a solution to unique problems and accelerate it exponentially. This is essential as a number of important problems still exist (such as biological simulations, molecular simulations and behavior, queries about deep space, state space explosion and many more) whose solution is either indeterminable by conventional computers or would require thousands of years or even infinite time. But finding solutions to such problem is imperative to both nature and survival of mankind. For computing solution to a given problem, a quantum computer relies on theory of computation, physics and mathematics. Together, heuristics derived from their principles optimize the search or state space to output most probable solution or solutions (Yes, there may be more than one correct answer.). But why are so many major players are leaning toward quantum computing technology all of a sudden, even if it has large number of constraints and may not output a definite solution? (We will answer the reason for the same in the upcoming sections).

3. Need and foundation for quantum hardware accelerator design

CPU's are designed to be able to assess and solve any problem. Therefore, they require a volatile software architecture along with a general purpose hardware setup. One might believe that this hybrid nature is extremely useful (Might be true some years ago). In today's date, CPU functions as a control unit in major computing facilities and less as a processing unit. However, todays enormous data specific loads require undivided attention from processing unit (making CPU a bad choice) [20,21]. Therefore, hardware accelerators are designed in a rigid non-volatile manner. This practice may have worked for some time, but simply adding accelerators every time some unique large data sets appear is not the solution (Just as adding multiple cores in CPU is not a good solution).

Above mentioned implications clearly state the fact that there is a need for a powerful generic hardware accelerator [22] unit. Primary aim of this chapter is to lead the scientists and industrialists towards the same goal. Furthermore, we reliable make the case that a Generic Quantum Hardware Accelerator (GQHA) is the solution to the posed query. Next, considering current academia and industry research standpoint, consider the following layered structured for a quantum hardware accelerator [23].

- Algorithms

- Programming a quantum accelerator
- Q-Arithmetic
- Compiler
- Runtime
- Q-ISA
- QEC
- Q-Microarchitecture
- Quantum to classical
- Quantum chip or wafer

Here, considering that major readers only have experience with conventional accelerator structure, we will follow a top to down approach for given quantum stack structure. For each layer we'll describe its function and its relevance or significance with respect to quantum mechanics. In addition, while this design is currently the accepted convention, it will be ages before one may imagine a personal quantum chip in personal electronics. Therefore, after brushing up on general principles and functional behavior for this design, we will present the design for a hybrid quantum processor and an overview of current commercially available quantum systems.

3.1 Algorithms

Designing any processing unit involves multitude of steps ranging from system specification, specifications, state machine design and many more. However, these form merely the crust of the unit. Actual deliverable specifications are determined on the basis of applied algorithms. It is to be noted, unlike CPU (utilizes generic differentiating algorithms), hardware accelerators are utilize specialized algorithms. The term specialized algorithms highlights the fact that hardware accelerators require minimal or no data pre-processing. The algorithms are designed in a manner to increase net throughput per step or in terms of CPU, though CPI is less, performance is high. In addition, these algorithms frame the hardware specifications for designing the actual hardware. Therefore, they play a crucial role in determining the overall performance of the hardware accelerator.

As you're already aware, an algorithm defines a series of precise finite steps to determine the solution of a given task. But, what is the designer supposed to do if the given problem is undecidable (i.e., a decision problem for which it is impossible to construct an algorithm that always leads to a correct solution) has never been solved (i.e., general transfer function is unknown). Consider another scenario where some problem requiring a large amount time to solve on a CPU say $t/8$ unit time. The designer designs multiple algorithms for a hardware accelerator, but finds even best possible case would deliver solution in $t/9$ unit time. A speedup of such a small order in comparison to very high cost is not suitable. In such scenarios, adding a hardware accelerator to System on Chip (SoC) would be simply waste of wafer space as Return of Investment (ROI) is very low.

Quantum algorithms [24,25] are ideal answer to such scenarios. An algorithm defining a series of finite steps which are inherently quantum in nature or utilizing any principle or feature of quantum computation is termed as a *quantum algorithm*. It is to be further noted that while classical algorithms can be run by quantum computers and that too exponentially faster, vive-versa is not true. Let us shed some light on the previously considered scenarios.

First, for an undecidable problem, as stated earlier, it is impossible to construct an algorithm which always gives a correct solution. However, quantum computation is not about correct or incorrect answer in such scenarios, it's about the most probable answer or rather the most recurring correct solution for the particular case. Let us simplify the stated case using the famous "Halting Problem" [26] by Alan Turing which proved it as an undecidable problem. It states that it is impossible to decide form an arbitrary description of a computer program and an input, whether the program will finish running or continue running forever. For a conventional computer, the halting problem statement holds true, however, not for a quantum accelerator. Reason being, a classical computer has a defined set run procedure and path, whereas, for a quantum processing unit, there are multiple runs for highly condensed data, furthermore, solution is depicted by the most probable solution as per histogram analysis [27–29]. Therefore, a quantum processing unit reliably proves that halting undecidable problems are true for conventional deterministic model based processing units (Fig. 1).

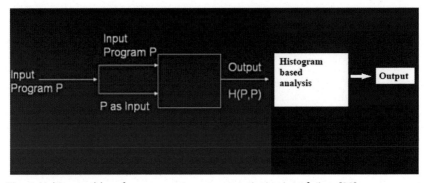

Fig. 1 Halting problem from a quantum computation point of view [30].

Second, as per the second scenario, we can consider a simple linear search problem in unordered data. Now, while a classical algorithm would require, say time *t*, the inherent quantum natured Grover's [31] algorithm would be faster by a factor of four or greater, i.e., it is quadratically faster. Further adding to that, a hybrid variant of Shor's algorithm [32] would be exponentially faster compared to classical version. Now, the question arises, what are factors leading to such results?

Prime factors responsible for exponential performance of quantum algorithms include its ability to exploit the physical nature of the entity itself rather than changing it entirely similar to classical mechanics. For example, consider the concept of parallelism. No matter the programming language or operating system used for increased parallelism, at some point, those threads or processes are being executed sequentially in classical systems. However, in case of quantum natured systems, we truly begin to grasp or operate on concept of parallelism as each individual entity is assigned a separate physical state. Furthermore, any operation deemed as parallel by the user is applied to the all associated states at the same instant or in a parallel manner. More importantly, depending on number of possible solutions, number of states increase or decrease. Let us illustrate this process using a simulation of the Grover's algorithm below (Fig. 2).

Similar to classical logic gates operating on binary bits in conventional electronics, above circuit depicts the quantum circuit, made using quantum logic gates operating on quantum bits or qubits. The above circuit particularly depicts a two qubit system (Notice the q[0] and q[n] notation in left corner). Just as classical mechanics depicts the logic behind logic gates, quantum mechanics depicts the logic behind quantum gates. But, contrary to classical computation, a quantum gate applies the operation associated with the gate to all the associated states with a quantum bit at once. This behavior enables parallel behavior at the lowest level of logic in quantum computation.

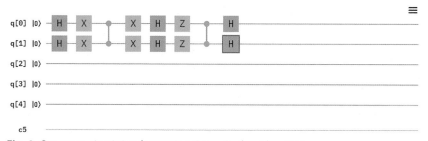

Fig. 2 Quantum circuit implementing Grover's algorithm [33].

Using algorithmic analysis [34], it is evident that for searching an unstructured database with N entries, for a marked entry, classical system would require $O(N)$ queries whereas Grover's algorithm would pose only $O(\sqrt{N})$ queries. Below Fig. 3 depict the state vector and its corresponding amplitude. Attention is to be given to the number of states (Depicted by count of states on x-axis) and their corresponding phase (Depicted by color of particular state).

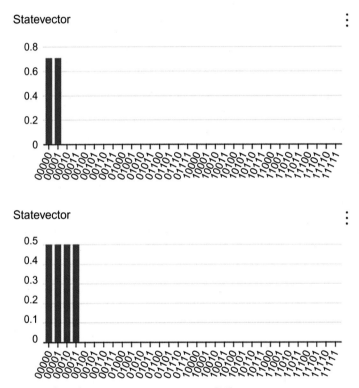

Fig. 3 Grover's algorithm stage-1, 2 state vectors [33].

In above Fig. 3, (i) depicts the initial state of the system where the two states (00000 and 00001) simply hold equal probability of occurrence. As we move to next stage of the circuit, we find system has devised two additional states (But only 2 qubits were used than how are additional states created?) by differentiating and dense packing of two qubits (More light would be shed on dense packing of qubits shortly below). It is to be noted that ideally, a single qubit can hold infinite amount of information [17]. In addition, as one may notice, amplitude value of each state has also decreased.

This highlights the fact that quantum systems tend to maintain equilibrium on their own. From the perspective of hardware acceleration, this enables true parallelism, a self-aware circuit model, accurate resource distribution and higher information carrying capacity per bit. These characteristics which are otherwise extremely hard to achieve even after tons of planning and optimization, comes easily due to nature in quantum natured accelerator unit.

Next, above figures (Fig. 4) depict stepping and parallel analysis on different states under different scenarios (i.e., individually and in entangled states). Here, the yellow bar depicts a phase change in the particular state. This highlights another features of quantum natured accelerators, namely their ability to operate use and operate on imaginary parts of state vector to accelerate the optimization process, while at the same time, retaining the integrity of data (i.e., data is not truncated or tampered).

Interestingly, as in Fig. 5, (v) is identical to (ii), we move forward to establish the fact that quantum natured processing units derive the most likely correct solution. In order to derive this solution, quantum systems

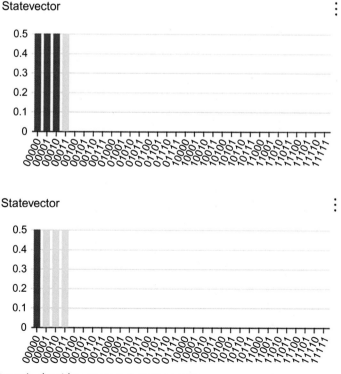

Fig. 4 Grover's algorithm stage-4, 8 state vectors.

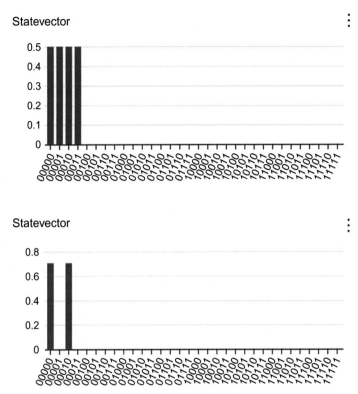

Fig. 5 Grover's algorithm stage-11, 12 state vectors.

conduct multiple runs, hence Fig. 5 (v) is identical to (ii). Furthermore, this reduces the error and provides fail-safe capability unlike conventional hardware accelerators, which base all solutions on single run which may or may not be correct and do not provide scope for failure (i.e., user will never know that system failed; system will output some random answer). Moving forward, after sufficient number of runs, the quantum system again begins to collapse non–solution states (Depicted by Fig. 5 (vi)). This sort of behavior can be termed as an intelligent virtual quantum mechanical pipeline.

Finally, Fig. 6 depicts the derived solution by the system. As is evident from the result, the quantum processing unit collapses [35,36] all the remaining states to determine the statistically correct solution.

This form of behavior is extremely helpful for considering ideal hardware accelerator characteristics as it enables optimum memory management and conducts multiple runs by itself for given data without any deliberate intimation from the control system (this eliminates a great number of

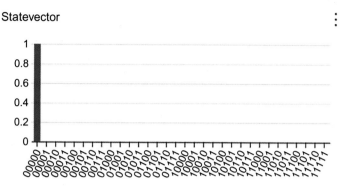

Fig. 6 Grover's algorithm stage-13 state vectors.

additional instructions which are otherwise required). Another aspect which is to be highlighted here is that *how does a quantum processing unit manage to retain such large amounts of data?* This question is answered by the data density per qubit. Ideally, a single qubit is enough to store infinite amount of data. However, currently in practice, data density per packet is still low, but enough to solve current major problems.

In terms of conventional electronics, we get exactly one bit worth of information from each bit. In realm of quantum computation, no such limit exists ideally. However, considering the current state of the quantum systems, it depends on the degree or order of superposition system is capable of performing on quantum bits. To illustrate this feature, let us consider the density matrices of state vectors involved in above example of Grover's algorithm.

To ease the technicality, for ordinate or vertical y-axis, simply consider each individual unit grid as a unique state. From qubit perspective, consider each vertically placed grid (i.e., state) to be superimposed with other vertically placed grids (i.e., states) but only for that particular state vector. Similarly, for abscissa or horizontal x-axis, each horizontally placed grid is entangled together. In layman terms, one can consider this as a graph depicting connectivity and coverage for the designed quantum system. In addition, it is to be noted, if required, quantum systems are also capable of entangling or superimposing the imaginary part of state vector besides the real part ($\text{Re}\,[\rho]$).

Here, Fig. 7 (i) depicts the density matrix for initial state of grovers algorithm state vector (refer to grovers (i)), and Fig. 7 (ii) depicts the corresponding density matrix for grovers second step state matrix (Refer to grovers (ii)). Form the point of initialization, one can see the superimposed quantum

Re[ρ]

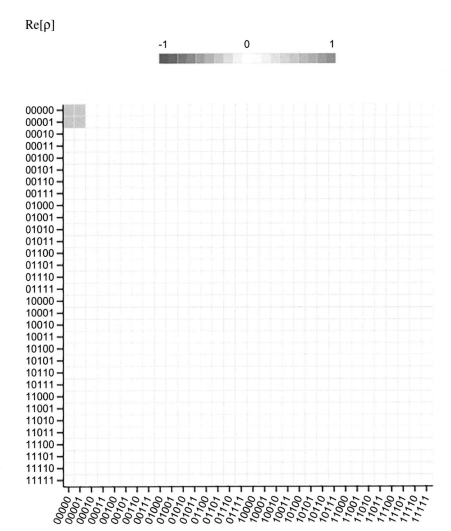

Fig. 7 Density matrix corresponding to Fig. 3 (i) [33].

states. From a binary perspective, one can interpret this as follow: Since the solution to problem is unknown initially, system considers both 0 and 1 as candidate solutions. It is to be noted here that the system is existing in such a state where the states have not been collapsed (i.e., the true state is still unknown and system merely travels via all possible scenarios for finding the solution.)

Next, as number of states increase to four, one may predict that system density matrix would fill 8 grids. But, as depicted in Fig. 8 (ii), density matrix

Re[ρ]

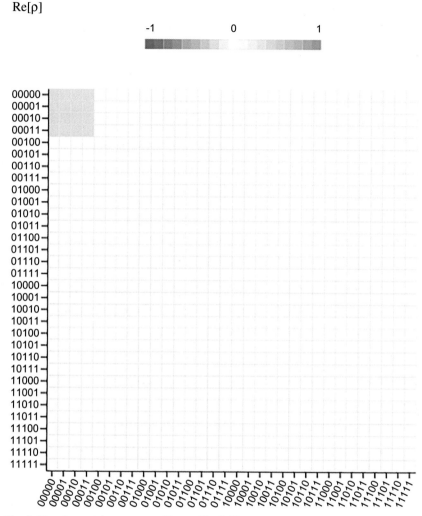

Fig. 8 Density matrix corresponding to Fig. 3 (ii).

shows 16 filled grids. Now question arises why? As is evident from our interpretation scheme, we see four superimposed states placed in a vertical fashion in each successive column. Here, the states start depicting true quantum characteristics as possible solution candidates now include 0, 1, 0–1, and 0 + 1 considering binary scale. Together these superimposed states form a highly dense packed unit or quantum bit is capable of holding greater amount of information compared to conventional bits. It is to be noted here that while quantum bits are existing in a superimposed or entangled states, they have not been collapsed yet (i.e., the solution is still unknown and system is merely considering the possibility of that state being a solution). Therefore, during run time, the quantum processing unit will travel via all possible states which it may deem as a possible solution. This highlights the fact that instead of merely calculating the solution like any general processing unit, quantum systems tend to question the very existence or reason of the derived solution for the particular problem. This form of behavior comes under the domain of Satisfiability Modulo Theory (SMT) under classical computation but even that is not entirely true to its nature. Considering the prime aspect of hardware accelerators, this property is of utmost usefulness as the quantum unit is not merely outputting a solution, but rather it questions the existence the calculated solution in terms of correctness, thereby, reducing CPU workload drastically and forming an intelligent system. This is one of the prime reasons, quantum accelerators are ideal for tasks concerning machine learning or deep learning neural nets. In artificial intelligence based applications, varying from neuromorphic or DNA computing, initial bias and weights are given by the user, but, with upcoming quantum era, one can hope for intelligent self-generative and corrective systems.

Moving forward, the density matrix for pre-final stage (refer to Fig. 5 (vi)) depicts the remaining states after collapsing 12 out of 16 states. This behavior of continuous state collapse leads system to statistically correct solution. It is to be noted, though collapsed, those states still present a probable solution. This form of design structure allows user the pick of pack depending on requirements of the hardware accelerator (Fig. 9).

Finally, below Fig. 10 depicts the derived solution for the given problem by narrowing it down to the statistically correct solution. The probability of each possible states being a perspective solution is mapped on the histogram. The state with the greatest probability in the end is declared as the probabilistically correct solution.

See carefully, the final solution is standalone in nature. This shows that when collapsed, qubits lose their associated properties. Therefore, final solution would be binary in nature.

While this example clearly guides the system to a specific solution, such is not the case with problems. If the designer fails to write an appropriate algorithm, the system will not give a single answer but rather an array of choices. Therefore, quantum algorithms or any principal algorithm responsible for system architecture, play a key role in system correctness, performance and characteristics.

Re[ρ]

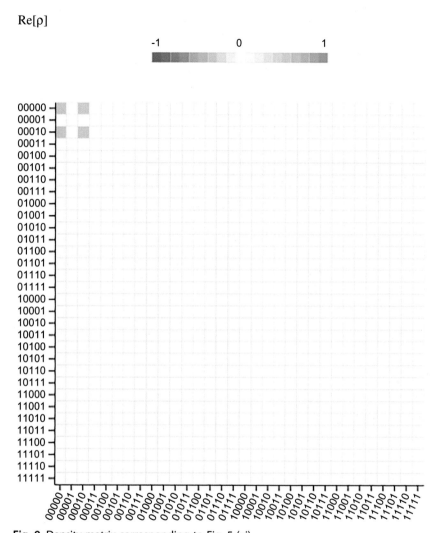

Fig. 9 Density matrix corresponding to Fig. 5 (vi).

3.2 Programming paradigm and languages

While quantum computing hardware does promise immense power, it would be useless if we cannot program it. Hence, just as during the earlier invention of conventional computers during the time of Alan Turing, we are in need of a programming paradigm in order to program and interpret the outcomes from a quantum processing system. However, since the technology is fairly young, each individual organizations are developing their own programming language.

Re[ρ]

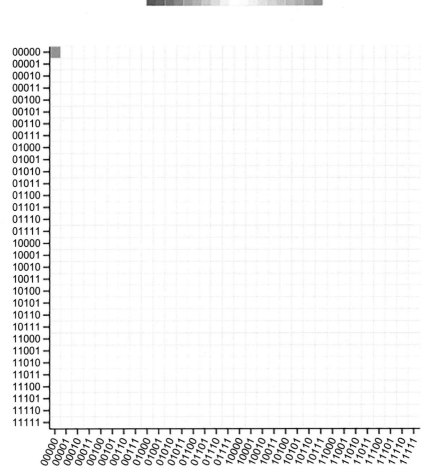

Fig. 10 Density matrix of derived solution (Refer to Fig. 6).

There is Qiskit, an open source SDK started by IBM for working with quantum computers at the level of pulses, circuits, and application modules. Other examples include Q#, QCL, pyquil, OpenQL, QASM, etc. Some languages like Quantum Computation Language (QCL) are bent towards framing this modern language similar to traditional programing languages like C and C++, etc. But, some such as pyquil believe a modern GUI based interface or Software Development Kit (SDK) (For example Oceans SDK from D-Wave) respectively. The user has to merely pick and drop the blocks for required operations, furthermore, it also allows a code interface if user is more comfortable in that particular format. While this format is adequate as per current requirements, it may not be the best fit for future. The said interface currently has no concept similar to conventional programming languages such as functions, classes or concept of reusability, etc. Whereas, a low level language such as the QCL, is similar in construct to the C-language. But, again it does not support concepts of object oriented programming and more but is more familiar to conventional programmers. Therefore, it is imperative from above facts and arguments that we still do require a generic standards, programming paradigm and language if we are to succeed in attracting a larger community of programmers and in turn solving the bigger problems. IBM offers both the drag-and-drop IBM Quantum Composer and the Jupyter notebook-based IBM Quantum Lab for working with the Qiskit [37].

Now, considering from a hardware acceleration perspective, it may not seem that relevant, but it is quite the opposite in practicality. Though the designer is able to setup the hardware architecture at will, if the data to be accelerated is not interpretable by the quantum accelerator, it's all worthless. Let us try to understand this statement with a practical example. Consider that you have assigned some humongous task to your conventional which is required to be accelerated by your quantum accelerator. It may seem that conventional data will fame the data as per requirement and pass it to the quantum processing unit, but in reality, conventional processing unit is not processing the input data. Next, the convention unit will simply check data size and nature and pass it to accelerator. Now the accelerator is responsible for actually framing and processing the data. As you may have guessed by know, for this process to be successful, the quantum processing unit is required to be programmed. For further clarification, let us consider a scenario from conventional computers. Suppose that you write any C++ program for adding two numbers present at some memory location. You may believe that it is your C++ program which directly interacts with data and gives the result, which could be nothing further from the truth. In reality, your program is used by the computer to form series of assembly line instructions from the instruction set architecture of the given processor. Next, these assembly code lines are executed to form a binary file. Finally, this binary file when run by the processor, interacts directly with data and outputs the result which follows a similar course of action to output the result. Therefore, here we are referring to encoding the very behavior a computer behaves with respect to given instructions. Hence, it is of utmost importance that we establish a generic programming paradigm and language for quantum processing systems. For current quantum systems, the said tasks are done using manual manipulations by utilizing microwaves, lasers and photonics, etc., in combination with conventional processors.

3.3 Compiler and runtime requirements

Figure 2.3 in [19] depicts a heterogeneous computational model. Initially, when a host program which may be conventional or quantum in nature is fed to the host CPU. The host CPU now feeds the decision of the given program is to be solved. But, identical to any programming language, a quantum program is a sequence of instructions. Early computers only understood machine language where the instructions are based on elementary logic (namely AND, OR, NOT, etc.). Any other programming language must be translated (i.e., compiled) into standard device language. The language of quantum systems is also a list of instructions. They depict the series of operations that must be applied to qubits in the required order.

Time runs from left to right and dictates the order of execution. When two gates are placed in the same left or right position, than their order of execution is not relevant and are rather executed in a parallel fashion.

Given a circuit, there are a number of its equivalents (i.e., they calculate the same computation) but with a different logic scheme and gate count. However, for any circuit (especially quantum systems where external noise is major factor), number of gates are advised to be kept minimal so as keep input intact, reduce Signal to Noise Ratio (SNR) and circuit depth (i.e., time taken for signal to travel from first to last gate), stating optimization is of core importance. This optimization task is termed as gate synthesis or compiling problem.

However, as the reader is already aware, one of the biggest challenges today is to build fault tolerant quantum computers and introduce better quantum error correction techniques. When building fault tolerant quantum systems, the quantum logic gates that make or break the performance form a very confined gate set that depends on the architecture of the hardware. Statistically, Clifford and T gate set comes as the usual outcome. These gates can be considered as the universal gates in quantum technology as a number of other gates exist, but they can be made using Clifford and T gates. Therefore, Clifford and T gates are often termed as the building blocks of quantum processing systems. Furthermore, since T-gates are implemented using magic state distillation, they are much more expensive in terms of cost function. Therefore, most of the current compilers are focused on reducing T gate count rather than total gate count.

Figure 3.2 in [19] depicts a quantum compilation infrastructure for a scenario where a quantum accelerator is deliberately added as an acceleration unit with a conventional processing unit (Intel i9 in depicted figure). This also highlights the fact that though quantum computers are extremely powerful, conventional quantum computers may not be replaced ever in the future considering their current role and the one in relation to quantum computing units.

From the aspect of hardware acceleration, quantum systems provides some unique options in terms of following: exact, inexact and random compiling. In exact synthesis, compiler implements the desired computational circuit with a finite size circuit. In an inexact circuit, the desired computation can only be achieved or approximated by using deeper and deeper circuits. While random synthesis or compiling is considered to be a part of inexact synthesis, it has proved to be extremely successful in solving some extraordinary problems and hence deserves a separate mention. For random synthesis, randomness is used to make decisions about which logic gates are to be implemented by simply rolling a dice. This process does not output a completely random circuit. Interestingly, a list of possible circuits that are all equivalent is constructed. This helps us combat noise processes and surprisingly enables us to use smaller circuits to achieve the same precision. This process can be further advanced based on histogram and performance analysis if necessary. Therefore, a quantum accelerator is capable of deliberately monitoring and controlling any aspect of any processing system, thus making it a true generic accelerator.

3.4 Quantum instruction set architecture (Q-ISA)

While many computer architects may interpret the function of instruction set architecture (ISA) as simply outputting instructions to be performed by the processing unit. In reality, the instruction set architecture is actually responsible for data abstraction. The better the ISA, more refine is the abstraction process. Here, the term better refers to a number of aspects of ISA including length of instruction (such as RISC or CISC), its nature of throughput (i.e., whether it is under domain of SIMD, SISD, etc.), number of instructions required for performing basic logical operations and many more. All these characteristics collectively decide the understanding of the given problem to your processing system. Depending on the required order of differentiation for basic logical operations, we are able to calculate the processing unit throughput. Classical logical gates and circuit model are designed to instruct and obtain definite and rigid results. This further indicates that even classical universal gates will not allow user to encounter each result as they are bound in terms of mathematics and classical mechanics. Conversely, quantum gates and circuit model are designed to instruct and obtain most probable result (While maintaining true universal nature). Therefore, nature of logical gates and circuit models tend to curb the concept of universality.

As is shown in previous sections, a number of high level languages exist in which quantum programming can be done. But compilers for all these languages convert them to some variant of Quantum Assembly Language (QASM) [38]. Similar to evolution of programming languages in domain of quantum computing, QASM is also required to be standardized for successful future course of action. Some Q-ISA currently available include Von-Neumann architecture based virtual instruction set architecture, Hierarchical QASM with loops (QASM-HL), Quil from Rigetti computing [39], Quantum Physical Operations Language (QPOL), and OpenQASM from IBM, etc. Since, their microarchitecture description varies greatly, these Q-ISA are framed without considering the lower level implications to interface with the quantum processor. It can be clearly established that given Q-ISA lack a generic explicit control structure and principles (This fact will be further illuminated under the section describing microarchitecture).

Currently, the Quil ISA by Rigetti computing presents one of the closest Q-ISA in process of being an ideal standard for reference. It would be compared against IBM's OpenQASM 3.0 coming in 2021. Rigetti computing has termed this underlying instruction architecture as Quantum Abstract Machine (QAM). The aim is to capture the essence of both quantum computation and conventional computing logic in a unique instruction set so as to support abstraction of quantum of entities and differentiate it to a classical level for user understanding. QAM can be considered as a Quantum Turing Machine (QTM) [40,41], yet more practical for accomplishing real world problems. For example, let us devise the circuit for Quantum Fourier Transform (QFT) (Fig. 11).

QASM Code:

```
// quantum Fourier transform
OPENQASM 2.0;
include "qelib1.inc";
qreg q [4];
creg c [4];
```

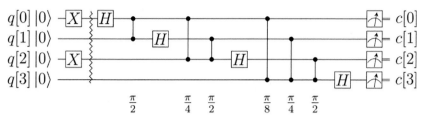

Fig. 11 4-qubit quantum Fourier transform circuit. The circuit applies the QFT to |1010> and measures in the computational basis. The output is read in reverse order c[3], c[2], c[1], c[0] [33].

```
x q[0];
x q [2];
barrier q;
h q[0];
cu1(pi/2) q [1],q[0];
h q [1];
cu1(pi/4) q [2],q[0];
cu1(pi/2) q [2],q [1];
h q [2];
cu1(pi/8) q [3],q[0];
cu1(pi/4) q [3],q [1];
cu1(pi/2) q [3],q [2];
h q [3];
measure q -> c;
```

3.5 Quantum microarchitecture

The microarchitecture of any processing unit plays major role in its performance, specifics (such as power consumption, etc.), and performance. It is responsible for acting as an explicit control unit for all internal and external data operations. In case of quantum hardware accelerators, microarchitecture [42,43] is even more important due to its non–volatile nature, and unlike CPU's they would be worthless as they cannot be exercised for any other operations. This also, highlights the fact that all the time a conventional accelerator unit is not operational, it serves as a waste of wafer space and power. Therefore, a generic natured hardware accelerator is necessary. Furthermore, as the VLSI industry progress to lower order nodes (i.e., 5 nm, 3 nm and so on), the role of microarchitecture becomes even more important because we are about to reach our physical limits and are therefore required to revisit the core microarchitecture for enhancement in processing power. Furthermore, as the conventional processors of today are unable to take the leap to quantum logic directly, designers should pave the way by implementing designs for reversible computing which enable implementation of some quantum principles using conventional logic gates. This may increase the wafer area requirement a little, but it compensates for itself in terms of performance, memory fetch operations and power requirements.

Note, other layers of the quantum stack are out of scope for an accelerator designer and of this book chapter. Reader should refer to any book explaining basics of quantum mechanics to understand the operation of remaining layers. In the next section, we will highlight the role of all layers mentioned until now and their application in designed QPU and industrial QPU's.

4. A generic quantum hardware accelerator (GQHA)

It is evident from the study of previous sections and general principles of quantum mechanics and computational theory, we derive following aspects about quantum systems:

- They are currently incapable of performing every task (i.e., it can only accelerate task which are possible on a classical quantum computer).
- As is evident from their physical structure of gold, they are expensive.
- Their current size depicts that perhaps making a personal quantum computer for mass production may never be possible.
- Additional apparatus is required for setting up the control system.
- Special conditions are required for proper functioning.
- More prone to noise compared to general computers.

From the above standpoint, quantum computing may seem more of a curse than boon. Some might suggest that being a particularly new technology, these development barriers are reasonable and may or may not be overcome over time.

However, nobody is paying attention to the fundamentals. Attention, if you compare the quantum systems present in today's date, all are creating quantum technology from scratch but not many are actually implementing the found results and fundamentals of quantum computation to current technology.

Therefore, here we present the design of a Hybrid Quantum Asynchronous Processor (HQAP) [44]. Parsing the each term of title, Hybrid refers to its semi-classical nature, next, quantum refers to its base principle of constructions based on quantum mechanics and computation, followed by its data operation nature which is asynchronous. The HQAP is designed from on from the classical tools of VLSI, but is based on and implements principles of quantum technology. This form of design enables it to act as a Generic Quantum Hardware Accelerator (GQHA).

The aim behind designing such a hardware acceleration unit is not merely data acceleration and generic data handling. The true purpose behind designing such hybrid units is to add an additional unit which is able to explore all the possibilities and outcomes concerning the given problem. Next, if necessary it should propose how to counter or bypass the undesired outcomes. Finally, and most importantly, it should accelerate the thinking process such as algorithms or artificial intelligence based units where the human mind may have stuck or reached a limit. Therefore, such a hybrid quantum accelerator unit will not only accelerate the present rate of technology development and problem solving but will also set the course for the future.

4.1 Deciphering HQAP as a GQHA

This section explains the core factors determining the nature of construction and operation in turn, for the HQAP. Initially, it is already proven that classical computers are important in their own respect, therefore, they are necessary for data handling and other operations in a generic natured accelerator unit. The classical part of the HQAP acts as the control unit for the entire accelerator and later as the interpreter. Compared to current quantum systems, this makes the control process much more precise, allows better synchronization, higher integration of the system, lowers the cost and eases the task to a much elementary level.

Next, following to the core principle of construction and operation, HQAP enables implementation of multitude of quantum natured operations such as superposition, entanglement, teleportation, etc. Together, this combination enables construction of the unique and first of its kind, an Accelerated Universal Quantum Turing Machine (AUQTM). A Universal Turing Machine or UTM, Accelerated Turing Machine or ATM and Quantum Turing Machine or QTM already exist. While they are capable of simulating any computational system, an accelerated Turing machine and a quantum computer respectively, none can simulate an AUQTM. The UTM has no concept of accelerating units, while ATM deem that though any task can be accelerated, for problems such as the halting problem, ATM will only accelerate recursion process with no net result and finally, a QTM may solve the problem but in unknown time.

4.2 Deconstructing GQHA

As described in the above sub-sections, HQAP is a hybrid unit consisting of both conventional and quantum units. Fig.12 denotes the standard units for the designed hybrid quantum asynchronous processor unit.

As you may have noticed already, unlike the present stack based quantum hardware accelerator design prevailing in the industry, HQAP is made multi-directional stacks (i.e., both vertical and horizontal) forming a structure in the form of an interconnected cube.

As depicted in Fig. 12, HQAP consists of following major units:

- QC-Asynchronous Unit
    It forms the control unit of HQAP. It operates on the principles of asynchronous computing.
- Q-Chip
- QEC Unit
- HTE Unit
- UHTT Unit
- QCEP
- VU Unit

The HQAP incorporates features of both conventional and quantum computation simultaneously and hence, builds a universal processing system.

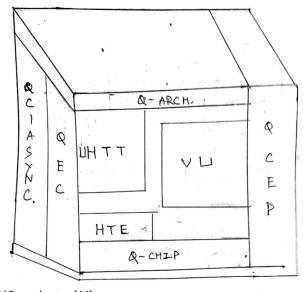

Fig. 12 HQAP accelerator [44].

Initially, quantum bits are synthesized from the wafer of semiconductor, which is termed as Q-chip in HQAP. The Q-chip or quantum chip refers to the wafer (i.e., silicon, germanium, etc.). The wafer is manipulated by microwaves for creating isolated particles (Quantum accelerators of today use a variety of apparatus including microwaves, lasers, photonics, etc., to manipulate the wafer.). It is to be highlighted here that while quantum accelerators in today's date isolate and operate on quantum bits separately, in actual truthfulness, quantum computing refers to in memory computing. In addition, if necessary, depending on analysis and requirement from the control unit, quantum bits can be superimposed or entangled by the Hyper Threading Entanglement (HTE) unit. The reader should note that while quantum bits are the necessary data carriers, it is the process of superposition and entanglement which enables true parallelism by allowing conduction of same operation on all threaded states together at once. Therefore, the HTE unit is of utmost importance.

Next, after creating the necessary amount and type of qubits, these qubits are measured or collapsed in mathematical terms. But, rather interestingly, in conventional computational terms, their values are tagged or stamped on asynchronous data tokens. As per the principles of asynchronous computing, data travels in the form of data tokens (each token can be considered as a generic coin which is stamped with a data value) and communicated via certain specific protocols. This behavior of asynchronous systems allows us to mimic behavior of quantum bits via batches of asynchronous bits. This also highlights the fact that the true computational power of quantum computers does not reside in specific hardware but rather in its principles along with quantum mechanics. In addition, due to such a mechanism of data handling, HQAP is enabled to assess and accelerate any type of data, thus formic a generic hardware accelerator. It is to be noted here that while asynchronous systems does allow us to mimic the behavior of quantum systems, the order of speed up provided via this process is still not comparable to a true quantum computer but significantly greater than current conventional computers.

Further in the process, these entangled semi-quantum asynchronous bits are to be directed by the control unit. The control unit conducts a time and space complexity analysis using one the present conventional cores. This time and space complexity analysis helps the control unit thread together the required resources. Once the required number of qubits are obtained and processing units are set in the initial state, HQAP begins forming sets of probable initial states and process them respectively in a parallel manner.

While this operation looks similar to initialization process of any conventional accelerator like a GPU, it differs greatly due to its earlier space, algorithmic and probabilistic analysis. The probabilistic analysis ensures minimum memory consumption instead of randomly allotting memory resources to every candidate. Apart from this factor, conventional accelerators have no aspect for algorithmic or time analysis, this leaves the control unit or CPU in conventional accelerators in an undeterminable or risky state where it simply awaits conformation form the accelerator unit. Therefore, contrary to conventional behavior of present accelerators, HQAP is ideal for superscalar architecture based computers or even as a standalone generic processing unit.

In order to ensure constant supply of error free quantum bits or qubits, the Quantum Computation enabled Pipelining (QCEP) unit is connected to every unit via a direct pipeline which is in turn connected to the Quantum Error Correction (QEC) unit. While the QCEP is responsible for transfer and maintenance of asynchronous qubits throughout their lifetime, QEC (Depicted in Fig. 15) is responsible for checking the correctness of their state at particular intervals and correcting it if necessary. Current industrial quantum computers follow a scheme of surface codes. This form of behavior establishes a chain reaction which if required can be examined if necessary. This may seem like simple communication between units, but when observed on a quantum scale in true quantum computers, this behavior or instance is termed as quantum teleportation This refers to the fact unlike conventional hardware accelerators, which are solemnly dependent on memory or CPU (Fig. 13 depicts the QCEP for System on Chip or SoC based system, it is also termed as Connected or Real QCEP) or other sources (Fig. 14 depicts the QCEP for Network on Chip or NoC based system, it is also termed as Disconnected or Virtual QCEP) for instructions in case of an error (Memory fetch and write operations require the highest number of cycles in conventional systems), quantum hardware accelerators are much faster and capable of acting on their own will. Furthermore, their actions are self-verifiable nature.

However, current quantum accelerators are incapable of such complex operations, therefore their actions can be verified by this process of chain reaction. But, for future reference, quantum natured hardware accelerators will be capable of memory transfer via this process of memory cloning using quantum teleportation (Fig. 15).

In addition, if necessary, some functioning of quantum circuits can be replicated by conventional logic gates by implementing reversible logic.

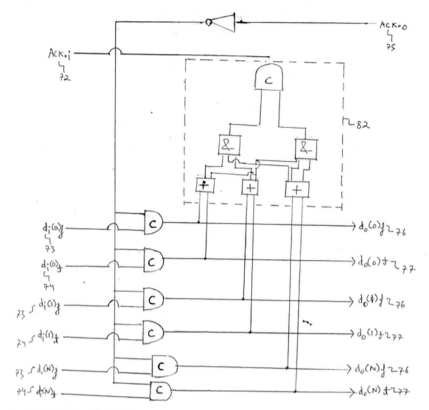

Fig. 13 QCEP for SoC [44].

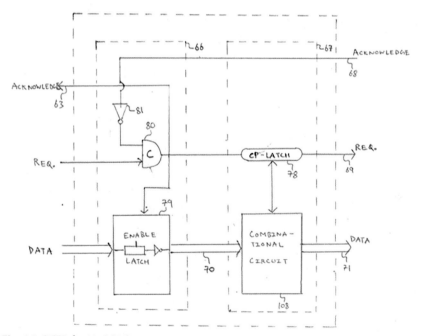

Fig. 14 QCEP for NoC [44].

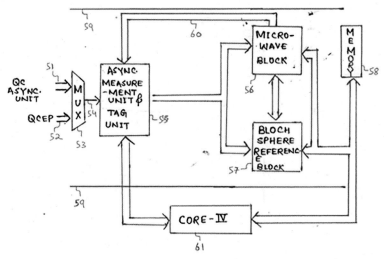

Fig. 15 Quantum error correction (QEC) Unit [44].

Reversible logic refers to the implementation of circuits whose output once given van be redirected into circuit to output earlier inputs. As per this convention, logical NOT gate is a hybrid quantum gate by nature when designed in a looped circuit. A not for conventional logic RTL designers, while reversible logic increases the gate count in some cases, it increases the overall efficiency by decreasing the memory fetch counts and memory requirements itself.

Finally, in order to ensure both classical and quantum natured capabilities during operations, the threaded resources together with Virtualizing Unit (VU) (Depicted in Fig. 16) map the Virtualized Digital Quantum Processing (VDQP) unit. For readers accustomed to conventional computation behavior, this process can be considered being similar to the process of FPGA mapping. Therefore, in all truthfulness, we only map the required part of the quantum circuit. For example, if the given fata is required to be subjected to Quantum Fourier Transform, the VDQP will only implement the circuit for QFT and nothing extra. But, if required, entire quantum processing unit can be mapped. The VDQP alone facilitates systems with quantum computation. On the other hand, VDQP along with the core-I to Core-IV can be used for conventional computation.

As is evident from the described behavior, quantum natured systems are ideal for acceleration purposes regardless of the data or operation type. This form of generic data acceleration will enable power saving, increase

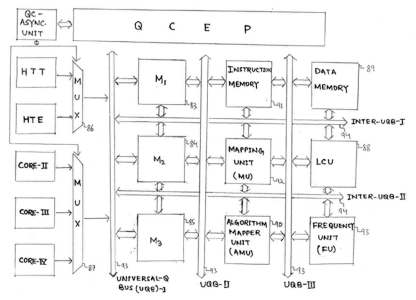

Fig. 16 Virtualization unit [44].

computational power, solve the problems associated with lower order nodes such as 3 and 5 nm, etc. But, most importantly, this form of hardware accelerators will enable us to find answers to some biggest and unsolvable problems. In addition, it would further highlight the reasoning behind the correctness or incorrectness of the obtained output.

5. Industrially available quantum hardware accelerators

In this section, we compare the current progress of quantum hardware accelerators from industrial firms and academic institutions. We compare them on the core principles of construction, computational power, primary source of control, error correction techniques and capability and correctness. Next, we compare and establish the pros, cons, similarities and dissimilarities between the designed hybrid quantum accelerator and industrial quantum accelerators. Finally, we will do an abstract comparison between conventional and quantum natured accelerators.

Currently, following major companies and academic institutions are involved in quantum computing research round the world:
- IBM
- Google

- Rigetti Computing
- D-Wave
- Qiskit from Delft University of Technology
- Cambridge Quantum Computing

5.1 IBM Quantum project

The IBM Quantum project [33] was one of the earliest companies along with Google to start quantum computation research. In fact, it was IBM fellow Charles Bennet who coined the phrase "quantum information science" in 1970. IBM was the first to put a quantum computer on the cloud for open access, doing so in May, 2016. IBM initially started with their fundamental "IBM Grid Framework". This framework (also known as 2 X Grid) [45], regardless of the device or medium of the user, grid provides enough structure and guidance so one can focus on their creative idea (Fig. 17).

While the grid framework was successful in providing the initialization for IBM Quantum project, it was not the perfect solution in the long run. Therefore, IBM adopted an optimized version of their grid framework

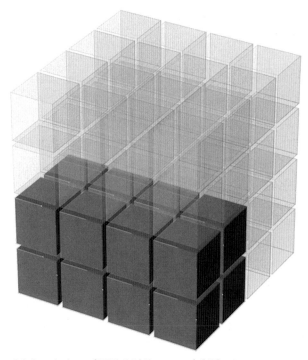

Fig. 17 Pictorial description of IBM Grid Framework [45].

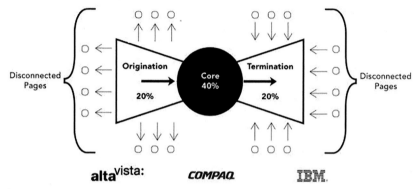

Fig. 18 IBM Bow Tie Layout [45].

to form a bow tie and lattice framework. Earlier 5-qubit designs namely, the IBM Q5 Yorktown and IBM Q5 Tenerife were based on bow tie layout (Figs. 18 and 19).

Now, while quantum mechanics and its principles are responsible for true power behind quantum computation, the actual credit for constructing such quantum hardware accelerators goes to designers. Torus Interconnect [46] can be considered as a three dimensional multi direction doubly linked list data structure. Most of the quantum accelerators todays are designed using the lattice framework and multiple instances of torus interconnect.

Apart, from these architectural details, another aspect of quantum bits is worrisome, i.e., quantum bits are extremely prone to noise. Therefore, it requires a dilution refrigerator for decreased impact of noise and maintain superconducting nature of qubits considering the energy requirements posed by Bose Einstein Condensate.

5.2 Google bristlecone

The other technological giant involved in quantum computing is Google. Google's 72 qubit quantum processor [47] named Bristlecone is a synchronous system. But, in actual hardware design, it is a square array of linear 9 Q-bits, arranged in 8 rows, which basically represents an array of mini quantum processing units rather than an integrated system. These individual processing centers can be either utilized in a standalone manner or in a sequence as a conventional synchronous system.

IBM's 10 Quantum Device Lineup

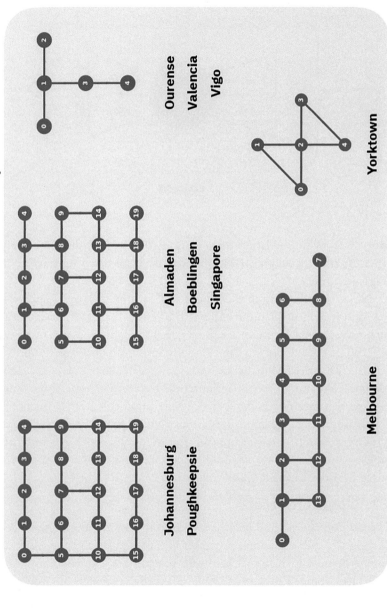

Fig. 19 IBM Quantum history as per qubit layout [33].

Recently, Google claimed that its quantum computer has achieved quantum supremacy. But, rather interestingly, its quantum computer solved a particular problem in approximately 3 s which would have taken over a decade or more in a conventional computer. In this sense, it becomes evident that bristlecone may have been the ideal choice for solving that particular problem but perhaps not every problem.

The processor is fabricated using aluminum for metallization and Josephson junction and indium for bump between two Si wafers. The states of all qubits can be read simultaneously using frequency multiplexing technique. The chip is wire bonded to a superconducting boards and cooled to 20 mK in a dilution refrigerator to reduce ambient thermal energy to well below the qubit energy.

5.3 D-Wave

The D-Wave [48] is the fastest growing and one of the most successful quantum computing based firms. The primary reason of its success is its quantum annealing based quantum computing system. In terms of qubit count, D-Wave currently supersedes all, its last quantum processing unit D-Wave 2000-Q was a 2048 qubit computer. Furthermore, D-Wave released Advantage version of 2020 which has 5000 qubits and 15-way qubit connectivity. D-Wave has been the only quantum technology firm to solve and greatest number of conventional problems from complicated simulations to verification or discovering of complex and novel substances (Fig. 20).

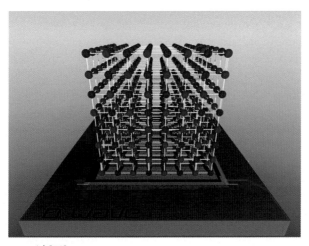

Fig. 20 D-wave grid [48].

Similar to other architectures, D-wave is using hybrid computing, leveraging quantum and classical systems in tandem to more efficiently solve problems. This blend between the quantum mechanics and conventional computation is what has enabled D-wave to greater performance.

5.4 Some other development areas and derived inference

However, quantum computing supremacy cannot be established merely on the basis of hardware as mentioned earlier in the algorithmic section. Therefore, technological firms like Rigetti Computing and Microsoft are making strides in the area of quantum natured software development as well. Rigetti's python SDK has enabled development of quantum natured circuits via a cloud based python interface. Microsoft is developing the compiler technology to back the various quantum programming languages currently being developed round the world.

Besides software and algorithms, major concern for quantum computing industry is the noise and coherence of the qubits. Therefore, Intel Labs [49] is developing state of the art quantum natured control units besides the general quantum computation devices.

6. Conclusion and future work

In conclusion, we have reliable evidence booth algorithmically and mathematically that quantum hardware accelerators are a necessity. If electronics industry and world in general wishes to move forward in solving the unsolvable problems, and find solutions to biggest problems of humanity, whether it is a cure to a pandemic situation or exploring the deep space, quantum computing is a necessity. It is highly recommended that future mathematicians, physicists, and engineers are proficient in concepts of quantum mechanics and computing. Furthermore, problems corresponding to lower order technology nodes, power consumption, and excessive core count in CPUs and other conventional accelerators, all can be successfully solved using quantum accelerators. However, world still requires numerous innovative minds who are able to peek inside the quantum world and interpret them to the conventional one, whether it is in form of algorithms or hardware itself.

References

[1] A. Mehrabian, M. Miscuglio, Y. Alkabani, V.J. Sorger, T. El-Ghazawi, A Winograd-based integrated photonics accelerator for convolutional neural networks, IEEE J. Sel. Top. Quantum Electron. 26 (1) (2020) 1–12, https://doi.org/10.1109/JSTQE.2019.2957443.

[2] C. Guo, et al., Control and readout software for superconducting quantum computing, IEEE Trans. Nucl. Sci. 66 (7) (2019) 1222–1227, https://doi.org/10.1109/TNS.2019.2920337.

[3] R. Weber, A. Gothandaraman, R.J. Hinde, G.D. Peterson, Comparing hardware accelerators in scientific applications: a case study, IEEE Trans. Parallel Distrib. Syst. 23 (1) (2011) 58–68, https://doi.org/10.1109/TPDS.2010.125.

[4] S.M. Shajedul Hasan, Y.W. Kang, M.K. Howlader, Development of an RF conditioning system for charged-particle accelerators, IEEE Trans. Instrum. Meas. 57 (4) (2008) 743–750, https://doi.org/10.1109/TIM.2007.911638.

[5] S. Stratigraphic, I. Of, S. Data, I. N. The, T. Region, and S. Basins, 1 1 1 I, 8641 (93) (1991) 26–32.

[6] M.P. Frank, R.M. Lewis, N.A. Missert, M.A. Wolak, M.D. Henry, Asynchronous ballistic reversible fluxon logic, IEEE Trans. Appl. Supercond. 29 (5) (2019), https://doi.org/10.1109/TASC.2019.2904962.

[7] E. Miranda, A. Mehonic, W.H. Ng, A.J. Kenyon, Simulation of cycle-to-cycle instabilities in SiOx-based ReRAM devices using a self-correlated process with long-term variation, IEEE Electron Device Lett. 40 (1) (2019) 28–31, https://doi.org/10.1109/LED.2018.2883620.

[8] C.D. Zuluaga, M.A. Alvarez, Bayesian probabilistic power flow analysis using Jacobian approximate Bayesian computation, IEEE Trans. Power Syst. 33 (5) (2018) 5217–5225, https://doi.org/10.1109/TPWRS.2018.2810641.

[9] I. Glasser, N. Pancotti, J. Ignacio Cirac, From probabilistic graphical models to generalized tensor networks for supervised learning, IEEE Access 8 (2020) 68169–68182, https://doi.org/10.1109/ACCESS.2020.2986279.

[10] F. Bin Muslim, L. Ma, M. Roozmeh, L. Lavagno, Efficient FPGA implementation of open CL high-performance computing applications via high-level synthesis, IEEE Access 5 (2017) 2747–2762, https://doi.org/10.1109/ACCESS.2017.2671881.

[11] M. Shen, G. Luo, N. Xiao, Exploring GPU-accelerated routing for FPGAs, IEEE Trans. Parallel Distrib. Syst. 30 (6) (2019) 1331–1345, https://doi.org/10.1109/TPDS.2018.2885745.

[12] S. Li, Y. Luo, K. Sun, N. Yadav, K.K. Choi, A novel FPGA accelerator design for real-time and ultra-low power deep convolutional neural networks compared with titan X GPU, IEEE Access 8 (2020) 105455–105471, https://doi.org/10.1109/access.2020.3000009.

[13] A. Ilic, L. Sousa, Efficient multilevel load balancing on heterogeneous CPU + GPU systems, in: High-Performance Computing on Complex Environments, vol. 9781118712, 2014, pp. 261–282, https://doi.org/10.1002/9781118711897.ch14.

[14] J.I. Agulleiro, F. Vazquez, E.M. Garzon, J.J. Fernandez, Real-time tomographic reconstruction through CPU + GPU coprocessing, in: High-Performance Computing on Complex Environments, vol. 9781118712, 2014, pp. 451–466, https://doi.org/10.1002/9781118711897.ch23.

[15] Y.S. Shao, D. Brooks, Research infrastructures for hardware accelerators, Synth. Lect. Comput. Archit. 34 (2015) 1–97, https://doi.org/10.2200/S00677ED1V01Y201511CAC034.

[16] D. Qiu, H. Wang, A probabilistic model of computing with words, J. Comput. Syst. Sci. 70 (2) (2005) 176–200, https://doi.org/10.1016/j.jcss.2004.08.006.

[17] M.A. Nielsen, I. Chuang, L.K. Grover, Quantum Computation and Quantum Information, vol. 70, 2002. no. 5.

[18] I. Ashraf, N. Khammassi, M. Taouil, K. Bertels, Memory and communication profiling for accelerator-based platforms, IEEE Trans. Comput. 67 (7) (2018) 934–948, https://doi.org/10.1109/TC.2017.2785225.

[19] M. D. Serr, n.d. "QuTech Central Controller: A Quantum Control Architecture for a Surface-17 Logical Qubit.".

[20] L. Yu, CCHybrid: CPU co-scheduling in virtualization environment, Concurr. Comput. Pract. Exp. 32 (3) (2020) 1–11, https://doi.org/10.1002/cpe.4213.

[21] Y. Kochura, S. Stirenko, O. Alienin, M. Novotarskiy, Y. Gordienko, Comparative analysis of open source frameworks for machine learning with use case in single-threaded and multi-threaded modes, in: Proc. 12th Int. Sci. Tech. Conf. Comput. Sci. Inf. Technol. CSIT 2017, vol. 1, 2017, pp. 373–376, https://doi.org/10.1109/STC-CSIT.2017. 8098808.

[22] P. Bir, S.V. Karatangi, A. Rai, Design and implementation of an elastic processor with hyperthreading technology and virtualization for elastic server models, J. Supercomput. (2020) 1–22.

[23] https://online-learning.tudelft.nl/courses/the-hardware-of-a-quantum-computer/.

[24] R.J. Lipton, K.W. Regan, Quantum Algorithms Via Linear Algebra: A Primer, vol. 52, 2015. no. 11.

[25] M. Benslama, A. Benslama, S. Aris, Quantum communications in new telecommunications systems, in: Quantum Communications in New Telecommunications Systems, 2017, pp. 1–179, https://doi.org/10.1002/9781119332510.

[26] G.T. Herman, The uniform halting problem for generalized one state turing machines, in: 9th Annual Symposium on Switching and Automata Theory (swat 1968), Oct. 1968, pp. 368–372, https://doi.org/10.1109/SWAT.1968.36.

[27] D.G. Marangon, et al., Long-term test of a fast and compact quantum random number generator, J. Lightwave Technol. 36 (17) (2018) 3778–3784, https://doi.org/10.1109/ JLT.2018.2841773.

[28] H. Li, H. Li, L. Zhang, Quaternion-based multiscale analysis for feature extraction of hyperspectral images, IEEE Trans. Signal Process. 67 (6) (2019) 1418–1430, https://doi. org/10.1109/TSP.2019.2892020.

[29] A. Dieguez, J. Canals, N. Franch, J. Dieguez, O. Alonso, A. Vila, A compact analog histogramming SPAD-based CMOS chip for time-resolved fluorescence, IEEE Trans. Biomed. Circuits Syst. 13 (2) (2019) 343–351, https://doi.org/10.1109/TBCAS.2019. 2892825.

[30] https://cs.stackexchange.com/questions/65401/proof-of-the-undecidability-of-the-halting-problem.

[31] S. Du, Y. Yan, Y. Ma, Quantum-accelerated fractal image compression: an interdisciplinary approach, IEEE Signal Process. Lett. 22 (4) (2015) 499–503, https://doi.org/10. 1109/LSP.2014.2363689.

[32] E. Martin-Lopez, A. Laing, T. Lawson, R. Alvarez, X.Q. Zhou, J.L. O'Brien, Experimental realisation of Shor's quantum factoring algorithm using qubit recycling, in: 2013 Conf. Lasers Electro-Optics Eur. Int. Quantum Electron. Conf. CLEO/Europe-IQEC 2013, vol. 1221, 2013, p. 4799, https://doi.org/10.1109/CLEOE-IQEC.2013. 6801701. no. 2009.

[33] https://www.ibm.com/quantum-computing/technology/experience/.

[34] Y. Wang, D. Dong, B. Qi, J. Zhang, I.R. Petersen, H. Yonezawa, A quantum Hamiltonian identification algorithm: computational complexity and error analysis, IEEE Trans. Autom. Contr. 63 (5) (2018) 1388–1403, https://doi.org/10.1109/TAC. 2017.2747507.

[35] N. Nayak, R.K. Bullough, B.V. Thompson, G.S. Agarwal, Quantum collapse and revival of Rydberg atoms in cavities of arbitrary Q at finite temperature, IEEE J. Quantum Electron. 24 (7) (1988) 1331–1337.

[36] J.S. Cohen, D. Lenstra, Spectral properties of the coherence collapsed state of a semiconductor laser with delayed optical feedback, IEEE J. Quantum Electron. 25 (6) (1989) 1143–1151.

[37] https://www.ibm.com/quantum-computing/tools.

[38] K.O.E.N. Bertels, A. Sarkar, T. Hubregtsen, M. Serrao, A.A. Mouedenne, A. Yadav, I. Ashraf, Quantum computer architecture: towards full-stack quantum accelerators, in: 2020 Design, Automation & Test in Europe Conference & Exhibition (DATE), IEEE, 2020, pp. 1–6.
[39] https://rigetti.com/.
[40] https://en.wikipedia.org/wiki/Quantum_Turing_machine.
[41] M. Muller, Strongly universal quantum turing machines and invariance of Kolmogorov complexity, IEEE Trans. Inf. Theory 54 (2) (2008) 763–780.
[42] X. Fu, M.A. Rol, C.C. Bultink, J. Van Someren, N. Khammassi, I. Ashraf, C.G. Almudever, An experimental microarchitecture for a superconducting quantum processor, in: Proceedings of the 50th Annual IEEE/ACM International Symposium on Microarchitecture, 2017, October, pp. 813–825.
[43] X. Zou, S.P. Premaratne, M.A. Rol, S. Johri, V. Ostroukh, D.J. Michalak, A.Y. Matsuura, Enhancing a near-term quantum accelerator's instruction set architecture for materials science applications, IEEE Trans. Quantum Eng. 1 (2020) 1–7.
[44] Hybrid Quantum Asynchronous Processor (HQAP) with QCEP, P. Bir, S.P. Singh, B.K. Singh, Lakshmanan M, N. Kumari. (2019, Dec. 20). Patent 201911031043. Accessed on: Dec. 21, 2019. [Online]. Available: https://ipindiaservices.gov.in/PublicSearch/PublicationSearch.
[45] https://www.ibm.com/design/language/2x-grid/.
[46] https://en.wikipedia.org/wiki/Torus_interconnect.
[47] https://research.google/teams/applied-science/quantum/.
[48] https://www.dwavesys.com/quantum-computing.
[49] https://www.intel.in/content/www/in/en/research/quantum-computing.html.

About the author

Parth Bir is currently working with Oneirix Labs as a Research & Development Engineer. His research domains include VLSI, High Performance Computing, Computer Architecture, Quantum Computing and Neuromorphic Computing. In the past, he has worked with Argonne National Laboratory, IIT Madras and Cadence Design Systems. In addition, he is also a reviewer for the Journal of Supercomputing, Springer Nature.

CHAPTER FIVE

FPGA based neural network accelerators

Joo-Young Kim
KAIST, Daejeon, South Korea

Contents

Advances in Computers, Volume 122
ISSN 0065-2458
https://doi.org/10.1016/bs.adcom.2020.11.002

135

Abstract

Machine learning (ML) and artificial intelligence (AI) technology are revolutionizing many different fields of study in computer science as well as a wide range of industry sectors such as information technology, mobile communication, automotive, and manufacturing. As more people are using the technology in their everyday life, the demand for new hardware that enables faster and more energy-efficient AI processing is ever increasing. Over the last few years, traditional hardware makers such as Intel and Nvidia as well as start-up companies such as Graphcore and Habana Labs were trying to offer the best computing platform for complex AI workloads. Although GPU still remains the most preferred platform due to its generic programming interface, it is certainly not suitable for mobile/edge applications due to its low hardware utilization and huge power consumption. On the other hand, FPGA is a promising hardware platform for accelerating deep neural networks (DNNs) thanks to its re-programmability and power efficiency. In this chapter, we review essential computations in latest DNN models and their algorithmic optimizations. We then investigate various accelerator architectures based on FPGAs and design automation frameworks. Finally, we discuss the device's strengths and weaknesses over other types of hardware platforms and conclude with future research directions.

1. Introduction

Machine learning (ML) and artificial intelligence (AI) technology have revolutionized how computers run cognitive tasks based on a massive amount of observed data. As more industries are adopting the technology, we are facing a fast-growing demand for new hardware that enables faster and more energy-efficient processing in AI workloads. Recently, traditional hardware makers such as Intel and Nvidia as well as new start-up companies such as Graphcore, Wave Computing, and Habana Labs have tried to offer the best computing platform for complex ML algorithms. Although GPU still remains the most preferred computing platform due to its large userbase and well-established programming interface, its top spot is not forever safe due to its low hardware utilization and energy efficiency. On top of energy efficiency and programming easiness, how to adapt fast-changing ML algorithms is another hot topic in AI hardware. Field-programmable gate array (FPGA) has a clear benefit on this point, as it can reprogram or amend its processing quickly with a low power budget.

In this chapter, we review essential computations in popular deep neural network (DNN) models and their algorithmic optimizations. We then investigate various accelerator architectures based on FPGAs and automation frameworks that leverage the platform' flexibility. Finally, we discuss the device's strengths and weaknesses over other types of hardware accelerators.

2. Background

2.1 Deep neural network models and computations

Before moving into FPGA based ML systems, we first introduce the basic models of deep neural networks and their major computations. As shown in Fig. 1, a deep neural network (DNN) model is composed of multiple layers of artificial neurons called perceptron [1]. Based on network connection, the most popular models are feedforward neural networks (FNNs, or often called fully-connected), convolutional neural networks (CNNs) [2], and recurrent neural networks (RNNs) [3]. Computation-wise, each CNN layer involves 3-dimensional convolution operations between an input feature map and filter weights. On the other hand, each layer of a FNN or a RNN model is converted into a matrix-vector operation, without and with time steps, respectively. The CNN models are usually computationally intensive with small-sized filter weights while the FNN/RNN models require large memory footprints for network weights.

2.2 Field programmable gate array

FPGA is a semiconductor device designed to be reprogrammed to a desired application or a functionality by a customer after manufacturing. This feature distinguishes FPGA from another semiconductor device type called Application Specific Integrated Circuits (ASICs), which are manufactured only for a specific application and cannot be reprogrammed after manufacturing. Fig. 2 shows the basic FPGA structure that consists of an array of programmable logic cells called configurable logic blocks (CLBs) and programmable interconnection networks, and a set of programmable input and output pads around the device. They also have a rich set of embedded components such as block RAMs (BRAMs), digital signal processing (DSP) blocks for compute-intensive operations such as multiply-and-accumulate (MAC), look-up tables (LUTs), and flip-flops (FFs). FPGAs provide a variety

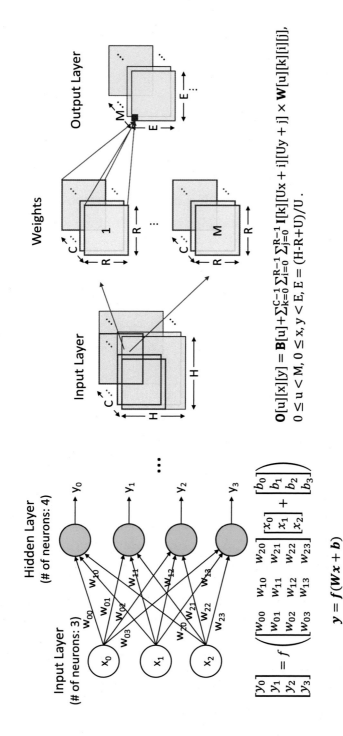

$$\mathbf{O}[u][x][y] = \mathbf{B}[u] + \sum_{k=0}^{C-1} \sum_{i=0}^{R-1} \sum_{j=0}^{R-1} \mathbf{I}[k][Ux+i][Uy+j] \times \mathbf{W}[u][k][i][j],$$
$$0 \leq u < M, 0 \leq x, y < E, E = (H-R+U)/U.$$

$$y = f(Wx + b)$$

Fig. 1 Deep neural network computations.

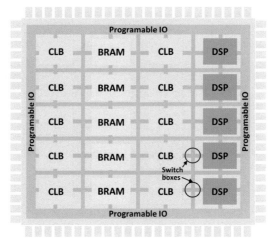

Fig. 2 Basic FPGA structure.

of IP blocks such as clock management units, high throughput memory controllers, high speed I/O links to allow users to design a whole system more easily.

Due to their low logic capacities and low operating frequencies compared to ASICs, FPGAs used to be chosen for verification platform or low-speed and low-complexity designs in the past. However, today's FPGAs can push their operating frequency up to 500 MHz with unprecedented logic density based on advanced silicon manufacturing technologies such as 14 and 7 nm. With a host of appealing features such as embedded processors, DSP blocks, and high-speed IOs on top of genuine re-programmability and agile development cycles, FPGAs are becoming a popular hardware platform for almost any types of hardware design.

2.3 FPGA based acceleration systems

Recently, FPGA draws a lot of attention as a compelling ML acceleration platform due to its adaptability to frequent algorithm changes and relatively good power efficiency compared to conventional CPU/GPU based systems. Fig. 3 shows three different architectures of FPGA based acceleration systems. The first architecture is the most typical architecture, which interconnects the host CPU and FPGA accelerator through PCIe bus. In this setting, the host CPU offloads a bulk of data to the FPGA accelerator and the FPGA accelerator performs coarse-grained tasks just like GPU. The second architecture is a network attached architecture, which is also used in

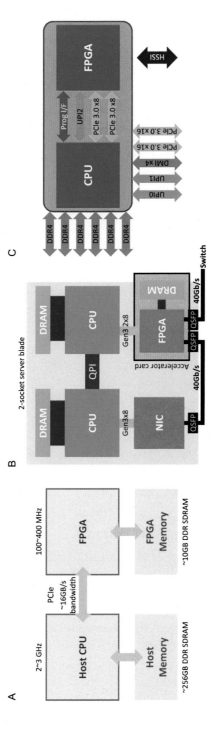

Fig. 3 FPGA based acceleration systems. (A) Local accelerator (B) Network attached (C) On-die integration.

Microsoft's Catapult project [4]. As the FPGA accelerator resides between the NIC card and a network switch, it always has the network traffic passing through and can apply in-line processing if needed. At the same time, the FPGA accelerator is connected to the host CPU through the PCIe bus like in the first case, it can be used as a local compute accelerator as well. In both architectures, the FPGA accelerator usually has its own local DRAMs on the board aside from the main memory of the host CPU. Third architecture integrates the FPGA accelerator much closer to the traditional CPU systems. In Intel's Xeon-FPGA hybrid chipset named Purley [5], recently developed for datacenter market, FPGA is connected to Intel Xeon processor through Ultra Path Interconnect (UPI) and 2 PCIe Gen 3 channels on the same die while the Xeon processor and FPGA logic share the main memory in a coherent way.

2.4 Challenges of FPGA based neural network acceleration

One of the challenges of designing a neural network accelerator on FPGA is that the device has a limited on-chip memory capacity. Although the high-end FPGA devices such as Xilinx's Virtex UltraScale + family offer up to 50 MB on-chip SRAM, most FPGA chips have less than 10 MB on-chip memory size. Fig. 4 shows the model accuracy, the number of required operations, and the number of parameters of various neural network models for image classification [6]. Their sizes range from a few millions to 150 million parameters, which will be up to several hundred MBs in data size. Therefore, for most neural network models, FPGA design needs to consider how to use on-chip memory wisely and how to reduce off-chip DRAM accesses. Another well-known challenge for the FPGA based design is a lower maximum operating frequency compared to the one of ASIC based design (usually 5–10 times slower). To overcome this disadvantage, FPGA design should be carefully optimized for its spatial architecture and should utilize the hardened specialized modules like DSPs efficiently.

Although FPGA based design has major challenges over ASIC based design, FPGAs do have good characteristics for ML applications. First, data-intensive but regular processing flow of the ML applications is suitable for FPGA's array structure. Second, FPGA's development turnaround time is much shorter than ASIC's yearly long development time, thanks to its re-programmability. This is actually a huge advantage in adapting quickly changing algorithmic advances from ML community. Third, FPGA's hardware/software co-design framework with design automation makes system-level performance improvement easier.

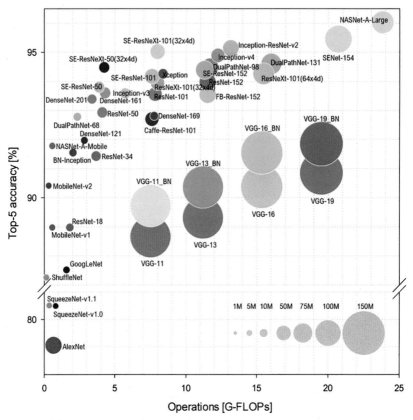

Fig. 4 Model accuracy, complexity, and size.

3. Algorithmic optimization

For the early phase of deep learning development after AlexNet was published in 2012 [2], most neural network models used the single-precision floating-point data format for both inference and training. In 2015, deep compression [7] proved that most neural network models are over-parametrized and the size of a model can be reduced with negligible accuracy loss or even with no any accuracy loss by utilizing techniques such as network pruning, data quantization, and weight sharing. Since then many researchers try to optimize the neural network models for various applications. In other words, they try to find the minimal model size and data format for target application. As FPGAs have relatively small on-chip logic and memory capacity, FPGA community also has been actively researching on model compression schemes as well as hardware-friendly algorithmic optimizations. In this section, we will review the algorithmic optimizations for efficient neural network processing on FPGAs.

3.1 Pruning

Pruning reduces the size of a neural network model by getting rid of redundant weights on the networks as shown in Fig. 5. One of the most famous and commonly used methods is to remove weights with small absolute values by making them to zero. The challenge in model pruning is how to redeem the accuracy drop caused by removing the weights. Han et al. [7] addressed this issue with weight retraining, which adjusts the weights for remaining connections after pruning. By iteratively applying pruning and retraining, they managed to reduce 89% of model weights without losing any accuracy for AlexNet. SqueezeNet [8] further reduces AlexNet by replacing 3×3 filters to 1×1 filters based on the proposed fire model. ZyncNet [9] implemented a SqueezeNet on a Kintex-7 FPGA using only its on-chip memory by removing pooling layers and resizing layers to power of 2. Energy-aware pruning [10] augmented an energy estimation model into its model pruning process. By optimizing energy efficiency with a minimal accuracy loss, the proposed pruning method achieved 1.6–3.7 times better energy efficiency in popular CNN models such as AlexNet, GoogleNet, and SqueezeNet.

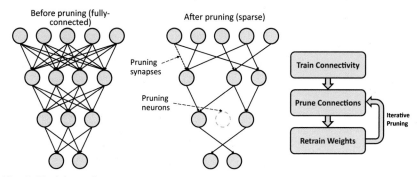

Fig. 5 Model pruning.

3.2 Data quantization

Data quantization is another popular method to reduce the size of neural networks, which replaces standard 32-bit floating-point weights and activation data to less complicated representations such as lower-bit floating-point numbers or fixed-point numbers. Usually combining with pruning that reduces the number of model parameters, data quantization minimizes the size of model by using a smaller bit for each parameter. Data quantization is also helpful for efficient hardware implementation. It not only reduces the storage and bandwidth requirement of a neural network accelerator but also allows it to use simpler arithmetic units with less logic gate counts.

Data format is essential to use different data representations. As shown in Fig. 6, IEEE 754 single-precision floating-point number format (FP32) composed of a single sign bit, 8-bit exponent, and 23-bit mantissa, where the data ranges from $\sim 1e^{-38}$ to $\sim 3e^{38}$. Its half-precision version (FP16) has 1-bit sign, 5-bit exponent, and 10-bit mantissa and ranges from $\sim 5.96e^{-8}$ to $\sim 6.5e^{4}$. Google invented its own floating-point format named brain floating-point (BFloat16) with 1-bit sign, 8-bit exponent, and 7-bit mantissa for accelerating machine learning applications [11]. BFloat16 preserves the data's dynamic range from $\sim 1e^{-38}$ to $\sim 3e^{38}$ like in FP32 to address the data truncation issue caused in training process with FP16. On the other hand, a fixed-point number is basically same as an integer except that the decimal point is located somewhere in the number. For example, a b-bit fixed-point number is composed of 1-bit sign, m-bit integer, and n-bit fraction where b equals to $1 + m + n$. The main problem of converting a floating-point data to a fixed-point data is that the dynamic range of former far exceeds that of latter. Many weights and activations will suffer from overflow/underflow problem and this may cause an accuracy loss of the model.

For fixed-point data representations, there have been studies on how much a model can lower its bit-precision without causing a significant accuracy degradation. Qiu et al. [12] chose an optimal fractional bit-width for

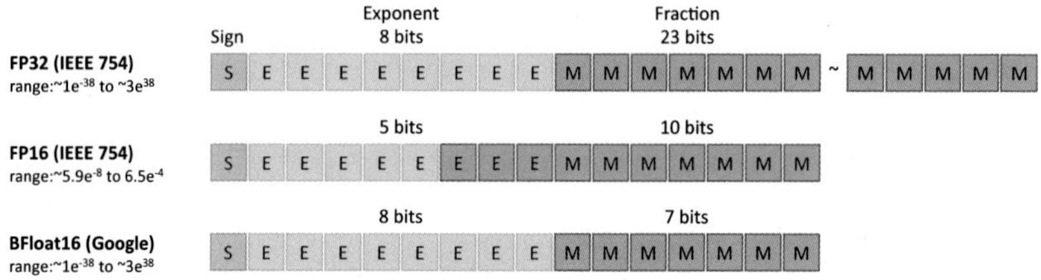

Fig. 6 Floating-point data representation.

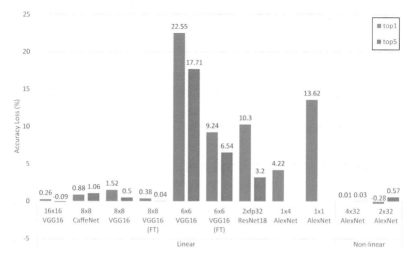

Fig. 7 Accuracy loss of various data quantization methods.

each layer to accommodate different bit-precision requirement of each layer to reduce the average bit-width of the model. Angel-Eye [13] fixed the fractional bit-width for all layers first and then applied a fine-tuning on the model. Fig. 7 shows the accuracy loss of various data quantization methods (captured from [14]) where the quantization configuration is represented as weight bit-width x activation bit-width (FT means fine-tuning). In the graph, you can see that 8-bit weight and 8-bit activation generally works robustly with negligible accuracy loss.

3.3 Data encoding and sharing

To further reduce the bit-width for each data representation, several works utilized code books that assign values to binary codes. Deep compression [7] clustered the weights of the model to reduce the number of weights represented and assigned each of them to a short binary code for further reduction. Although the work greatly reduced the number of bits used per weight, it changes read data flow with additional look-up table accesses. HashedNets [15] used a low-cost hash function to randomly group connection weights into hash buckets and shared a single representative value for all weights within the same hash bucket. CodeX [16] is an end-to-end framework that facilitates non-linear encoding, per-layer bit-width customization, and fine-tuning on a neural network accelerator on FPGA.

3.4 Fast convolution algorithms

A convolution layer in CNNs iteratively performs 3-D convolution operations on an activation volume for each of different kernels in the layer. Therefore, how to process the repeated convolution operations more efficient becomes one of the most important design issues in CNN accelerators. As the CNN accelerators on FPGA usually run out of logic resources, there have been several efforts to reduce the complexity of convolution computations using fast convolution algorithms such as FFT convolution and Winograd convolution. Basically, these algorithms transform the activation and kernel data in spatial domain to frequency domain and perform only a single element-wise multiplication for the entire convolution based on the convolution theorem. This method greatly reduces the number of multiplications and accumulations for convolution layers, but requires additional domain conversions and memory storages. Mathieu et al. [17] showed that the conversion overhead amortizes as the kernel size increases so that the FFT based convolution can reduce the total number of computations for CNN models with the kernel sizes larger than 5×5. However, as the state-of-the-art CNN models tend to use small kernel sizes such as 1×1 or 3×3, significant computation and memory overhead are expected for FFT convolutions. To address this issue, Zhang et al. [18] proposed the Overlap-and-Add method that tiles the input map into smaller blocks and processes them with zero-padding to reduce the overhead significantly.

Winograd convolution is another fast convolution algorithm developed by Winograd in 1980 [19]. The convolution of a 2-D feature map F_{in} with a kernel K using Winograd algorithm is expressed by the following equation.

$$F_{out} = \mathrm{A}^T \left[\left(GF_{in}G^T \odot \left(BKB^T \right) \right) \right] A$$

where G, B, and A are transformation matrices which are only related to the size of kernel K and feature map F_{in}. For example, if the size of feature map and kernel are 4×4 and 3×3, respectively, the transformation matrices are given like following.

$$G = \begin{bmatrix} 1 & 0 & 0 \\ \dfrac{1}{2} & \dfrac{1}{2} & \dfrac{1}{2} \\ \dfrac{1}{2} & -\dfrac{1}{2} & \dfrac{1}{2} \\ 0 & 0 & 1 \end{bmatrix} \quad B = \begin{bmatrix} 1 & 0 & -1 & 0 \\ 0 & 1 & 1 & 0 \\ 0 & -1 & 1 & 0 \\ 0 & 1 & 0 & -1 \end{bmatrix} \quad A = \begin{bmatrix} 1 & 0 \\ 1 & 1 \\ 1 & -1 \\ 0 & -1 \end{bmatrix}$$

Since the transformation matrices mostly consist of 0, 1, ± 1, and ± 0.5, Winograd based convolution operation greatly reduces the number of multiplications but it increases the number of additions. Lavin and Gray [20] showed that Winograd algorithm can achieve speed-up over 2.25 times compared to direct convolution computation. Meng and Brothers [21] proposed a new class of Winograd algorithm for an integer-based convolution for further speed up.

4. Accelerator architecture

4.1 Processing element

The most important compute component of neural network accelerators is multiply-and-add (MAC) unit as it is the basic computation for both matrix multiplication and convolution. How to pack MAC units densely and how to utilize them efficiently are main design goals for neural network accelerators. To this end, they set a basic compute unit called processing element (PE), which usually consists of a MAC unit and register/scratchpad memories, and instantiate multiple of it for high throughput neural network processing. As modern FPGAs offer a number of performance-optimized DSPs with hardened circuitry, FPGA based neural network accelerator designs rather focus on how to fully utilize the given DSPs on chip.

4.1.1 DSP architecture

For efficient PE design, we first need to understand the DSP architectures that FPGA vendors provide. Fig. 8 shows the block diagram of Xilinx's DSP unit named DSP48E2 [22] and Intel's Variable Precision DSP [23]. The DSP48E2 slice contains a 27-bit pre-adder, a 27-bit × 18-bit multiplier, and a flexible 48-bit ALU, which supports various functions including multiply, multiply-add, and multiply-accumulate up to four inputs. On the other hand, the Variable Precision DSP offers a variety of data processing modes such as four-way 9 × 9 fixed, two-way 18 × 19 fixed, 27 × 27 fixed, and floating-point modes in a single unit. For higher compute density in floating-point operation, it supports multi-way half-precision mode as well.

4.1.2 PE architectures based on dataflows

Chen [24] proposed the dataflow taxonomy that classifies the existing deep neural network accelerators into three categories: weight stationary, output stationary, and no local reuse. Weight stationary method has weight data

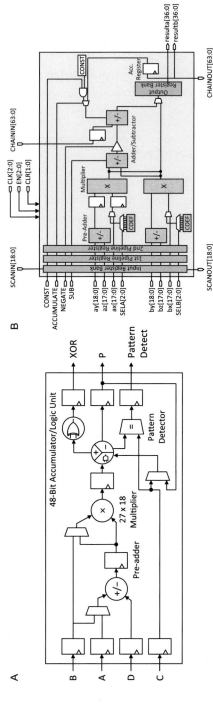

Fig. 8 DSP Units on FPGAs (A) Xilinx DSP48E2 (B) Intel Variable Precision DSP.

pinned in PE's local memories, while input activation and partial sums are moving among PEs for MAC computations. Since the weight data stay in the same location, their reusability and energy efficiency are maximized. NeuFlow [25] designed for general-purpose vision workloads is an example of weight stationary dataflow on FPGA. On the other hand, output stationary method focuses on computation output side with reducing the number of partial sum fetch/store accesses. ShiDianNao [26] is an example of output stationary because each PE is responsible for processing of an output activation by fetching the corresponding input activations from neighboring PEs. No local reuse is a more generic approach than the previous two. It tries to utilize a global buffer efficiently without fixating a specific type of data on a certain memory location. Zhang et al. [27] read the filter weights and input activations from the global buffer and processes them using MAC units and then stored the resulting output activations back to the global buffer. Fig. 9 shows the typical PE architecture in DNN accelerators. It contains buffers for input activations and filter weights, a multiplier and an accumulator to accumulate the products between activations and weights or the partial sums from another PE. For output, the PE may have a dedicated buffer for output partial sums or send the results to another PE or global buffer. Lu et al. [28] proposed a flexible dataflow architecture named FlexFlow to support various types of data paths in different layers efficiently with slight modification in PE architecture.

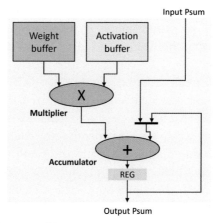

Fig. 9 Processing element architecture.

4.2 Vector architecture

While a scalar processor is a processor whose instructions operate on a single data, a vector processor refers to a processor that implements instructions operating on an array of data. Vector processors used to be very popular in supercomputer design in 1970s and 1980s as they greatly improved the computing performance for data-intensive numerical simulations. In CPU design, Intel's SIMD extensions such as MMX, SSE, and AVX, AMD's Fused Multiply-Add, and ARM's Scalable Vector Extension are all vector processing architectures. VESPA [29] is a soft vector processor specifically designed for FPGAs. It is portable to any FPGA device, scalable to larger designs, and flexible to add custom functions. Yu et al. [30] proposed a vector processing unit that can be added to a scalar processor and can collaborate together. VectorBlox [31], whose block diagram is shown in Fig. 10, is another good example of vector processor on FPGA. It targets an embedded supercomputing market with performance-oriented features like parallel scratchpads, scatter/gather DMA engine, custom vector instructions, and hybrid vector/SIMD execution.

4.3 Array architecture

Although the term "vector architecture" can be broadly used for multiple processing units controlled by a single control unit, it usually refers to the processor architecture that has a vectorized execution unit extended in a single dimension. On the other hand, an array architecture refers to the processor architecture that has a two-dimensional array of PEs at the core. Losing quite a lot of programmability, array architecture achieves much higher computational performance and data bandwidth for array-type processing.

Microsoft's CNN accelerator [32] (Fig. 11A) used a 2-D spatial mapping, which the input volume is segmented into multiple rows while each column is responsible for a different kernel weight, to maximize data parallelism for both input and output activations. In order to reuse data between neighbor layers, a network-on-chip module performs tensor reshaping when it sends the data from the output buffers to input buffers. Ma et al. [33] also chose a 2-D array architecture for their loop optimization techniques such as loop tiling, loop unrolling, and loop interchanges.

Another popular array architecture is systolic array (Fig. 11B) used in Google's TPU processor [34]. For the case of matrix multiplication, the systolic array feeds the input activations from the top row to the bottom row

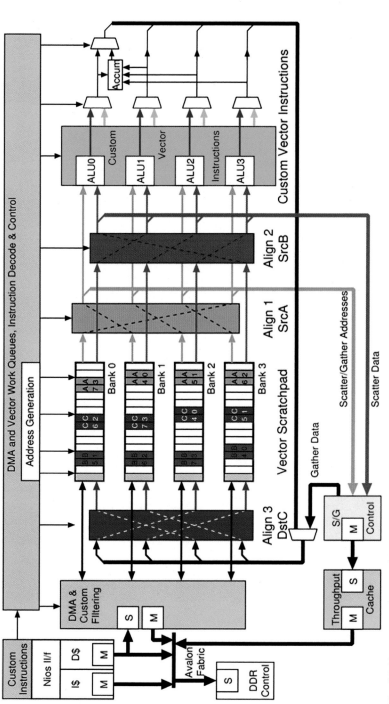

Fig. 10 Vector processor architecture.

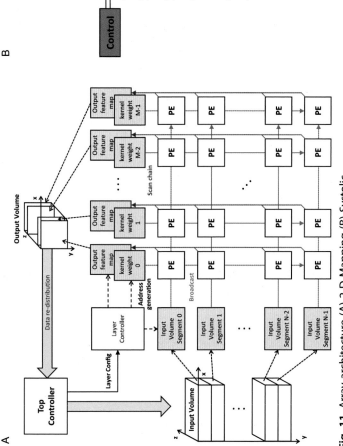

Fig. 11 Array architecture (A) 2-D Mapping (B) Systolic.

with a cycle difference while the partial sums between the incoming activations and pre-loaded weights are accumulated through the PEs in vertical direction with forming a diagonal wave-front. Once the entire NxN systolic array is fully loaded and whole PEs start to work, N final sums will be produced every cycle out of all columns. Therefore, the systolic array architecture is a very throughput-oriented and energy-efficient dataflow architecture with little control overhead. Wei et al. [35] leveraged OpenCL language to generate a 2-D systolic array that optimizes the clock speed and resource usage for a given CNN on FPGA. Moss et al. [36] used a systolic array architecture to implement high performance binarized neural networks on an Intel's Xeon + FPGA hybrid platform.

4.4 Multi-FPGA architecture

As the model size increases, it is getting harder to map the whole neural networks on a single FPGA device. With the increased demand for large-scale ML inference and training systems, there have been growing number of works using multiple FPGAs. Geng et al. [37] used multiple FPGAs for CNN training with load balance techniques. Jiang et al. [38] split a CNN model by layers and scheduled them to the heterogeneous FPGAs with different capacities for higher throughput and higher utilization. Shen et al. [39] used multiple FPGAs to accelerate lung nodule segmentation. Zhang et al. [40] and Zhang et al. [41] employed deeply pipelined multi-FPGA architectures to expand the design space of neural network acceleration and to find the optimal design for performance and energy efficiency. They also proposed scheduling algorithms based on dynamic programming in mapping CNN layers on multiple FPGAs. At datacenter scale, Microsoft BrainWave platform [42] distributed large-scale RNN models on multiple FPGAs and employed a massively parallel SIMD architecture on each FPGA to satisfy a low latency requirement for a real-time inference even with a single batch or small batch sizes.

4.5 Narrow bit-precision architecture

Along with quantization algorithms, there exists many DNN accelerators on FPGA which use narrow bit-precision architectures for compute and memory efficiency. Since FPGAs are reprogrammable, they are suitable to use custom computations for various bit precisions. Colangelo et al. [43] created a framework that explores the throughput and accuracy of various networks with different bit precisions including 8-bit, 4-bit, 3-bit, 2-bit and 1-bit.

They also showed that extremely narrow bit-precision cases such as ternary and binary can be efficiently mapped to FPGA by utilizing the logic blocks and DSP blocks. Zhao et al. [44] explained computational benefits of binarized neural networks (BNNs) and designed an BNN accelerator synthesized from C++ to FPGA-targeted Verilog. Wang et al. [45] developed a design flow to accelerate extremely low bit-width neural networks on embedded FPGAs. They allowed to use different quantization schemes for different layers and created an end-to-end automation tool that deploys the quantized networks on FPGA easily. Prost-Boucle et al. [46] proposed a scalable architecture for ternary CNNs on FPGA.

5. Design methodology

Unlike ASIC development takes a long time due to the rigorous verification process for manufacturing, FPGA development is agile and flexible because its final product is ready after a relatively short logic synthesis time. Therefore, FPGA development tends to be more open for hardware/ software co-design approaches, algorithmic changes, and design automation. Utilizing high-level languages such as OpenCL and HLS for generating custom logic is another distinctive design methodology of FPGA.

5.1 Hardware/software co-design

As deep learning requires more data-intensive computations with growing size of model parameters, hardware/software co-design methodology gains an attraction to break the performance wall caused by separated hardware and software processing on traditional computing. Han et al. [47] built the EIE processor that accelerates irregular data processing caused by model pruning and data quantization. Caffeine [48] created a hardware/software co-designed library based on a uniformed convolutional matrix-multiplication representation for efficient CNN acceleration on FPGA. Leveraging model compression techniques, a start-up company Deephi built neural network accelerators for CNN and RNN models [49].

5.2 High level synthesis

High level synthesis or HLS is a software tool that generates Verilog logic gates from C-like high-level code. It allows the users who are not familiar with logic design to develop hardware accelerators for complex ML algorithms on FPGA. Xilinx included VivadoHLS [50] on their development framework so that any user can use HLS for custom hardware development. Although HLS enables rapid prototyping for any random algorithms, there exists limitations on its achievable performance, memory bandwidth, and logic count compared to the ones from manual designs by domain experts. For ML applications, which usually have regular compute and memory access patterns, HLS provides a decent performance, short development time, and easier debugging environment. Mentor Graphic tries to use their own HLS tool named Catapult HLS to reduce the gap between deep learning frameworks such as TensorFlow and custom hardware accelerators [51]. HLS4ML [52] is a recent work from CERN collaborating with other US universities. It automatically converts open-source machine learning package models such as Tensorflow and PyTorch to digital circuits on FPGA through HLS.

5.3 OpenCL

Open Computing Language, also known as OpenCL, is another high-level language developed for parallel programming on multiple computing platforms including CPUs, GPUs, DSPs, and FPGAs. A key benefit of OpenCL is that it is a portable, open, and royalty-free standard maintained by a nonprofit technology consortium Khronous Group. Based on standard ANSI C (C99), OpenCL extends the programming language to extract parallelism explicitly. Both Intel and Xilinx actively use OpenCL for their board support packages and software tools such as Intel FPGA SDK for OpenCL and Xilinx SDAccel. Since OpenCL is a synthesizable language for parallel programming, there have been many acceleration works on FPGA that utilizes the language. Suda et el. [53] designed a throughput-optimized OpenCL-based FPGA accelerator for large-scale CNNs. Intel published an architecture named Deep Learning Accelerator (DLA) written in OpenCL which maximizes data reuse and minimizes external memory bandwidth [54]. Wang et al. [55] devised a performance analysis framework to improve the bottlenecks of the OpenCL applications targeted on FPGA. Mu et al. [56] developed a framework that performs automatic FPGA code generation, coarse-grained modeling, and fine-grained modeling at the same time using OpenCL.

5.4 Design automation framework

Although it is not correct to say that FPGA has a complete tool flow from software application to hardware implementation, it is one of the most

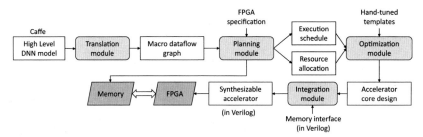

Fig. 12 Design automation from neural network model to FPGA.

available and affordable platforms for optimizing a whole computing stack if needed. There have been several approaches to create automation frameworks that generate custom hardware design from target applications with minimal user inputs.

Fig. 12 illustrates the overall block diagram of the automation framework from Sharma et al. [57], which generates synthesizable accelerators for multiple FPGA platforms based on the specification of a neural network model written in Caffe framework. First, the translation module transforms the specified neural network model into macro dataflow graphs such as input, output, convolution, and pool. This specifies data dependency among nodes and allows the framework to achieve layer-specific optimizations using a wide range of accelerators. Second, the planning module receives the target FPGA's specification and optimizes the hardware templates for the target device. It partitions computations of each layer and generates a static execution scheduling that maximizes the data reuse and concurrency under the budget of on-chip memories. Then, the optimization module generates the final accelerator core using multiple hand-tuned design templates and customizes them based on the execution scheduling by the planner module. It also produces state machines and micro-codes for the accelerator if necessary. Finally, the integration module glues a memory interface for the accelerator and generates synthesizable Verilog code for the target FPGA. Wang et al. [58] also presented a design automation tool that builds an accelerator for a custom CNN configuration based on an RTL generator and a compiler that generates a dataflow under the user-specified constraints. Venieris et al. [59] surveyed existing tool flows for mapping CNNs on FPGAs and suggested recent research trends such as sparsity handling, low-precision data types, training, and integration to high-level deep learning frameworks.

6. Applications

6.1 Image recognition

Since the ImageNet competition ignited deep learning revolution, CNN based image recognition has been one of the most popular ML applications with huge demand from auto and manufacturing industry. FPGA has been a good platform for image recognition as CNNs have moderate model sizes with abundant data reuse opportunities. Recently, there have been many FPGA based accelerators for real-time object detection based on YOLO and SSD algorithm [60–63].

6.2 Speech recognition

Speech recognition is another significant ML application and there have been several accelerators on FPGAs that run RNN/LSTM models [64–66]. As the sizes of state-of-the-art speech recognition models are quite large, most of the previous works tried to compress the target model to fit a single FPGA. Kwon et al. [67] implemented a BrainWave NPU and used multiple of it on multiple FPGAs to scale-up the RNNs.

6.3 Autonomous vehicle

In automotive industry, self-driving car and driving-assistance systems are at the center of future innovations. Tesla recently developed a new ASIC named full self-driving (FSD) [68] to meet the computational requirements of intensive sensor data processing as well as machine learning algorithm processing. It also needs to be complaint with the safety and security requirements for autonomous vehicle. On FPGA side, there have been several prototype accelerators focusing on real-time road scene analysis using camera and Lidar sensors [69–71].

6.4 Cloud computing

FPGAs used to be a second-class citizen in large-scale computing systems such as cloud computing and high-performance computing, where the main performance-critical computing was done by CPUs, GPUs, and other ASICs. However, since Microsoft proved that FPGAs can perform at cloud scale with improving the system performance and reducing the total cost of operation on the Bing search ranking case in 2014 [72], global IT companies started to use FPGAs as a major hardware acceleration platform. Chinese search giant Baidu has been deploying Xilinx FPGA boards in their datacenters for image and speech recognition since 2016. Another Chinese IT company Alibaba is using FPGAs for their personalized recommendation systems. Although Microsoft used FPGAs for their infrastructure and internal services including network and storage for the first time, Amazon Web Services (AWS) is the first company which offered FPGAs to the public. On AWS's EC2, a user can instantiate F1 instances and use FPGA development tools and services. Likewise, AliCloud provides FPGAs-as-a-Searvice (FaaS) to general users.

7. Evaluation

As FPGAs become a major hardware acceleration platform, there have been approaches to compare FPGAs against other hardware platforms such as multi-core CPUs, GPUs, and ASICs. Most of the works evaluated the performance metrics of each platform on compute-intensive domains such as matrix-vector multiplication, deep neural networks, computer visions, etc. Although FPGA platform hardly beats GPU in an absolute performance, many works found that FPGA is promising as it shows comparable performances for many useful cases with much higher energy efficiencies.

7.1 Matrix-vector multiplication

Even before the rise of deep learning, matrix-vector multiplication and general matrix multiplication (GEMM) were one of the most important compute kernels in high performance computing (HPC) domain. Kestur et al. [73] evaluated Basic Linear Algebra Subroutines (BLAS) library performance on CPU and GPU, and compared them against their accelerator's performance implemented on a FPGA platform. As a result, the FPGA platform managed to offer comparable performances against the CPU and GPU platform as well as it showed 2.7–293 times better energy efficiency for some test scenarios. Dorrance et al. [74] presented a highly utilized sparse matrix-vector multiplication (SpMxV) engine that outperforms the CPU and GPU with 38–50 times better energy efficiency. Vestias and Neto [75] reported peak performances of latest CPUs, GPUs, and FPGAs and surveyed on their application performances and power efficiencies. They suggested that FPGAs become competitive with custom floating-point or fixed-point representations but they have more trouble running software-based HPC applications than the other two platforms.

7.2 Deep neural networks

Nurvitadhi et al. [76] compared Intel's two latest FPGAs, Arria 10 and Stratix 10, against Titan X GPU. They showed that the Stratix 10 FPGA can perform 10%, 50%, and 5.4 times better in performance than the Titan X GPU on GEMM operations for pruned, integer 6-bits, and binarized DNNs, respectively. In [77,78], Nurvitadhi and the authors compared an Arria 10 FPGA against a 14-nm ASIC as well as CPU and GPU for an advanced variant of RNN and binarized neural networks. They proved that the FPGA and ASIC can be much better in performance and energy efficiency than the state-of-the-art CPU and GPU system. They also observed that the FPGA is less efficient than the ASIC, but the gap between the two can be reduced if the FPGA design heavily utilizes hardened DSPs and BRAMs. Cong et al. [79] ported 15 benchmark kernels on a Virtex 7 FPGA using HLS and compared the FPGA's performances against Nvidia K40 GPU's to analyze which device performs better under various workloads. They found that the FPGA usually achieves a higher number of operations per cycle in customized deep pipelines, but its parallelization factor is limited due to a low off-chip memory bandwidth. As a result, the FPGA showed comparable or higher performance than the GPU for 6 out of the 15 kernels while consuming less than one third of the GPU power. A FPGA acceleration company InAccel recently published a report that FPGA's performance for a logistic regression training is comparable to GPU's performance but it can provide a much better total cost of operation in cloud computing [80].

7.3 Vision kernels

FPGAs have been used for embedded vision applications due to their customizable real-time image processing and good energy efficiency. Recently, Xilinx Research Labs with Iowa State University published a paper on comparing energy efficiency of CPU, GPU, and FPGA for vision kernels [81]. They benchmarked popular vision libraries such as OpenCV, VisionWorks, and xfOpenCV on three representative embedded hardware platforms: ARM57 CPU, Jetson TX2 GPU, and ZCU102 FPGA. Although the GPU achieved the lowest energy per frame results for simple kernels, the FPGA outperformed the other two for more complicated kernels and complete vision applications with a factor of 1.2–22.3. They also claimed that the FPGA tend to perform better if the pipeline complexity of a target vision application grows.

8. Future research directions

We have reviewed many FPGA based deep neural network accelerators from a few important aspects of hardware design such as algorithmic optimization, hardware architecture, design methodology, and application. The FPGA device gained re-programmability and a shorter development cycle with the sacrifice of logic speed and memory capacity. Because of that, ML acceleration researches on FPGA platform have been different from the case of ASIC platform. However, thanks to recent advances in device speed and capacity, FPGA is gaining traction in a wide range of fields such as high-performance computing, cloud computing, networking, automotive vision systems, and image/sound recognition. I firmly believe that the FPGA will play a key role in hardware acceleration for the post Moore's law era.

For future research, I suggest the following four directions. First, I expect that FPGA based neural network accelerator design will have two mainstreams. One stream will be more customized design with aggressive software co-design for maximizing energy efficiency while the other is more generalized design that can support a variety of network models. Second, there will be approaches to integrate FPGA accelerators into existing or new software frameworks. As a result, developers will have complete tool flows for ML acceleration which penetrate from application software to hardware implementation. Third, FPGA based neural network accelerator research will be towards deep learning training, not inference. Lastly, there will be more researches on multi-FPGA systems than standalone systems.

References

[1] D.E. Rumelhart, G.E. Hinton, R.J. Williams, Learning representations by back-propagating errors, Nature 323 (1986) 533–536.
[2] A. Krizhevsky, I. Sutskever, G. Hinton, Imagenet classification with deep convolutional neural networks, in: Conference on Neural Information Processing Systems (NIPS), 2012.
[3] A. Graves, et al., A novel connectionist system for unconstrained handwriting recognition, IEEE Trans. Pattern Anal. Mach. Intell. 31 (5) (2009) 855–868.
[4] A. Caulfield, et al., A cloud-scale acceleration architecture, in: IEEE/ACM International Symposium on Microarchitecture (MICRO), 2016.
[5] https://www.intel.com/content/www/us/en/design/products-and-solutions/processors-and-chipsets/purley/intel-xeon-scalable-processors.html.
[6] S. Bianco, R. Cadene, L. Celona, P. Napoletano, Benchmark analysis of representative deep neural network architectures, IEEE Access 6 (2018) 64270–64277.
[7] S. Han, H. Mao, W.J. Dally, Deep Compression: Compressing Deep Neural Networks with Pruning, Trained Quantization and Huffman Coding, arXiv:1510.00149, 2015.

[8] F.N. Iandola, et al., Squeezenet: AlexNet-level Accuracy with 50x Fewer Parameters and <0.5MB Model Size, arXiv:1602.07360, 2016.

[9] D. Gschwend, Zynqnet: An FPGA-accelerated Embedded Convolutional Neural Network, arXiv:2005.06892, 2020.

[10] T.J. Yang, Y.H. Chen, V. Sze, Designing energy-efficient convolutional neural networks using energy-aware pruning, in: IEEE Conference on Computer Vision and Pattern Recognition (CVPR), 2017.

[11] https://en.wikipedia.org/wiki/Bfloat16_floating-point_format.

[12] J. Qiu, et al., Going deeper with embedded FPGA platform for convolutional neural network, in: ACM/SIGDA International Symposium on Field-Programmable Gate Arrays (FPGA), 2016.

[13] K. Guo, et al., Angel-eye: a complete design flow for mapping CNN onto embedded FPGA, IEEE Trans. Comput. Aided Des. Integr. Circuits Syst. 37 (1) (2018) 35–47.

[14] K. Guo, et al., A survey of FPGA-based neural network inference accelerator, ACM Trans. Reconfig. Technol. Syst. 12 (1) (2019) 1–26.

[15] W. Chen, et al., Compressing neural networks with the hashing trick, in: International Conference on Machine Learning (ICML), 2015.

[16] M. Samragh, M. Javaheripi, F. Koushanfar, CodeX: Bit-Flexible Encoding for Streaming Based FPGA Acceleration of DNNs, Arxiv:1901.05582, 2019.

[17] M. Mathieu, M. Henaff, Y. LeCun, Fast Training of Convolutional Networks Through FFTs, arXiv:1312.5851, 2013.

[18] C. Zhang, V. Prasanna, Frequency domain acceleration of convolutional neural networks on CPU-FPGA shared memory system, in: ACM/SIGDA International Symposium on Field-Programmable Gate Arrays (FPGA), 2017.

[19] S. Winograd, Arithmetic Complexity of Computations, vol. 33, SIAM, 1980.

[20] A. Lavin, S. Gray, Fast Algorithms for Convolutional Neural Networks, arXiv:1509.09308, 2015.

[21] L. Meng, J. Brothers, Efficient Winograd Convolution via Integer Arithmetic, arXiv:1901.01965, 2019.

[22] https://www.xilinx.com/support/documentation/user_guides/ug579-ultrascale-dsp.pdf.

[23] https://www.intel.com/content/dam/www/programmable/us/en/pdfs/literature/hb/stratix-10/ug-s10-dsp.pdf.

[24] Y.-H. Chen, J. Emer, V. Sze, An energy-efficient reconfigurable accelerator for deep convolutional neural networks, in: ACM/IEEE International Symposium on Computer Architecture (ISCA), 2016.

[25] C. Farabet, et al., NeuFlow: a runtime reconfigurable dataflow processor for vision, in: IEEE Workshop on Embedded Computer Vision, 2011.

[26] Z. Du, et al., ShiDianNao: shifting vision processing closer to the sensor, in: ACM/IEEE International Symposium on Computer Architecture (ISCA), 2015.

[27] C. Zhang, et al., Optimizing FPGA-based accelerator design for deep convolutional neural networks, in: ACM/SIGDA International Symposium on Field-Programmable Gate Arrays (FPGA), 2015.

[28] W. Lu, et al., FlexFlow: a flexible dataflow accelerator architecture for convolutional neural networks, in: IEEE International Symposium on High Performance Computer Architecture (HPCA), 2017.

[29] P. Yiannacouras, J.G. Steffan, J. Rose, VESPA: portable, scalable, and flexible FPGA-based vector processors, in: ACM International Conference on Compilers, Architecture and Synthesis for Embedded Systems (CASES), 2008.

[30] J. Yu, G. Lemieux, C. Eagleston, Vector processing as a soft-core CPU accelerator, in: ACM/SIGDA International Symposium on Field Programmable Gate Arrays (FPGA), 2008.

[31] A. Severance, G.G. Lemieux, Embedded supercomputing in FPGAs with the VectorBlox MXP matrix processor, in: IEEE International Conference on Hardware/Software Codesign and System Synthesis (CODES + ISSS), 2013.

[32] K. Ovtcharov, et al., Accelerating deep convolutional neural networks using specialized hardware, in: Microsoft Research Whitepaper, 2015.

[33] Y. Ma, Y. Cao, S. Vrudhula, J.S. Seo, Optimizing the convolution operation to accelerate deep neural networks on FPGA, IEEE Trans. VLSI Syst. 26 (7) (2018) 1354–1367.

[34] N.P. Jouppi, et al., In-datacenter performance analysis of a tensor processing unit, in: ACM/IEEE International Symposium on Computer Architecture (ISCA), 2017.

[35] X. Wei, et al., Automated systolic array architecture synthesis for high throughput CNN inference on FPGAs, in: ACM/IEEE Design Automation Conference (DAC), 2017.

[36] D.J.M. Moss, et al., High performance binary neural networks on the Xeon + FPGA™ platform, in: IEEE International Conference on Field Programmable Logic and Applications (FPL), 2017.

[37] T. Geng, et al., FPDeep: acceleration and load balancing of CNN training on FPGA clusters, in: IEEE International Symposium on Field-Programmable Custom Computing Machines (FCCM), 2018.

[38] W. Jiang, et al., Heterogeneous FPGA-based cost-optimal design for timing-constrained CNNs, IEEE Trans. Comput. Aided Des. Integr. Circuits Syst. 37 (11) (2018) 2542–2554.

[39] J. Shen, et al., Scale-out acceleration for 3D CNN-based lung nodule segmentation on a multi-FPGA system, in: ACM/IEEE Design Automation Conference (DAC), 2019.

[40] C. Zhang, et al., Energy-efficient CNN implementation on a deeply pipelined FPGA cluster, in: International Symposium on Low Power Electronics and Design (ISLPED), 2016.

[41] W. Zhang, et al., "An efficient mapping approach to large-scale DNNs on multi-FPGA architectures," IEEE Design, Automation & Test in Europe Conference & Exhibition (DATE), 2019.

[42] J. Fowers, et al., A configurable cloud-scale DNN processor for real-time AI, in: ACM/IEEE International Symposium on Computer Architecture (ISCA), 2018.

[43] P. Colangelo, et al., Exploration of low numeric precision deep learning inference using intel FPGAs, in: IEEE International Symposium on Field-Programmable Custom Computing Machines (FCCM), 2018.

[44] R. Zhao, et al., Accelerating binarized convolutional neural networks with software-programmable FPGAs, in: ACM/SIGDA International Symposium on Field-Programmable Gate Arrays (FPGA), 2017.

[45] J. Wang, et al., Design flow of accelerating hybrid extremely low bit-width neural network in embedded FPGA, in: IEEE International Conference on Field Programmable Logic and Applications (FPL), 2018.

[46] A. Prost-Boucle, et al., Scalable high-performance architecture for convolutional ternary neural networks on FPGA, in: IEEE International Conference on Field Programmable Logic and Applications (FPL), 2017.

[47] S. Han, et al., EIE: efficient inference engine on compressed deep neural network, in: ACM/IEEE International Symposium on Computer Architecture (ISCA), 2016.

[48] C. Zhang, Z. Fang, P. Zhou, J. Cong, Caffeine: towards uniformed representation and acceleration for deep convolutional neural networks, in: International Conference on Computer-Aided Design (ICCAD), 2016.

[49] K. Guo, et al., Software-hardware codesign for efficient neural network acceleration, IEEE Micro 37 (2) (2017) 18–25.

[50] https://www.xilinx.com/products/design-tools/vivado/integration/esl-design.html.

[51] https://embeddedvisionsummit.com/2019summit/session/using-high-level-synthesis-to-bridge-the-gap-between-deep-learning-frameworks-and-custom-hardware-accelerators/
.

[52] G.D. Guglielmo, et al., Compressing Deep Neural Networks on FPGAs to Binary and Ternary Precision with HLS4ML, arXiv:2003.06308, 2020.

[53] N. Suda, et al., Throughput-optimized OpenCL-based FPGA accelerator for large-scale convolutional neural networks, in: ACM/SIGDA International Symposium on Field-Programmable Gate Arrays (FPGA), 2016.

[54] U. Aydonat, et al., An OpenCL deep learning accelerator on Arria 10, in: ACM/SIGDA International Symposium on Field-Programmable Gate Arrays (FPGA), 2017.

[55] Z. Wang, et al., A performance analysis framework for optimizing OpenCL applications on FPGAs, in: IEEE International Symposium on High Performance Computer Architecture (HPCA), 2017.

[56] J. Mu, et al., Optimizing OpenCL-based CNN design on FPGA with comprehensive design space exploration and collaborative performance modeling, ACM Trans. Reconfig. Technol. Syst. 13 (3) (2020) 13–28.

[57] H. Sharma, et al., From high-level deep neural models to FPGAs, in: IEEE/ACM International Symposium on Microarchitecture (MICRO), 2016.

[58] Y. Wang, et al., DeepBurning: automatic generation of FPGA-based learning accelerators for the neural network family, in: ACM/IEEE Design Automation Conference (DAC), 2016.

[59] S.I. Venieris, A. Kouris, C.-S. Bouganis, Toolflows for mapping convolutional neural networks on FPGAs: a survey and future directions, ACM Comput. Surv. 51 (3) (2018) 56.

[60] L. Lu, et al., Evaluating fast algorithms for convolutional neural networks on FPGAs, in: IEEE International Symposium on Field-Programmable Custom Computing Machines (FCCM), 2017.

[61] H. Nakahara, H. Yonekawa, T. Fujii, S. Sato, A lightweight YOLOv2: a binarized CNN with a parallel support vector regression for an FPGA, in: ACM/SIGDA International Symposium on Field-Programmable Gate Arrays (FPGA), 2018.

[62] D.T. Nguyen, T.N. Nguyen, H. Kim, H.J. Lee, A high-throughput and power-efficient FPGA implementation of YOLO CNN for object detection, IEEE Trans. Very Large Scale Integr. VLSI Syst. 27 (8) (2019) 1861–1873.

[63] D. Caiwen, et al., REQ-YOLO: a resource-aware, efficient quantization framework for object detection on FPGAs, in: ACM/SIGDA International Symposium on Field-Programmable Gate Arrays (FPGA), 2019.

[64] S. Li, et al., FPGA acceleration of recurrent neural network based language model, in: IEEE International Symposium on Field-Programmable Custom Computing Machines (FCCM), 2015.

[65] S. Han, et al., ESE: efficient speech recognition engine with sparse LSTM on FPGA, in: ACM/SIGDA International Symposium on Field-Programmable Gate Arrays (FPGA), 2017.

[66] M. Lee, et al., FPGA-based low-power speech recognition with recurrent neural networks, in: IEEE International Workshop on Signal Processing Systems, 2016.

[67] D. Kwon, et al., Scalable multi-FPGA acceleration for large RNNs with full parallelism levels, in: ACM/IEEE Design Automation Conference (DAC), 2020.

[68] P. Bannon, et al., Tesla compute and redundancy solution for the full self-driving computer, in: IEEE Hot Chips Symposium, 2019.

[69] Y. Lyu, L. Bai, X. Huang, ChipNet: real-time LiDAR processing for drivable region segmentation on an FPGA, IEEE Trans. Circuits Syst. Regul. Pap. 66 (5) (2019) 1769–1779.

[70] J. Peng, et al., Multi-task ADAS system on FPGA, in: IEEE International Conference on Artificial Intelligence Circuits and Systems (AICAS), 2019.

[71] A. Kojima, Y. Nose, Development of an autonomous driving robot car using FPGA, in: IEEE International Conference on Field-Programmable Technology (FPT), 2018.

[72] A. Putnam, et al., A reconfigurable fabric for accelerating large-scale datacenter services, in: ACM/IEEE International Symposium on Computer Architecture (ISCA), 2014.

[73] S. Kestur, J.D. Davis, O. Williams, BLAS comparison on FPGA, CPU and GPU, in: IEEE Computer Society Annual Symposium on VLSI, 2010.

[74] R. Dorrance, F. Ren, D. Markovic, A scalable sparse matrix-vector multiplication kernel for energy-efficient sparse-blas on FPGAs, in: ACM/SIGDA International Symposium on Field-Programmable Gate Arrays (FPGA), 2014.

[75] M. Vestias, H. Neto, Trends of CPU, GPU and FPGA for high-performance computing, in: IEEE International Conference on Field Programmable Logic and Applications (FPL), 2014.

[76] E. Nurvitadhi, et al., Can FPGAs beat GPUs in accelerating next-generation deep neural networks? in: ACM/SIGDA International Symposium on Field-Programmable Gate Arrays (FPGA), 2017.

[77] E. Nurvitadhi, et al., Accelerating recurrent neural networks in analytics servers: comparison of FPGA, CPU, GPU, and ASIC, in: IEEE International Conference on Field Programmable Logic and Applications (FPL), 2016.

[78] E. Nurvitadhi, et al., Accelerating binarized neural networks: comparison of FPGA, CPU, GPU, and ASIC, in: IEEE International Conference on Field-Programmable Technology (FPT), 2016.

[79] J. Cong, et al., Understanding performance differences of FPGAs and GPUs, in: ACM/SIGDA International Symposium on Field-Programmable Gate Arrays (FPGA), 2018.

[80] https://inaccel.com/cpu-gpu-or-fpga-performance-evaluation-of-cloud-computing-platforms-for-machine-learning-training/.

[81] M. Qasaimeh, et al., Comparing energy efficiency of CPU, GPU and FPGA implementations for vision kernels, in: IEEE International Conference on Embedded Software and Systems (ICESS), 2019.

About the author

Joo-Young Kim received the B.S., M.S., and Ph. D degree in Electrical Engineering from Korea Advanced Institute of Science and Technology (KAIST), in 2005, 2007, and 2010, respectively. He is currently an assistant professor in the School of Electrical Engineering at KAIST since September 2019. His research interests span various aspects of hardware design including VLSI design, computer architecture, FPGA, domain specific accelerators, hardware/software co-design, and agile hardware development. Before joining KAIST, Joo-Young was a Senior Hardware Engineering Lead at Microsoft Azure working on hardware acceleration for its hyper-scale big data analytics platform named Azure Data Lake. Before that, he was one of the initial members of Catapult project at Microsoft Research, where he deployed a fabric of FPGAs in datacenters to accelerate critical cloud services such as machine learning, data storage, and networking.

Joo-Young is a recipient of the 2016 IEEE Micro Top Picks Award, the 2014 IEEE Micro Top Picks Award, the 2010 DAC/ISSCC Student Design Contest Award, the 2008 DAC/ISSCC Student Design Contest Award, and the 2006 A-SSCC Student Design Contest Award. He serves as Associate Editor for the IEEE Transactions on Circuits and Systems I: Regular Papers (2020–2021).

This research was supported by the MSIT (Ministry of Science and ICT), Korea, under the ITRC (Information Technology Research Center) support program (IITP-2020-0-01847) supervised by the IITP (Institute of Information & Communications Technology Planning & Evaluation).

CHAPTER SIX

Deep learning with GPUs

Won Jeon, Gun Ko, Jiwon Lee, Hyunwuk Lee, Dongho Ha, and Won Woo Ro
School of Electrical and Electronic Engineering, Yonsei University, Seoul, South Korea

Contents

Abstract

Deep learning has been extensively researched in various areas and scales up very fast in the last decade. It has deeply permeated into our daily life, such as image classification, video synthesis, autonomous driving, voice recognition, and personalized recommendation systems. The main challenge for most deep learning models, including convolutional neural networks, recurrent neural networks, and recommendation models, is

167

their large amount of computation. Fortunately, most computations in deep learning applications are parallelizable, therefore they can be effectively handled by throughput processors, such as Graphics Processing Units (GPUs). GPUs support high throughput, parallel processing performance, and high memory bandwidth and becomes the most popularly adopted device for deep learning. As a matter of fact, many deep learning workloads from mobile devices to data centers are performed by GPUs. In particular, modern GPU systems provide specialized hardware modules and software stacks for deep learning workloads. In this chapter, we present detailed analysis on the evolution of GPU architectures and the recent hardware and software supports for more efficient acceleration of deep learning in GPUs. Furthermore, we introduce leading-edge researches, challenges, and opportunities of running deep learning workloads on GPUs.

1. Deep learning applications using GPU as accelerator

Graphics Processing Units (GPUs) have been widely used to improve execution speed of various deep learning applications. With hundreds of thousands running threads and a massive amount of hardware computation units, the GPUs can provide strong computing power and performance. In addition, the wide support of deep learning frameworks including PyTorch [1], Caffe [2], and TensorFlow [3] enable rapid deployment of GPUs for modern high level AI algorithms, both for inferencing and training. In fact, GPUs are heavily used in the various areas and devices ranging from the large-scaled computing systems such as the servers in data centers to the small-scaled mobile devices running AI applications.

One of the biggest merits using GPUs in the deep learning application is the high programmability and API support for AI. Although GPUs spend a large area of silicon with a heavy power consumption compared to the other accelerators, the portability and programmability of GPUs provided with a help of rich software support makes GPUs popularly used in the AI business. Additionally, GPUs show higher flexibility than the other AI accelerators and take advantage of running many kinds of applications on it. With those advantages, GPUs can be used for acceleration in various ways. In each case, the GPU hardware configuration or usage method varies depending on the applications and the device characteristics. For example, Internet of Things (IoT) systems [4], smart city [5], health care [6], robotics [7], autonomous driving [8], and super-resolution [9] are accelerating AI applications with GPUs, as shown in Fig. 1. In each case, there is a similar pattern and difference in how GPUs are used for AI acceleration. In this introductory section, we will first look through the applications using GPUs for accelerating AI and how those AI applications use GPU for machine learning acceleration.

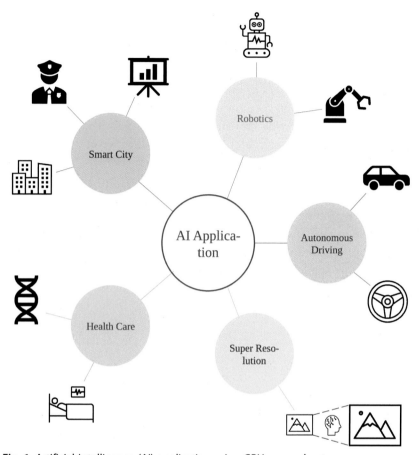

Fig. 1 Artificial Intelligence (AI) application using GPU as accelerator.

Firstly, we look into the deep learning applications combined with the IoT technologies and GPU acceleration. The IoT applications are using AI for algorithms such as vision recognition and speech recognition to provide fast and efficient results [10,11]. Smart city and health care are the representative areas that use IoT applications with AI. These applications provide services to users through edge devices using various sensors and manage data through a centralized server. GPUs contribute to the training phase of AI on the centralized servers in these applications and are also used for AI acceleration in an embedded SoC (System on a Chip) developed for edge devices. Looking at the case of a smart city, the most common edge device in a smart city is a camera. In the smart city system, the camera performs various AI powered operations such as object detection and face recognition and is used for security and traffics management. The execution information of these AI applications is collected in a centralized data center, used in the smart city through additional inference steps depending on the situation, and also used for periodic training in the data center. The GPU can be used in both the data center and edge devices of the smart city system. In the data center, a multi-GPU system suitable for a large amount of computation is used, and in the case of edge devices, the limitation of area and power consumption is relatively small due to the characteristics of the camera used in the smart city, so an SoC solution with multiple GPUs can be used.

In the field of health care, deep learning is used for gene analysis [12] and patient care [13,14]. The gene analysis is an application that consumes a large amount of computing power and is performed on a server with multiple GPUs. On contrary, the patient care applications usually operate on wearable devices, for example a patient bed for each user. Similar to the smart city system, the application training process is performed on the server of the service provider with multi-GPU systems, and only the inference process for each user occurs on the edge device. In the health care area, edge devices need to process multiple AI applications for users, and in such an environment, GPUs that can execute AI applications more flexibly and have an advantage as an accelerator.

Another area where GPU supports the acceleration of AI applications through SoCs format is autonomous robotics. AI applications such as visual recognition and route finding can be used in manufacturing robots [15]. Since robotics need high responsiveness, many robotics use SoCs to accelerate AI applications. GPUs in the SoCs accelerate inference of AI applications at real-time robotics operates. Due to the characteristics of robotics that run various AI applications in one machine with visual processing, it can provide advantages to employ GPUs to SoCs.

It has been also well known that autonomous driving can be one of the main AI applications accelerated by GPUs. Various steps for autonomous driving runs through GPU systems in various reasons. First, the training step of AI for autonomous driving runs in a server-scale environment due to its large scale of resources requirement; the server systems usually accelerate AI training with multi-GPU system. The inference step of AI for autonomous driving runs when an autonomous vehicle is driving, and many companies design special embedded boards to their vehicle for accelerating the inference step. Unlike other edge devices introduced above, autonomous vehicles are relatively free in terms of area and power. For these reasons, the embedding boards consist of a combination of various processors for

AI acceleration. In the multi-processor board, GPUs are used for graphics processing and cooperate with other accelerators to accelerate AI applications using high flexibility of GPUs.

Through the following parts of this chapter, we will look into the details of the reasons why the GPU takes an important position in the deep learning through the characteristics of GPUs. We will look at the overall characteristics and history of GPUs, and what kind of the hardware and software supports GPU vendors provide for deep learning applications. Moreover, we will introduce studies on AI acceleration using GPU proposed in the academic field and summarize the pros and cons of GPU as AI accelerators.

2. Overview of graphics processing unit

In this section, we present overview of the GPU architectures and designs. We also introduce the differences between the conventional CPUs and GPUs. We summarize the evolution of GPUs over the last decade toward the end of the section.

2.1 History and overview of GPU architecture

The first GPUs (as a separate device) were introduced in the late 1990s [16]. In fact, the early GPUs were developed to mainly process real-time graphics rendering operations. Since the graphics computations are highly parallelizable, the GPU hardware has been developed toward higher throughput. For that, GPUs employ a larger number of computing units compared to CPUs. As shown in Fig. 2, most of the hardware resources in CPUs are occupied by the control logic and cache memory. With the resources, CPUs provide

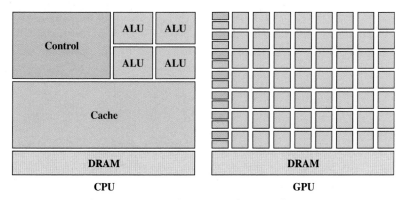

Fig. 2 Comparison between microarchitectures of CPU and GPU.

complicated out-of-order execution, branch prediction, and higher cache locality, thus achieve higher single-core performance. On the other hand, computing units occupy most of the hardware resources in GPUs, as shown in the figure. With the massive number of computing units, GPUs are able to achieve higher Thread-Level Parallelism (TLP) and throughput. In details, computing units in GPUs are architected as Single Instruction, Multiple Data (SIMD) structures and with the SIMD units, a GPU efficiently performs parallel execution of the massive number of threads. In the 2000s, the concept of General-Purpose computing on Graphics Processing Units (GPGPU) has firstly been introduced to take advantage of the high throughput of GPUs in areas other than graphics. Nowadays, with the software and hardware supports for GPGPU, the field of GPU applications has been expanded to various high-level applications including machine learning, blockchain, weather forecast, molecular dynamics, and many High-Performance Computing (HPC) applications. Furthermore, the recent NVIDIA GPUs include domain-specific processors, such as *Tensor Core*, other than the conventional SIMD processors [17]. With the various computing capability and applicable areas, GPUs are widely used in edge devices, desktops, laptops, servers, and high-performance data centers.

In this section, we provide detailed hardware architecture of a GPU and its operation flow. In addition, we briefly introduce the evolution of NVIDIA GPUs and the state-of-the-art techniques for GPUs.

2.2 Structure of GPGPU applications

A key approach to fully utilize the high throughput of GPU devices relies on the loop unrolling of a loop-based sequential code. When a CPU performs operations on non-scalar data (e.g., vectors, matrices, and tensors), it requires serialized loop execution, as shown in Fig. 3. In the figure, a total of 10 operations are taken to perform simple vector addition operation with the single-core CPU since the vector size is 10. In case there are no

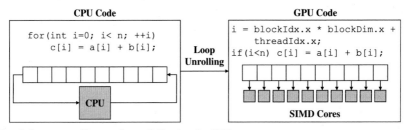

Fig. 3 Loop unrolling and parallelization in GPUs.

dependencies among the data, the loop execution can be unrolled and parallelized. Furthermore, if a processor has N or more computing cores for N-sized non-scalar data, the loop can be processed in a single cycle. As presented in Fig. 3, a GPU code first calculates vector indexes using thread IDs and performs an identical operation on multiple data in parallel. In the figure, the GPU completes the same operation as the CPU in a single cycle. GPGPU applications often show such data parallelism, thus can be accelerated with the loop unrolling approach.

Due to the loop unrolling and massive parallel thread execution, the programming model for GPUs is different from CPUs. In particular, to maximize the throughput of GPGPU applications, a programmer needs to understand SIMD hardware and its execution model. To alleviate the burden of the programmer and improve the programmability, GPUs support sets of programming Application Programming Interfaces (APIs) such as Computed Unified Device Architecture (CUDA) and Open Computing Language (OpenCL).

The CUDA programming language is developed based on a parallel programming model. It is designed and provided by NVIDIA in order to support the general-purpose processing on GPUs. NVIDIA has released an initial version of CUDA and its software development kit (CUDA SDK) in 2006 [18]. CUDA works on all NVIDIA GPUs with compute capability from GeForce G8x series and upwards. CUDA has become popular among many researchers and programmers due to having extensions to industry-standard programming languages such as C, C++, and Fortran [19]. Once a programmer launches a large number of threads, the APIs deliver the thread executions to the GPU hardware as shown in Fig. 4. The start of a GPGPU application is handled by a host CPU processor. In the CPU part, the program receives input variables, reads data from file systems, and prepares program libraries. Then, using *cudaMalloc* and *cudaMemcpy* APIs, the program allocates memory space in the GPU device and transfers input data from the host processor to the GPU device. Lastly, the host processor launches kernel execution on the discrete GPU device, and the parallel thread execution is processed by the SIMD cores.

For efficient management of the massive number of threads, NVIDIA GPUs handle the threads in hierarchical structure of grid, thread block (or Cooperative Thread Array, CTA), warp, and thread as shown in Fig. 5. A CUDA application can consist of multiple GPU kernels, and a grid is directly matched with a kernel. A grid contains all threads in a kernel and can be partitioned into multiple thread blocks. A thread block contains

Host Side

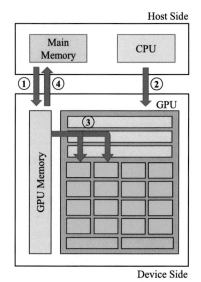

① Copy data from main memory to GPU memory

② CPU instructs GPU for computing

③ GPU Proceeds to parallel execution of instructions

④ GPU sends back the result to the main memory

Device Side

Fig. 4 Basic CUDA processing on GPUs.

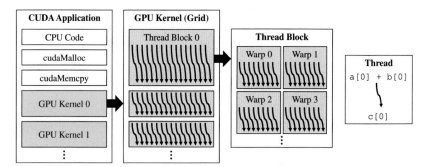

Fig. 5 The hierarchical structure of a CUDA GPU application.

multiple warps, which are the smallest units from the GPU hardware perspective. Each warp consists of multiple threads (32 for NVIDIA GPUs) and executes the identical instruction on the different data in parallel. In GPU hardware, a warp is executed on a *Streaming Multiprocessor (SM)*. An SM can execute maximum 2048 threads (or 64 warps) at a time, thus a recent NVIDIA GPUs can execute hundreds of thousands of threads in parallel. Using the massive number of concurrently executing threads and the supports of APIs, GPUs effectively accelerate the unrolled loop executions for vector data in GPGPU applications.

Similar to NVIDIA CUDA applications, AMD OpenCL applications have a hierarchical structure of *NDRange* (*N-Dimensional Range*), *work group*, *wavefront*, and *work item*. Table 1 describes the corresponding terminologies between CUDA and OpenCL. Despite some numerical specifications that can be different (a *warp* contains 32 threads while a *wavefront* contains 64 threads), the same hierarchy has a similar role in the structure of GPGPU applications.

2.3 GPU microarchitecture

A recent GPU device has thousands of CUDA cores to maximize throughput and execute hundreds of thousands of threads in parallel. In NVIDIA GPUs, to efficiently manage a large number of cores, multiple CUDA cores are grouped in an SM (*Compute Unit, CU* in AMD GPUs). For example, an NVIDIA Ampere A100 GPU has 108 SMs, and an SM has 64 CUDA cores, thus the GPU device has 6912 CUDA cores [20]. In the recent NVIDIA and AMD GPUs, each computing core can execute independent threads (or *work items* in AMD GPUs), thus can follow independent execution paths. This concept of execution model is called Single Instruction, Multiple Threads (SIMT), by distinguishing it from the SIMD model [21]. Similar to an instruction in a CPU, the smallest unit of a GPU program is called a *warp instruction*. A warp instruction consists of up to 32 threads (64 in a wavefront) and executed on an SM.

SM or CU is the basic unit that composes a GPU hardware, thus a GPU device can be viewed as a set of several SMs or CUs. Fig. 6 describes the architecture of an SM in the recent NVIDIA GPUs. In general, the number of 32-bit single-precision floating-point cores is called CUDA core count, thus the SM consists of 64 CUDA cores. In the figure, the CUDA cores are split into 4 groups with 16 CUDA cores, and each group has dedicated

Table 1 GPU terminologies for CUDA and OpenCL.

NVIDIA CUDA	AMD OpenCL
Grid	NDRange (N-Dimensional Range)
Thread Block	Work Group
Warp	Wavefront
Thread	Work Item
Streaming Multiprocessor	Compute Unit

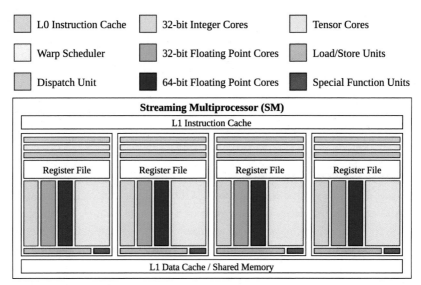

	L0 Instruction Cache		32-bit Integer Cores		Tensor Cores
	Warp Scheduler		32-bit Floating Point Cores		Load/Store Units
	Dispatch Unit		64-bit Floating Point Cores		Special Function Units

Fig. 6 Architecture of a streaming multiprocessor [20].

structures such as L0 instruction cache, warp scheduler, dispatch unit, register file, tensor cores, load/store units, and special function units. In addition, L1 instruction cache and L1 data cache/shared memory are shared by all CUDA cores in an SM.

Warp instructions are delivered from the L1 and L0 instruction caches to the warp scheduler. Then the warp scheduler determines the processing order of warp instructions. In detail, the warp scheduler tries to maximize the pipeline utilization and hide operation stalls caused by long latency instructions such as global memory access. Several warp scheduling techniques have been proposed, and *Loosed Round Robin* (LRR), *Greedy-Than-Oldest* (GTO), and *Two-Level* (TL) warp scheduling policies are widely used in academia [22,23]. Once the warp scheduler determines the warp instruction to be executed, the dispatch unit receives instruction information and sends the instruction to an available functional unit. The dispatch unit monitors the status of the functional unit that has an identical data type (e.g. 32-bit integer, 32-bit floating-point, or 64-bit floating-point). If the target functional unit is busy, the dispatch unit queues the warp instruction until the current operation of the functional unit ends.

The functional units in GPUs operate as SIMT cores and provide high parallel performance. In general, 32-bit integer, 32-bit floating-point, and 64-bit floating-point functional units are implemented inside of an SM and perform vector arithmetic operations such as addition, subtraction, multiplication, and division. The load/store unit generates memory requests for the functional units. The Special Function Units (SFUs) performs transcendental operations such as sin, cosine, and square root. In the recent NVIDIA Ampere GPUs, more various data types such as 16-bit half-precision floating-point and 16-bit brain floating-point (bfloat16) are supported by the functional units. Furthermore, the tensor core performs tensor operations on 4-bit integer, 8-bit integer, and 32-bit tensor floating-point (TensorFloat32), including conventional data types.

The high throughput of GPUs is achieved by a large number of parallel processing cores. To provide sufficient data to a large number of cores, a large amount of register file is required. Furthermore, fast context switching between multiple warps is important to maximize the utilization of an SM. If an SM can only execute 1 warp at a time, long latency instructions such as global memory access will disrupt the entire SM and lead to performance degradation. To avoid that, each SM has architectural state of multiple warps (64 warps in recent NVIDIA GPUs) and each warp scheduler in SMs finds an ready warp to execute. In summary, the register file takes a substantial amount of GPU resources for two reasons: parallel data supply and fast context switching [24,25]. Fig. 7 shows the size of the on-chip memory structures of NVIDIA GPUs over the last decade. Before the Volta architecture (Fermi ∼ Pascal), the register file clearly takes most of the on-chip memory. From the Volta architecture, NVIDIA GPUs focused on the support of deep learning and HPC applications and introduced tensor core. These applications use large dataset which does not fit into conventional on-chip memory. As a result, the size of L1 data cache and L2 cache has been significantly increased to store the large dataset.

In NVIDIA GPUs and the CUDA programming model, the shared memory (or scratchpad memory) is a fast memory space that is accessible by all threads in a thread block. In particular, the contents of the shared memory space are defined by a programmer and cannot be flushed by a hardware control logic, unlike conventional cache memory. From the hardware perspective, the shared memory is implemented with an SRAM and located inside of an SM, thus provide much faster memory performance compared to the DRAM-based global memory. Despite programming difficulties, fine-grained optimization or extra implementation using shared memory can further improve the performance of GPU applications [26]. The L1 data cache in GPUs operates similarly to a conventional on-chip cache memory. It contains a subset of the global memory space and transparent to the programmer. When a functional unit issues a memory request

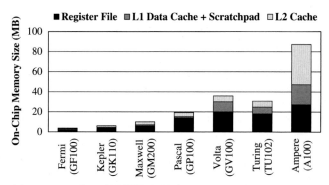

Fig. 7 On-chip memory size of NVIDIA GPUs by generations.

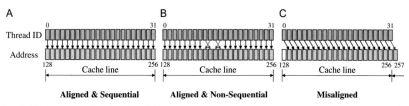

Fig. 8 Memory coalescing in GPUs.

using the load/store unit, the requested data in the global memory space is delivered through the L2 cache, the L1 data cache, and the register file. In NVIDIA GPUs, the L1 data cache is implemented along with the shared memory, and the L1 data cache and the shared memory are often interchangeable. The programmer can configure the size of the L1 data cache and shared memory using CUDA APIs.

When a warp accesses the memory address space, the memory accesses from the 32 threads are merged into a single memory transaction for efficient memory access. This merged memory access is called *memory coalescing*. Fig. 8 shows three different cases of memory coalescing. When all memory accesses are aligned and sequential as presented in (A), all threads in the warp can obtain data with a single cache line access. Before NVIDIA Kepler architecture (2012), aligned but non-sequential memory accesses such as (B) required multiple memory requests. Recent GPUs from Kepler architecture support coalesced memory access for such cases. Lastly, when a warp accesses misaligned address space as presented in (C), it requires two cache line accesses. In the example, thread 31 accesses memory address 257, which is the next cache line. The additional cache line access of misaligned memory access leads to the performance degradation of GPUs [27]. In summary, a programmer can improve GPU performance by reducing the number of cache line accesses with the consideration on the memory coalescing and memory misalignment.

Scaling up tens of the SMs, a GPU device can be formed as shown in Fig. 9. In the latest NVIDIA GPUs, 2 SMs are grouped into one Texture Processing Cluster (TPC) and 8 TPCs are grouped into one Graphics Processing Cluster (GPC). The L2 cache is located outside of GPCs and shared among all GPCs, thus accessible from all SMs. The L2 cache is connected with memory controllers that manage and access the main memory modules. To provide sufficient memory bandwidth to computing cores, GPUs employ Graphics DDR DRAM (GDDR) or High Bandwidth Memory (HBM). When multiple GPU devices are employed in a computer system, NVLink provides fast GPU-to-GPU interconnection for data transfer.

2.4 Evolution of GPUs

Over the last decade, the floating point operations per second (FLOPS) performance of a single GPU device has improved significantly. In addition, domain-specific computation unit such as tensor core were introduced to accelerate popular workloads beyond conventional vector operations. Table 2 shows the hardware specifications of NVIDIA GPUs from Fermi architecture (2010) to Ampere architecture (2020). During the period, FLOPS increased 20-fold, memory bandwidth increased 10-fold, while power consumption increased only 1.8-fold. Single GPU devices can have around 17.5 times more transistors owing to the advance of manufacturing process technology and the increased die size. Around the appearance of Pascal architecture (2016), deep learning applications began to be widely

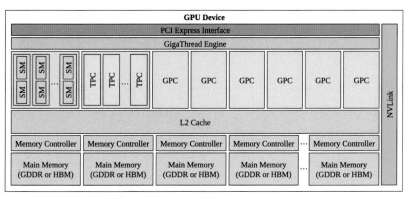

Fig. 9 The architecture of an NVIDIA GPU device [20].

Table 2 NVIDIA GPU in the recent decade [17,20,28–32].

Specifications	Fermi (GF100)	Kepler (GK110)	Maxwell (GM200)	Pascal (GP100)	Volta (GV100)	Turing (TU102)	Ampere (A100)
FP32 CUDA Core Count	448	2880	3072	3584	5120	4608	6912
FP32 Tera FLOPS	1.0	5.2	6.8	10.6	15.7	16.3	19.5
Memory Interface	384-bit GDDR5	384-bit GDDR5	384-bit GDDR5	4096-bit HBM2	4096-bit HBM2	384-bit GDDR6	5120-bit HBM2
Memory Size (GB)	6	12	12	16	32 or 16	24	40
Memory Bandwidth (GB/s)	144	288	317	720	900	672	1555
Thermal Design Power (W)	225	225	250	300	300	260	400
Transistor Count (Billion)	3.1	7.0	8.0	15.3	21.1	18.6	54.2
GPU Die Size (mm^2)	529	561	601	610	815	754	826
Technology (nm)	40	28	28	16	12	12	7

used in various areas. That led to the introduction of the tensor core in the Volta architecture. Tensor core performs much faster and more efficient operations on tensor data types. Detailed operations and usages of the tensor core is given in Sections 2 and 3.

3. Deep learning acceleration in GPU hardware perspective

As stated earlier, GPU has become one of the widely used hardware solutions for deep learning applications and helps improve the execution speed of the AI applications. In this section, we will present architectural details of the advanced core technologies of commercial GPUs, ranging from the NVIDIA Pascal to the Ampere architectures.

3.1 NVIDIA tensor core: Deep learning application-specific core

NVIDIA presented a new computing unit for deep learning operations named Tensor Core which came with the Volta architecture in 2017. Ever since programmers started to use GPUs as deep learning hardware solutions, they faced with the situation that GPU's various SIMD computing units do not provide enough computing power; as a matter of fact, those units by nature are developed to accelerate graphics operations. As a result, using the conventional CUDA cores for certain deep learning operations such as a large amount of neural-network training suffers from performance degradation. Many researchers observed that the degradation is caused by tremendous Matrix Multiplication and Accumulation (MMA) operations, illustrated in Fig. 10, which make up a large portion of neural network applications. Thus, in order to accelerate the MMA operation, NVIDIA presented Tensor Core which helps conduct General Matrix-Matrix

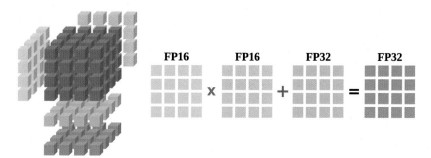

Fig. 10 Tensor Core 4 × 4 matrix multiply and accumulate.

Multiplication (GEMM) operations efficiently. In the original Tensor Core, the input matrix A and B are FP16 matrices, and the result of multiply A and B is added to C which is represented in either 4 by 4 FP16 or FP32 matrix. As a result, the newer GPUs with hundreds of Tensor Core produce a massive increase in throughput and execution efficiency for deep learning applications compared to the previous generation GPUs[17,20,32].

The initial version of Tensor Core presented in the V100 architecture supports only two data types of inputs (FP32 or FP16; FP32 by default) and only FP32 accumulation is supported [17]. It has later been discovered that reducing precision will bring further increase in performance and energy efficiency with little impact in accuracy. In fact, the Tensor Core design in Turing architecture supports FP16 accumulate, INT4, and INT8. In $D = A \times B + C$ where input matrix A and B are in 16FP, matrix C can include 16FP data. Whereas if A and B are in INT4 and INT8 meaning matrix A and B include 4-bit integer and 8-bit integer data, matrix C has to be in 32-bit integer [32]. Table 3 describes sign, exponent, and mantissa bits of precision mode in the latest A100 architecture. In the A100 architecture, TF32, BF16 and FP64 precision mode (not in IEEE standard) is added on the Tensor Core design. TF32 which is the default precision of A100's Tensor Core includes combination of 8-bit exponent such as FP32 and 10-bit mantissa in FP16. Also, BF16 includes 8-bit exponent and 7-bit mantissa as an alternative to IEEE FP16 which includes 5-bit exponent and 10-bit mantissa. These newly added precisions bring a performance benefit by reducing representation although representation error of data might increase. In order to support HPC computing, A100's Tensor Core also supports IEEE-compliant FP64 computations [20].

Fig. 11 shows Fine-Grained Structured Sparsity developed in order to double compute throughput for deep neural networks in A100 architecture. Utilizing sparsity is possible in deep learning since only a subset of weights serves as a meaningful purpose in determining the learned output and the remaining weights could be no longer needed. For example, one of the input prunes trained weight with a 2:4 non-zero pattern and the weight is compressed to 1/2 size of non-zero data and non-zero indices. Also, non-zero indices are used to indicate computation that needs to be performed. This computation selection can be made by a simple mux architecture in Tensor Core[20].

According to the NVIDIA's publicly released whitepaper [17], Tesla V100 GPU contains 80 SMs where each SM comes with eight Tensor Cores, totaling 640 Tensor Cores per GPU. This brings computation performance of

Table 3 Sign, exponent, and mantissa bits of precisions that A100 supports.

Input operands	Precision
FP32 (1, 8, 23)	s e e e e e e e e m m m m m m m m m m m m m m m m m m m m m m m
TF32 (1, 8, 10)	s e e e e e e e e m m m m m m m m m m
FP16 (1, 5, 10)	s e e e e e m m m m m m m m m m
BF16 (1, 8, 7)	s e e e e e e e e m m m m m m m
INT8	b b b b b b b b
INT4	b b b b
Binary	b
IEEE FP64 (1, 11, 52)	s e e e e e e e e e e e m m m m m m m m m m m … m m

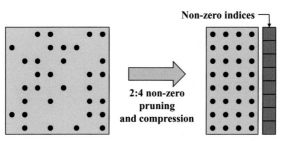

Fig. 11 A100 fine-grained sparsity [20].

15.7 TFLOPs for FP32 and 125 TFLOPs for FP16. Also, a Turing TU102 GPU contains 576 Tensor Cores, eight per SM, and 72 SMs per GPU. In Turing architecture, NVIDIA has stated to support various precision computations, such as INT4, INT8, etc. Furthermore, for their latest A100 GPU, it contains 432 third-generation Tensor Cores, four Tensor Cores per SM, and 108 SMs per GPU. This increases the performance up to 19.5 TFLOPs for FP32, 312 TFLOPs for FP16 computation and supports TF16, TF32 and sparse matrix computation [17,20,32].

3.2 High-bandwidth memory

As today's AI applications handle a larger amount of data, the demand on more memory capacity and bandwidth is continuously requested. In a discrete GPU computation model, all the data required for the computation need to be copied from the host memory to GPU DRAM; however, the data size cannot exceed the device DRAM capacity. When the data becomes larger than the GPU DRAM capacity, programmers need to separate it into the smaller size. For example, if we want to run a CNN application which needs 40 GB memory space on a GPU device with 4 GB DRAM, we need to slice the application kernel into multiple smaller kernels. This leads to poor programmability and decreases performance due to the overhead from kernel launch and memory copy. For this reason, the GPUs have been integrated with bigger and powerful (faster access time and higher communication bandwidth) DRAMs over time and NVIDIA includes HBM2 and GDDR6 for the memory device in their GPUs from Pascal to Ampere architectures.

The world's first GPU that came to the market with the high-bandwidth memory (HBM2) is NVIDIA Tesla P100. HBM2 is a 3D-stacked SDRAM that stacks multiple dies located on the same physical package reducing power consumption and area compared to previous generation GDDR5 designs. Therefore, it enables data centers and servers to be built with increase in performance while maintaining the cost range. A memory device integrated in Tesla P100 provides a maximum 16 GB of memory size and 732 GB/s of peak memory bandwidth. Also, an HBM2 in Tesla V100 comes with four memory dies with four stacks providing a maximum 16 GB of memory and 900 GB/s of peak memory bandwidth. The latest A100 GPU contains HBM2 DRAM with increased memory capacity and bandwidth up to 40 GB and 1555 GB/s accordingly [17,20,31].

Furthermore, to reduce the pressure on L2 caches from increased memory size and bandwidth, A100 GPUs come with 40 MB L2 cache which is nearly 7 times larger than V100 GPUs. Also, NVIDIA has added a new partitioned crossbar structure to provide 2.3 times higher bandwidth between L2 and memory compared to V100 GPUs [20].

3.3 Multi-GPU system

NVLink is the NVIDIA high-speed interconnect technology which was first introduced in 2016 with the Pascal GP100 GPU. NVLink provides higher bandwidth for GPU-to-GPU and GPU-to-CPU communications and better performance compared to the conventional inter-device interconnect using PCIe. NVLink enables connections of up to eight GPUs together where each of four GPUs makes up a quad. A quad is a fully NVLink-connected GPU bundle and one master CPU is dedicated to manage each quad. Even when there are not enough GPUs to make up a quad, NVLink or PCIe can still be used to connect any number of GPUs. However, the advantage of utilizing quad in using NVLink is that each GPU in a quad can access device memory using DMA and the connection between the two quads with NVLink can mitigate the pressure on PCIe. Fig. 12 shows NVIDIA's DGX-1 server that is recommended to users who want to

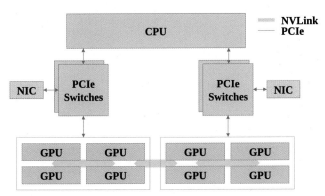

Fig. 12 Hybrid cube mesh NVLink topology as used in DGX-1 [31].

accelerate computation of general-purpose Deep Learning applications. As illustrated, each of four GPUs is connected with NVLink that provides 40 GB/s and 80 GB/s bidirectional bandwidths for single and double links correspondingly. Furthermore, NVLink can also be used for CPU-to-GPU interconnect. For instance, Tesla P100 supports connection with IBM's POWER8 through NVLink with up to 160 GB/s bidirectional bandwidths for a single GPU connection which provides 5 × higher bandwidth than PCIe [17,20,31,32].

The second generation NVLink allows direct load/store/atomic access from CPU to each GPU's HBM2 memory. Also, it adds supports for atomics initiation from either GPU or CPU and Address Translation Services (ATS) to enhance CPU performance by enabling GPUs to access CPU page tables. This CPU-to-GPU technique further increases overall system performance by improving efficiency of memory copy. Overall, the increased number of links, faster link speed, and enhanced functionality help accelerating deep learning performance in multi-GPU systems [17,32].

The third generation NVLink interconnect, shown in Fig. 13, is introduced for the latest NVIDIA A100 GPUs with increased number of links, speed, and bandwidth. The new NVLink provides data rate of 50 Gbit/s for each link which brings the total bandwidth up to 600 GB/s with 12 links (NVLink in Tesla V100 supports 300 GB/s of total bandwidth). Also, NVIDIA has introduced NVSwitch with the third generation NVLink, which is a networking fabric similar to Mellanox InfiniBand and Ethernet. In fact, NVSwitch is developed to be used in scaling data centers.

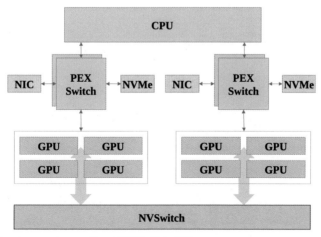

Fig. 13 Structure and operation of NVSwitch [17].

Furthermore, all writes in the third generation NVLink are not posted enabling synchronization at the requester. These new features help to improve efficiency of NVLinks and data center platforms such as NVIDIA DGX POD and NVIDIA DGX SuperPOD include NVSwitch and multiple A100 GPUs [20].

3.4 Multiple-instance GPU

Data center applications often come in various scales, for example, early-stage development or simple inferencing requires smaller memory size and a fewer computing units compared to deep learning training and HPC simulation. Generally, data centers are expected to maintain resource utilization high with providing better efficiency. For this, earlier functionalities such as Volta Multi-Process Service (MPS) were developed to improve Quality of Service (QoS) in data centers by providing higher throughput and lower latency. However, since system resources such as L2 cache and memory are shared across all the processes, the coherency problem could result in a performance slowdown.

Multiple-Instance GPU (MIG) which is newly added in the A100 GPU allows a single GPU to be managed and utilized as multiple virtual GPUs. Each virtual GPU is called an instance and each instance's SMs occupy separate and isolated data/control paths and memory areas. As illustrated in Fig. 14, a GPU is divided into four GPU instances each with varying portion

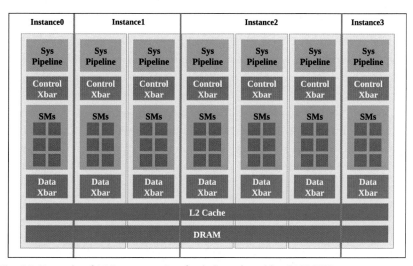

Fig. 14 Example of MIG computation firstly introduced by A100 [20].

of compute and memory resources. In MIG, every GPU comprises seven GPU slices, and a single GPU instance takes at least one GPU slice. When given a GPU instance, certain amount of resources such as system pipeline, interconnect, SMs, caches, and memory is solely dedicated to the user for running applications. This feature enables individual user's kernel to run with predictable throughput, latency, and stability, which are the key factors for improving QoS [20].

4. GPU software for accelerating deep learning

With the ever-increasing popularity of deep learning applications, users demand a better optimized solution of GPU software along with its high-performance hardware. As discussed previously, traditional GPUs were originally built and utilized for graphical applications. However, modern-day GPUs with general-purpose computing capability allow users in various fields of profession to run applications such as scientific computing, image processing, deep learning, etc. To meet consumers' needs for high performance, GPU hardware architecture has evolved with increasing complexity. Along with it, GPU vendors (such as NVIDIA and AMD) developed software solutions specialized for high-performance computing. Since we mainly discuss accelerating deep learning applications for the scope of this chapter, we will look at how GPU software for executing deep learning applications have been developed in this section.

4.1 Deep learning framework for GPU

For deep learning on GPUs, many developers and institutions have presented frameworks to support easier implementation of DNN models. There are a number of publicly available deep learning frameworks such as TensorFlow, PyTorch, Caffe, etc. Each of them has its own unique features, and we will discuss some of the key aspects of using each framework. We picked three well-known and popular frameworks, TensorFlow, PyTorch, and Caffe, to go through a brief introduction of each framework.

4.1.1 TensorFlow

TensorFlow, developed by Google Brain team, is a framework designed for deep learning in servers/cloud-scale systems, CPUs, GPUs, and Google Tensor Processing Units (TPUs). One important feature of TensorFlow is that it represents computations and algorithms with a dataflow graph. Dataflow graph elements include tensors and operations, which are input/output data in n-dimensional arrays and computation processes, respectively. Also, TensorFlow enables distributed execution on multiple GPUs for training. Being one of the most widely known frameworks for having many included libraries, extension, tools, and guidelines, TensorFlow is efficient to implement while providing high performance for running deep learning applications [3].

4.1.2 PyTorch

PyTorch is developed mainly by Facebook's AI Research lab to achieve both the usability and performance of deep learning applications. By using Python style for programming, PyTorch aims to provide a usability-centric deep learning framework. Although it runs on Python, the core library, libtorch, of PyTorch is written in C++ to achieve high performance. With core library implementations of tensor data structure, both the GPU and CPU operators and basic functions to support parallel programs, PyTorch shows strong support for GPUs especially when extending application execution to multiple GPUs. Also, PyTorch keeps control and data flow separately so that the CPU can run the code while GPU can perform tensor operations at the same time. Overall, PyTorch is a user-friendly deep learning framework with providing optimized performance on GPUs [1].

4.1.3 Caffe

Caffe is another well-known deep learning framework developed by the Berkeley Vision and Learning Center. Caffe is implemented in C++ with CUDA for GPU computation. One key feature of Caffe is that storing and communicating of data occurs in 4-dimensional arrays called blobs. The use of blobs enables reducing of synchronization overhead between the host CPU and the GPU. Also, Caffe is known for easy setup of neural network models and its high performance (processing speed of over 60 million images per day on a single K40 GPU) [2].

Since the initial version of CUDA, the newer release has been continuously launched with extended libraries to accelerate general-purpose computing on newer GPUs. Recent attention to deep learning applications brought the development of a group of features and libraries specialized for deep learning acceleration in the CUDA platform, commonly known as NVIDIA Deep Learning SDK. There are a number of core libraries we will discuss next, such as cuDNN, TensorRT, cuBLAS, cuSPARSE, and DALI. Also, significant scaling of GPU performance has been achieved through multi- or multi-node GPU systems. A key factor to further increase its performance is the utilization of optimized hardware and software solutions for efficient data communication between devices. We will also introduce a software library accelerating data communications in multi-GPU systems later.

4.2 Software support specialized for deep learning

Huge attention to deep learning has changed the technological advances of GPU in both hardware and software perspectives. We looked at how hardware support for deep learning acceleration has been expanding throughout recent years in the previous section. We will now see the details and examples of GPU software mainly designed for deep learning acceleration.

With introduction of Deep Learning SDK which is a tool and set of libraries created by NVIDIA, deploying deep learning applications onto GPUs have become more powerful and efficient. There are a number of libraries we will study included in the Deep Learning SDK. Training and inference of neural network models have become handy to many software developers.

4.2.1 cuDNN: NVIDIA CUDA deep neural network library

cuDNN is a GPU library designed to accelerate low-level primitives of deep neural networks. cuDNN not only supports for functions such as convolution, pooling, normalization, activation, etc. which are commonly used in DNN models but supports also newly added functions implemented in the latest models. Since cuDNN is a library in the CUDA platform, it provides flexibility to work with any standard programming language. Also, by providing portability to most commonly used deep learning frameworks such as Caffe, Torch, TensorFlow, cuDNN is widely used in deep learning on GPUs [33].

Since the initial release of cuDNN a few years back, NVIDIA has been continuously releasing newer versions of cuDNN to optimize performance on their later GPU architecture by adding new features. For example, cuDNN v7 is released to support the Volta architecture, which introduced Tensor Core. Its one key feature is that users are allowed to choose between normal operations and Tensor Core operations in convolution and RNN layers by calling `cudnnSetConvolutionMathType`, `cudnnSetRNNMatrixMathType` functions, respectively. Users can find added functions and their supported algorithms of each version of cuDNN through NVIDIA documentation [34].

With the launch of NVIDIA's latest Ampere architecture, cuDNN v8 has been announced to optimize deep learning performance on A100 GPUs with new compute mode, TensorFloat-32 (TF32). In cuDNN v8, the library is separated into a number of sub-libraries, as shown below. Also, Tensor Core performance is increased for 1D, 2D, 3D, and grouped convolutions [34].

cuDNN libraries in version 8 [34]:

- `cudnn_ops_infer`—enables cuDNN context, tensor operation management, and contains common functions used in the inference of neural network models, including batch normalization, softmax, dropout, etc.
- `cudnn_ops_train`—contains common training functions similar to `cudnn_ops_infer` and depends on `cudnn_ops_infer`.
- `cudnn_cnn_infer`—enables inference routines in convolutional neural networks, and it depends on `cudnn_ops_infer`.
- `cudnn_cnn_train`—enables training routines in convolutional neural networks, and it depends on all the above libraries.
- `cudnn_adv_infer`—includes all other features not contained in the above libraries and deep learning inference algorithms such as RNNs. This library also depends on `cudnn_ops_infer`.
- `cudnn_adv_train`—contains the training features and algorithms not included in previous libraries and depends on `cudnn_ops_infer`, `cudnn_ops_train`, and `cudnn_adv_infer`.
- `cudnn`—optional part that lies between application and cuDNN code to open the library for an API at runtime.

4.2.2 TensorRT: NVIDIA SDK for accelerating deep learning inference

TensorRT is a library that provides APIs to import pre-trained DNN models from deep learning frameworks and optimize them to have increased performance in deep learning inference on NVIDIA GPUs (Fig. 15). TensorRT is aimed to provide an optimal inference solution to embedded systems, datacenters, or automotive environments. In such conditions, throughput and latency are critical for cost-effective scaling and real-time services. Also, when hardware resource is limited, such as in automotive environments, achieving both efficiency and performance is difficult.

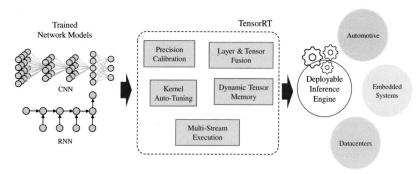

Fig. 15 Key features and workflow of TensorRT [35].

According to NVIDIA, ResNet-50 throughput on V100 GPU (DGX-2) achieved up to 7867 images per second for inference image classification on CNNs [36]. Instead of using the same frameworks used in training, developers can utilize TensorRT to increase throughput with low latency for DNN inference while maintaining efficiency [37].

There are five key optimizations of TensorRT: Weight & Activation Precision Calibration, Layer & Tensor Fusion, Kernel Auto-Tuning, Dynamic Tensor Memory, and Multi-Stream Execution. For Weight & Activation Precision Calibration, TensorRT automatically converts a trained network in FP32 to reduced INT8 precision to reduce inference latency. Then, TensorRT employs the calibration process in order to minimize accuracy loss that can occur from reducing precision. For Layer and Tensor Fusion, TensorRT accelerates performance by enabling the fusion of layers in neural network models. By reducing the network size, fusion optimization achieves a simpler form of the overall network. Also, TensorRT ensures the same behavior of layers before and after the fusion. There are several types of fusions supported with TensorRT, such as *ReLU ReLU Activation, Convolution and ReLU Activation*, etc. (Supported types of fusions can further be found from NVIDIA TensorRT Documentation). TensorRT selects kernels with the lowest runtime for optimizing execution time to target the GPU platform through the Kernel Auto-Tuning process. Also, through Dynamic Tensor Memory, TensorRT reduces memory footprint and dynamically allocates memory for each tensor efficiently. Lastly, Multi-Stream Execution allows scaling of multiple input streams in parallel [35,38].

4.2.3 cuBLAS: CUDA basic linear algebra subroutine library

cuBLAS is a library to allow users to access computational resources and accelerate the performance of basic linear algebra on GPUs. The library extends to batched operations, multi-GPU execution, and mixed and low precision by utilizing Tensor Cores. There are three sets of cuBLAS API, as shown below, that are currently available [19].

- cuBLAS API—For the use of cuBLAS API, users must allocate application data in the GPU memory space, call cuBLAS functions, and copy the results back to the host memory.
- cuBLASXt API—At application runtime, data may reside either in the host or other compute-related devices. Hence, the library automatically allocates memory, dispatches functions, and transfers data to single or multiple GPUs in the system. cuBLASXt API is primarily designed to support memory allocation and data communication of cuBLAS in multi-GPU systems.

- cuBLASLt API—Is a lightweight library newly designed in the latest cuBLAS version to perform General Matrix-to-matrix Multiply (GEMM) functions. Also, the API supports flexibility in matrix data layouts, input and compute types, and supports parameter programmability to select implementation of algorithms.

One important reason for using cuBLAS library in deep learning is that it allows fast matrix computation, which accounts for a huge portion of many DNN layers and functions. For matrix multiplication APIs in the latest cuBLAS release (cuBLAS 11.0.0), GEMM functions come with added support for bfloat16 data type to match the release of the latest Ampere architecture. Also, a new compute type TensorFloat32 has been added to accelerate matrix multiplication on Tensor Cores with higher precision than bfloat16. Furthermore, a new release of cuBLAS adds support for using Tensor Cores for matrix multiplication in half and mixed-precision by default and automatic use of double-precision Tensor Cores [19].

4.2.4 cuSPARSE: CUDA sparse matrix library

cuSPARSE is another GPU-accelerated library to perform linear algebra computation for sparse matrices. As with other CUDA SDK libraries, cuSPARSE is flexible to use C and C++ interface for optimized computation and data storage. The library comprises four major functions as described below. There are a number of functions that can be called in each level, and the details can be found in the CUDA Toolkit Documentation for use of cuSPARSE library. Helper functions and other functions are also available for use [19].

- cuSPARSE Level 1 function—supports sparse linear algebra functions performing operations between dense and sparse vectors.
- cuSPARSE Level 2 function—performs operations between sparse matrices and dense vectors.
- cuSPARSE Level 3 function—performs operations between sparse and dense matrices.
- Conversion—responsible for conversion operations between different matrix formats.

Similar to the standard cuBLAS library, users are responsible for allocating memory and copying data between GPU device memory and the host memory. However, support for auto-parallelization of resources in multi-GPU systems is not available. The latest release of cuSPARSE comes with improved performance for sparse matrix-sparse matrix multiplication (SpGEMM) with added APIs [19].

4.2.5 DALI: NVIDIA data loading library

DALI is NVIDIA's open-source library to support the decode and augmentations of images, videos, and speech for accelerating preprocessing input data of deep learning applications on GPUs. In systems with high GPU-to-CPU ratio, processing of large input data can slow down the overall deep learning performance scaling due to bottleneck in host CPU. To overcome this, DALI is designed to accelerate reading, labeling, and augmenting of input data, as shown in Fig. 16, to increase the utilization of available GPU compute capabilities[39].

One key feature of DALI is that it provides portability of data preprocessing pipelines between different deep learning frameworks. DALI also provides flexibility in the integration of frameworks such as MXNet, PyTorch, TensorFlow, and PaddlePaddle to enable easy implementation. Furthermore, DALI offers an option for utilizing CPU for processes other than data loading when GPU compute cycles are completely consumed [39,40].

The performance of input data processing for deep learning applications can significantly impact overall computation throughput. Due to large datasets and limited hardware resources (for data preprocessing), preparing input data in deep learning such as loading, decoding, resizing, and augmentation is becoming more challenging to scale performance. Therefore, employing an optimized software solution seems promising in such that it can achieve power efficiency while increasing performance.

4.3 Software to optimize data communications on multi-node GPU

To achieve scalable performance in deep learning, GPU vendors not only have started producing multi-node GPUs as a single platform such as

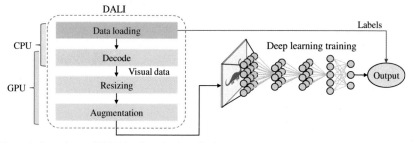

Fig. 16 Overview of DALI in the pipeline [39].

DGX but also developed a software solution for accelerating communication in multi-GPU and multi-node GPU systems. One such solution we will discuss in this section is the NVIDIA Collective Communication Library (NCCL). The latest release of NCCL provides basic collective communication functions in multi-GPU systems as described below [41].

- AllReduce—reduction operations on input data across multiple ranks (or CUDA devices) and writing the result to every rank. AllReduce operations are rank-agnostic, which means that the reordering of data does not change the outcome.
- Broadcast—copies data from one rank to all other ranks.
- Reduce—works in the same fashion as AllReduce but only writing the result to a single designated rank.
- AllGather—data in each rank is aggregated in the order of ranks and writes the result to every rank. Since the order affects the data layout of the outcome, the operation depends on the mapping of devices to ranks.
- ReduceScatter—works in the same fashion as Reduce, but each rank only receives a partial result based on its rank index.

NCCL library also provides group functions such as `ncclGroupStart` and `ncclGroupEnd` to allow merging multiple calls into one for three key purposes: managing multiple GPUs from one thread, aggregating operations to increase performance, and merging multiple point-to-point operations. For the management of multiple GPUs using a single thread, group functions must be used to avoid deadlocks. With aggregated operations, multiple collective operations can be performed from a single NCCL launch to reduce the latency. Lastly, point-to-point communications can be combined into complex communication patterns using group functions and they represent any communication pattern between ranks. Grouping of these communication patterns includes one-to-all (scatter), all-to-one (gather), all-to-all, and exchanging of data with neighbors. The details of use cases and examples of NCCL group features can be found in the NCCL user guide [41].

With the popular use of multi-GPU and multi-node GPU systems for scaling deep learning performance, optimized data communication across multiple GPUs is getting more attention in both industry and academia. Release of hardware solutions such as NVLink interconnect, and NVSwitch enables high-bandwidth data communication between devices. A software library specifically designed for data communication such as NCCL seems promising to further increase the performance of Multi-GPU systems.

5. Advanced techniques for optimizing deep learning models on GPUs

In the previous sections, we discussed how GPUs effectively handle a massive number of computations within their hardware components along with software support. GPUs efficiently compute massive data by executing a number of concurrent threads in parallel within their SIMD units. State-of-the-art GPUs are now shipped with tensor cores to accelerate matrix multiplication further. To provide high performance and programmability to GPUs, software support, such as cuDNN and CUTLASS, improves resource utilization efficiency as they are aware of the internal architecture of GPUs [42,43].

Running deep learning models on GPUs, however, suffers slowdowns even with the previously introduced hardware and software support. A common reason for the slowdowns is the underutilization of GPU hardware resources on running deep learning models since they are not fully aware of GPU architecture. Convolutional neural networks, for example, use 2D filters on convolution layers and pooling layers. Direct computation on these layers creates stride memory access patterns, causing extra memory requests. Moreover, optimization techniques for deep learning models conflict with GPU architecture. For instance, pruning skips multiplications with near-zero values by considering them as multiplying with zero and skipping them. However, this technique does not reduce computation on GPU SIMD units unless all threads in a warp can skip the computation. To efficiently accelerate deep learning models on GPUs, advanced techniques that are aware of GPU architecture is essential.

In this section, we focus on four advanced techniques shown in Table 4, including each topic in the table. Table 4 classifies advanced techniques on

Table 4 A summary of advanced approaches of GPUs for deep learning models.

Category	References
Increasing compute efficiency	[44–50]
Enhancing memory performance	[48,51–56]
Overcoming GPU memory capacity limit	[53,57–62]
Improving multi-GPU performance	[62–65]

GPUs for deep learning models. We refer readers to the original papers for details of each techniques. First, we introduce two techniques for improving compute efficiency in pruned deep learning models. Second, we present a case for improving memory efficiency of GPUs with data reuse. Lastly, we show an algorithm to overcome memory capacity limit of a GPU when training heavy deep learning models. Before introducing each approach, we first study the cause of inefficiency executing deep learning models on GPUs. Then, we introduce corresponding techniques to mitigate the cause of slowdowns on GPUs.

5.1 Accelerating pruned deep learning models in GPUs

Pruning is a technique to make deep learning models lighter by eliminating unnecessary values. Typically, near-zero values are eliminated and handled as zero. When computing a pruned deep learning models, the elimination can decrease compute intensity since multiplication with zero always produces zero.

Unfortunately, pruning does not contribute to accelerating deep learning models on GPUs and instead faces longer execution time [66–68]. The main reason of this phenomenon is the sparsity in pruned models. Generally, pruning does not selectively eliminate values; it erases any near-zero values. As a consequence, a sparse weight is mixed with zeros and non-zero values randomly. Even though multiplying zero can be skipped within a single ALU, skipping computation in a SIMD lane is possible only if the entire ALUs in the lane trying to compute pruned data. In extreme cases, only a single ALU computing non-zero value prevents skipping computation for an entire SIMD lane. In addition, accessing sparse data incurs uncoalesced memory accesses, further slowing down GPUs due to the inefficient memory accesses such as memory divergence. Therefore, computing a pruned model is not accelerated in GPUs.

Furthermore, a pruned deep learning model often use an encoding scheme for compressing sparse weights, which incurs an additional delay in decoding them. In the CUSPARSE library, sparse weight is encoded into the Compressed Sparse Row (CSR) format [69]. This format consists of three vectors. A vector stores non-zero values of a sparse matrix and the other vectors store row and column index for each non-zero values, respectively. A compressed data is decoded into a matrix when a GPU computes a pruned model. Decoding, however, causes additional overhead in data management, and accessing sparse data in a GPU creates uncoalesced memory requests. To sum up, executing a pruned deep learning model in a GPU suffers slowdowns from the inefficient use of both computation and memory resource.

To mitigate the inefficiency of computing pruned deep learning models in GPUs, researchers have proposed customized, SIMD-friendly pruning techniques. In common, they propose specialized sparse matrix layouts that can compute non–zero values fully utilizing GPU resources. In this section, we study two pruning techniques that are built for SIMD units and tensor cores, respectively.

5.1.1 Increasing compute efficiency of GPU SIMD lanes via synapse vector elimination

Discussed above, executing pruned deep learning models on GPUs suffer from inefficient computation on SIMD lanes due to the irregularity of sparse data structures. Fig. 17A and B show executing deep learning models on GPUs, which have a dense and a sparse data structure, respectively. Although a pruned model in Fig. 17B requires fewer memory accesses and computations, they offer little performance improvement of executing the model on GPUs. This is because computing sparse and irregular data on a GPU suffers slowdowns from both low SIMD lane efficiency and uncoalesced memory access.

Hill et al. proposed synapse vector elimination to accelerate pruned deep learning models in GPUs [45]. Synapse vector elimination reduces the number of redundant computation while achieving high GPU resource utilization rate. Fig. 17C shows the proposed technique reducing the execution time of computation by merging sparse vectors into dense ones. Two pairs of sparse vectors in Fig. 17B are transformed into a pair of dense vectors.

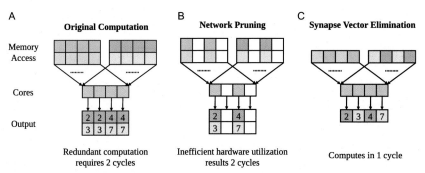

Fig. 17 Execution time comparison of a deep learning model with (A) no pruning, (B) network pruning, and (C) synapse vector elimination [45].

Then, a GPU computes the dense pair of vectors efficiently by fully utilizing a SIMD lane in a GPU and creating a coalesced memory request.

Synapse vector elimination improves creates dense vectors by reordering vectors in a pruned matrix. Fig. 18 compares matrix multiplication between a naive computation and the proposed synapse vector elimination. The proposed technique reduces the number of operations by decreasing the size of both input and weight matrices. Further, Fig. 18B depicts how it removes non-contributing vectors.

The two main steps of synapse vector elimination are synapse reordering and matrix truncation. First, synapse reordering simplifies removing redundant synapses while maintaining the uniformity of data structure. The number of non-contributing vectors, D, is found during training a deep learning model. Then, input and weight matrices are divided at row or column at K-D as to be retained and eliminated, respectively. After partitioning, the data is not arranged; some non-contributing vectors are in a partition to be retained and vice versa. Since the number of vectors to be unused and

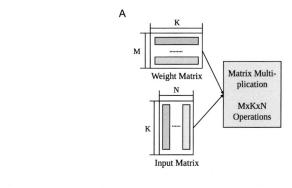

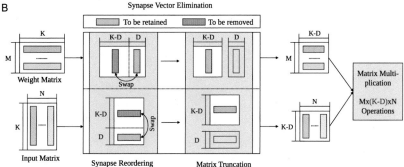

Fig. 18 Computing matrix multiplication of a deep learning model with (A) naïve computation and (B) synapse vector elimination [45].

retained is identical, swapping two vectors in a pair makes an input or weight matrix arranged. Next, matrix truncation reduces the dimension of input matrices by skipping computation at discarded partition.

5.1.2 Algorithm and hardware co-design for accelerating pruned deep learning models on tensor cores

Modern GPU architectures such as Volta and Turing include tensor cores, which are specialized hardware to accelerate matrix multiplication. When these GPUs execute pruned deep learning models, tensor cores suffer inefficiency computing sparse matrices since not all computing units are utilized. Previous customized pruning technique is SIMD-aware showing high efficiency in vectors. However, tensor cores can only accelerate matrix multiplication. Hence, the previous technique is not directly available to tensor cores.

To take advantage of both tensor cores and pruning, Zhu et al. proposed a novel algorithm and hardware co-design technique that efficiently computes sparse matrices on tensor cores [46]. First, the author observes the data placement of unified sparsifying shown in Fig. 19B is more SIMD-friendly. Then, the authors find when dividing a matrix into multiple L-dim vectors, less than 30% of 4-dim vectors have 75% or less sparsity and only less than

A

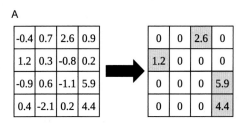

Generic Sparsifying

B

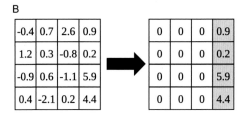

Unified Sparsifying

Fig. 19 Comparison between (A) Generic sparsifying and (B) Unified sparsifying. Both cases show a 4×4 matrix with a sparsity of 75% [46].

2% of 4-dim vectors have 25% or less sparsity. This means when a sparse matrix is divided into vectors, the number of non-zero values is likely to be balanced across the vectors. It motivates a balanced vector-wise encoding format.

Motivated from the above observation, Zhu et al. proposed VectorSparse: a methodology of iterative vector-wise sparsifying and retraining [46]. In this section, we introduce a new data layout and refer readers to the original paper for details on retraining in the customized pruning method. In VectorSparse, a sparse matrix is divided into multiple L-dim vectors and encoded with parameter K, which is the most significant number of non-zero values present among the vectors. During encoding, only non-zero values remain, and their position is stored in an offset matrix. Fig. 20 shows an example of a 4×8 matrix are divided into four 8-dim vectors, where parameter K is 2.

Matrix multiplication between a vector-wise matrix and a dense matrix is computed as SIMD-friendly vector operations. Fig. 21 shows an example of matrix multiplication with a vector-wise sparse matrix whose original size is 4×8. With naïve matrix multiplication, calculating the first row of matrix C requires $8 \times 8 \times 6$ multiplication and additional accumulation. In contrast, with a vector-wise sparse matrix, the computation becomes two vector multiplication with a constant. In Fig. 21, calculating the first row of matrix

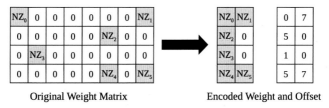

Fig. 20 An example of encoding a sparse matrix in VectorSparse [46].

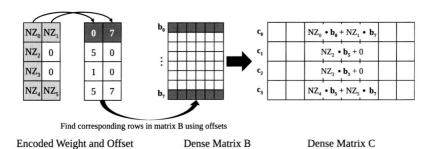

Fig. 21 Matrix multiplication between a vector-wise sparse matrix A and a dense matrix B [46].

C requires two multiplication between a vector and a scalar, NZ0, and b0 and NZ1 and b7, respectively. This achieves a computation reduction of 75%. Matrix multiplication with a vector-wise sparse matrix is SIMD-friendly since between a dense vector and a scalar fully utilizes a SIMD-lane.

Matrix multiplication with a vector-wise sparse matrix is SIMD-friendly, but some modifications are required to support this computation in a tensor core. Tensor cores accelerate matrix multiplication between dense matrices, so NVIDIA currently provides programming APIs for dense matrix multiplications. Similarly, new load and compute instructions enable accelerating a vector-wise sparse matrix multiplication with new instructions. The new load instruction loads a sparse matrix and its corresponding offset, and the another instruction computes matrix multiplication in tensor cores.

A new microarchitecture for tensor cores includes three hardware modifications. Firstly, the proposed microarchitecture adds a special-purpose register for an offset matrix, and the new load instruction fetches data to the register. Secondly, the size of a buffer for a dense matrix B is doubled, and an additional buffer called a ping-pong buffer is added to hide load latency. Increasing the buffer size for the matrix B doubles the latency to load data, but it enables computation between a constant and a vector while maintaining high throughput of GPUs. Lastly, the broadcasting of the operand buffer A to four dot product units is enabled along with a new load instruction with sparse matrix and computation instruction. Broadcasting data in the buffer A enables a multiplication between a vector and a scalar within a tensor core, accelerating sparse matrix multiplication.

Fig. 22 shows a timeline for comparing relative execution time of computing a dense matrix, a sparse matrix without ping-pong buffer, and with ping-pong buffer, respectively. Vector-wise sparse matrix multiplication reduces computation cycles by skipping redundant multiplication with zero in a tensor core. In the figure, we assume matrix A has a sparsity of 75%. Therefore, the number of computation cycles is decreased by 75%. In addition, ping-pong buffer hides latency of loading matrices, further improving the performance of tensor cores. In summary, simple hardware modifications on tensor cores enable accelerating sparse matrix multiplication and pruned neural networks.

5.2 Improving data reuse on CNN models in GPUs

GPUs compute convolution layers in CNN models with various methods. The most straightforward method is Direct Convolution which computes

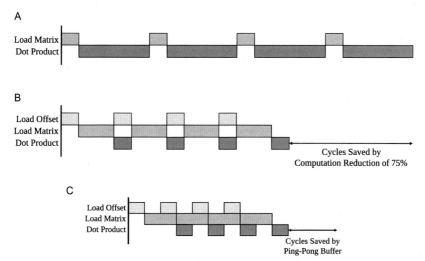

Fig. 22 Comparison of execution timeline among (A) a dense matrix computation, (B) vector-wise sparse mode without ping-pong buffer, and (C) vector-wise sparse mode with ping-pong buffer [46].

by sliding 2D-filters to input data, which is implemented in cuda-convnet2 [70]. Another technique is to convert the convolution into matrix multiplication with lowering technique, currently supported by Caffe [2] and cuDNN [43].

These two techniques are not always superior to the other. Li et al. found thatthe performance gap between the two approaches comes from memory efficiency caused by different data layout [52]. The data layout of a convolution is 4D. In this section, we define each dimension as N (the number of images), C (the number of feature maps), H (height of an image), and W (width of an image). In addition, the hierarchy of the data layout is ordered from highest to lowest. For instance, the dimension C is the highest dimension in CHWN layout. While cuda-convnet2 computes direct convolution with CHWN, Caffe, and cuDNN compute matrix multiplication with NCHW [70]. Li et al. executed AlexNet with NVIDIA GTX Titan Black using CHWN (cuda-convnet2) and NCHW (cuDNN) data layout, respectively [71]. Experimental results show CHWN layout outperforms on the first convolutional layer and all pooling layers. On the other hand, NCHW layout works well from second to the last convolutional layer. The results imply a single data layout do not always outperform the others.

Li et al. discovered CHWN layout performs well when N is higher. They executed the third convolution layer of ZFNet on a GTX Titan

Black GPU by changing the length of N and C, respectively [72]. Their results show that the CHWN layout in cuda-convnet outperforms the other when the length of N is greater or equal to 128, and it is insensitive to C. A GPU executes threads by grouping them as a warp, and a warp consists of 32 threads. Therefore, when N is greater or equal to 32, the memory accesses of threads in a warp can coalesce, achieving high memory efficiency in a GPU. In addition, when N is even higher, each thread can handle more than one image so that the image data can be reused in the register file. Coalesced memory requests and data reuse reduce the number of off-chip memory requests. As a consequence, the CHWN layout outperforms the other when N is high.

In contrast, NCHW layout outperforms the other when the length of C is higher. To figure out when the NCHW layout outperforms the CHWN layout, Li et al. also performed sensitivity tests on the size of N and C to the same convolution layer. Their results showed the NCHW layout performs well when C is greater or equal to 32 and is insensitive to N. The NCHW layout do not directly compute convolution layers; they transform 4D data layout into 2D arrays along H and W dimensions. Unrolling matrices, however, incurs additional overhead in data reorganization. When C is higher, creating 2D arrays becomes efficient as more data is reused [43].

A heuristic selects a more efficient data layout for each layer. First, Li et al. defined Nt and Ct to be thresholds for each data layout. Then, from the previous experiments, they proposed a heuristic to choose the CHWN layout when N is greater or equal to Nt or C is below Ct. If the two conditions do not fit, N is smaller than Nt, and C is higher than Ct, they choose the NCHW layout. With a GTX Titan Black GPU, the threshold becomes 128 and 32, respectively. However, the thresholds may vary with different GPUs. Additional experiments with a GTX Titan X GPU showed thresholds for the new GPU is 64 for Nt and 128 Ct. To find Nt and Ct for a GPU, one-time profiling with varying N and C is required.

Pooling layers, on the other hand, always perform well with the CHWN layout of cuda-convnet. As explained above, the CHWN layout achieves higher memory efficiency since memory requests are coalesced with a higher value of N. In contrast, computing a pooling layer with the NCHW layout shows lower performance due to uncoalesced memory request. In the NCHW layout, H and W are the two lowest dimensions; hence each thread computes a filter of a pooling layer. This results in memory requests with stride access patterns; therefore, the layout suffers low memory efficiency with uncoalesced memory requests. As a consequence, using the CHWN

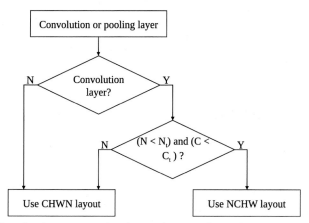

Fig. 23 A flow for determining data layout for convolution layers and pooling layers proposed by Li et al. [52].

layout improves GPU performance on pooling layers. The algorithm for selecting an appropriate data layout for each convolution layer and pooling layer is summarized in Fig. 23.

To take advantage of both data layouts, a fast data layout transformation technique is necessary. Li et al. noted that two layouts share C, H, and W have the same relative position and can be seen as a single dimension. Therefore, converting data layout becomes transposing a 2D array between [CHW][N] and [N][CHW]. With a shared memory-based tile optimization that achieves coalesced memory access, data layout transpose is easily implemented [73].

5.3 Overcoming GPU memory capacity limits with CPU memory space

Training deep learning models requires a substantial amount of memory, sometimes exceeding the memory capacity limit of a single GPU. For instance, training VGG-16, a heavy CNN model, with a batch size of 256 requires 28 GB of memory space, while high-end GPUs, such as Titan V is shipped with the memory capacity of 12 GB [17]. The disparity between the memory space requirement of deep learning models and insufficient memory capacity of a GPU forces to train these models with a smaller batch size or use multiple GPUs [71] or a new GPU shipped with higher memory capacity [20]. However, reducing batch size takes longer time to train a deep learning model, and building a new system with multiple GPUs or a new GPU is costly. Moreover, using GPUs with higher memory capacity is not a panacea. As deep learning models are including more layers and becoming more complex, even current state-of-the-art GPUs may not be able to train future models with higher batch size. To mitigate memory capacity limits of GPUs in training deep learning models, researchers have proposed memory management schemes by observing memory access patterns of deep learning models.

Rhu et al. noted that memory allocation scheme in deep learning frameworks does not effectively manage precious GPU memory space [57]. They observed DNN models are trained using a stochastic gradient descent (SGD) algorithm in layer-wise. To train a neural network model, a forward propagation is first computed from an input layer to an output layer creating intermediate values (feature maps) on each layer. Then, a loss function is calculated at the end of the forward propagation. Backward propagation trains each layer, with the result of the loss function and feature maps created from forward propagation, but this time, its computation is done from the output layer to an input layer. As a result, data in the layers on the front will not be accessed until the layers are trained in backward propagation. Furthermore, when a backward propagation is finished on a layer, data for computing the layer will not be further accessed. Therefore, the memory space for the data can be saved. Deep learning frameworks, however, allocate the entire memory space required for training. This policy wastes precious GPU memory space since a GPU must hold unused data in its memory.

To overcome the memory capacity limit of a single GPU, Rhu et al. proposed virtualized Deep Neural Networks (vDNN), a transparent runtime memory management solution virtualizing memory usage of both GPU and CPU [57]. Based on the previous observation that not all data is used in GPUs during training deep learning models, vDNN transparently manages data in the following three cases. First, vDNN offloads data of a layer to a CPU at forward propagation if a computation is finished. Migrating data is overlapped with layer computation in a GPU, so most of the offloading overhead is hidden. Fig. 24A shows unused data for computing forward propagation for a layer which can be migrated to the CPU. Second, vDNN prefetches the offloaded data to the GPU before the processor begins computation for backward propagation. For instance, Fig. 24B illustrates an example of prefetching data for computing backward propagation of a layer. Lastly, vDNN saves additional memory space by releasing data that will not be further accessed. With layer-wise memory management, vDNN achieves memory savings up to 92%, only requiring GPU memory of 4.2 GB when training a VGG-style network containing hundreds of layers.

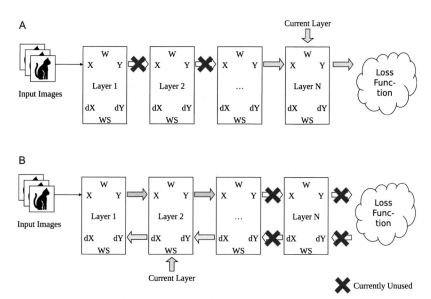

Fig. 24 An example of vDNN saving GPU memory in (A) forward propagation and (B) backward propagation [57].

6. Cons and pros of GPU accelerators

At the end of the chapter, we summarize the pros and cons of using the GPU as an AI accelerator. We will see how each of the characteristics and structures of the GPUs discussed above provide advantages or disadvantages when accelerating AI. Based on these pros and cons, we can see in which case the GPU can be used for AI acceleration.

The first advantage of the GPU for accelerating AI applications is high programmability. Unlike accelerators designed to use a specific AI, GPUs can accelerate a variety of applications depending on the user. High programmability of the GPU has advantages in AI acceleration on cloud computing and data center systems. This is because, in a situation such as a cloud computing, where various applications need to be accelerated for the user, the GPU can accelerate all the various AI applications and other applications that require parallel processing. The second advantage is the support of various AI API and libraries on the GPU. There are a various library of linear algebra algorithms supporting GPU, and API support for AI is also diverse. NVIDIA supports internal APIs such as CUDNN and CUBLAS for AI applications, and AMD also supports AI acceleration through OpenCL, an open-source library.

Through this support, many deep learning frameworks such as PyTorch, TensorFlow, and Caffe support AI acceleration using GPU. These supports also allows the GPU to have high user accessibility and lower application development costs, which are advantages of the GPU accelerators.

On the contrary, as the above advantages exist, there are also disadvantages in using the GPU for AI acceleration. First, to ensure high programmability, the power efficiency of AI acceleration is relatively low compared to other accelerators. Unlike other accelerators that have minimal control logic for accelerating AI, the GPU has control logic to support various parallel processing and the structure for graphic processing. This is because the GPU is designed not only for AI acceleration, which results in lower efficiency in AI acceleration. With complex control logic and structure, which is not needed for AI acceleration, the GPU consumes more power than others in accelerating AI. Second, as GPU is designed for general purposes, the GPU has disadvantages compared to other accelerators that can have specific computational logic to accelerate AI functions. This characteristic not only reduces efficiency but also makes the overall process throughput lower than other accelerators. Lastly, GPUs have less computing power per area than other accelerators, and the size of the form factor is also larger. Unlike other accelerators with minimal control logic to accelerate AI, GPUs have much more control logic and structure, so the area and the computing power per area are relatively less competitive than others due to the area occupied by those logics. The above disadvantages make it difficult to use the GPU on edge devices such as wearable devices that are sensitive to power and area.

Acknowledgment

This research was supported by the MOTIE (Ministry of Trade, Industry & Energy) (No. 10080674, Development of Reconfigurable Artificial Neural Network Accelerator and Instruction Set Architecture) and KSRC (Korea Semiconductor Research Consortium) support program for the development of the future semiconductor device and also supported by the Super Computer Development Leading Program of the National Research Foundation of Korea (NRF) funded by the Korean government (Ministry of Science and ICT (MSIT)) (NRF-2020M3H6A1084852).

Key terminology and definitions

Graphics Processing Unit Graphics Processing Unit is designed for accelerating graphic processing. GPU provides great parallelism through Single Instruction Multi-Thread (SIMT) support using many computing units, so it is used to accelerate various parallel programs. This characteristic is also an advantage for accelerating Deep Learning, so many server systems and users use GPUs as accelerators for their Deep Learning application.

Deep Learning Deep learning is an algorithm that calibrates an algorithm through learning and uses it to perform tasks with high accuracy. Deep Neural Network (DNN), widely used deep learning algorithm, use matrix type weights to modify algorithms, and perform tasks. As DNN is consists of matrix operations, high parallelism is required to accelerate them, and various hardware with high parallelism are studied and used for DNN acceleration.

Parallel Processing Parallel processing is a method used to run programs faster on multiprocessors. This method reduces execution time by executing code simultaneously, which does not have a dependency on each other using multiple computing devices. It is divided into instruction-level parallelism, thread-level parallelism, and more, depending on the unit of code to take parallelism.

Tensor Operation Tensor operation was first introduced in the NVIDIA Volta architecture and is used in the tensor core created to accelerate linear algebra computation on NVIDIA GPUs. In GPUs which have Tensor core, linear algebra operations used in Deep Neural Networks such as matrix multiplication are compiled into tensor operations and allocated to the tensor core for accelerating them.

Data Center Computing The data center is a system having a large scale of computing and memory devices. It is used in programs requiring a large amount of computation or memory resources or is used to execute programs of various users simultaneously. As a lot of computational and memory resources are required for the training stage of AI applications, it is executed in the data center, and the inference process of various users is also executed through the data center.

References

[1] A. Paszke, S. Gross, F. Massa, A. Lerer, J. Bradbury, G. Chanan, et al., PyTorch: an imperative style, high-performance deep learning library, in: Advances in Neural Information Processing Systems, 2019, pp. 8026–8037.

[2] Y. Jia, E. Shelhamer, J. Donahue, S. Karayev, J. Long, R. Girshick, et al., Caffe: convolutional architecture for fast feature embedding, in: Proceedings of the 22nd ACM International Conference on Multimedia, 2014, pp. 675–678.

[3] M. Abadi, P. Barham, J. Chen, Z. Chen, A. Davis, J. Dean, et al., Tensorflow: a system for large-scale machine learning, in: 12th USENIX Symposium on Operating Systems Design and Implementation (OSDI 16), 2016, pp. 265–283.

[4] NVIDIA. https://developer.nvidia.com/taxonomy/term/785.

[5] NVIDIA. https://www.nvidia.com/en-us/industries/smart-cities/.

[6] NVIDIA. https://www.nvidia.com/en-us/industries/healthcare-life-sciences/.

[7] NVIDIA. https://www.nvidia.com/en-us/deep-learning-ai/industries/robotics/.

[8] NVIDIA. https://www.nvidia.com/en-us/industries/transportation/.

[9] NVIDIA. https://developer.nvidia.com/rtx/ngx.

[10] M. Mohammadi, A. Al-Fuqaha, S. Sorour, M. Guizani, Deep learning for IoT big data and streaming analytics: a survey, IEEE Commun. Surv. Tutorials 20 (4) (2018) 2923–2960.

[11] H. Li, K. Ota, M. Dong, Learning IoT in edge: deep learning for the internet of things with edge computing, IEEE Netw. 32 (1) (2018) 96–101.

[12] P. Danaee, R. Ghaeini, D.A. Hendrix, A deep learning approach for cancer detection and relevant gene identification, in: Pacific Symposium on Biocomputing, 2017, pp. 219–229.

[13] J. Camps, A. Sama, M. Martin, D. Rodriguez-Martin, C. Perez-Lopez, J.M.M. Arostegui, A. Prats, Deep learning for freezing of gait detection in Parkinson's disease patients in their homes using a waist-worn inertial measurement unit, Knowl.-Based Syst. 139 (2018) 119–131.

[14] A. Davoudi, K.R. Malhotra, B. Shickel, S. Siegel, S. Williams, M. Ruppert, P. Rashidi, Intelligent ICU for autonomous patient monitoring using pervasive sensing and deep learning, Sci. Rep. 9 (1) (2019) 1–13.

[15] S. Levine, P. Pastor, A. Krizhevsky, J. Ibarz, D. Quillen, Learning hand-eye coordination for robotic grasping with deep learning and large-scale data collection, Int. J. Robot. Res. 37 (4–5) (2018) 421–436.

[16] NVIDIA, NVIDIA Launches the World's First Graphics Processing Unit: GeForce 256, 1999.

[17] NVIDIA, NVIDIA Tesla V100 GPU Architecture: The World's Most Advanced Data Center GPU, 2017.

[18] NVIDIA. https://www.developer.nvidia.com/about-cuda.

[19] NVIDIA, NVIDIA CUDA Toolkit Documentation v11.0.171, 2020.

[20] NVIDIA, NVIDIA A100 Tensor Core GPU Architecture: Unprecedented Acceleration at Every Scale, 2020.

[21] E. Lindholm, J. Nickolls, S. Oberman, J. Montrym, NVIDIA tesla: a unified graphics and computing architecture, IEEE Micro 28 (2008) 39–55.

[22] A. Jog, O. Kayiran, A.K. Mishra, M.T. Kandemir, O. Mutlu, R.R. Iyer, et al., Orchestrated scheduling and prefetching for GPGPUs, in: A. Mendelson (Ed.), Proceedings of the 40th Annual International Symposium on Computer Architecture, ACM, 2013, pp. 332–343.

[23] V. Narasiman, M. Shebanow, C.J. Lee, R. Miftakhutdinov, O. Mutlu, Y.N. Patt, Improving GPU performance via large warps and two-level warp scheduling, in: Proceedings of the 44th Annual IEEE/ACM International Symposium on Microarchitecture, 2011, pp. 308–317.

[24] S. Lee, K. Kim, G. Koo, H. Jeon, M. Annavaram, W.W. Ro, Warped-compression: enabling power efficient GPUs through register compression, in: Proceedings of the 42th Annual International Symposium on Computer Architecture, 2015.

[25] Y. Oh, M.K. Yoon, J.W. Song, W.W. Ro, FineReg: fine-grained register file management for augmenting GPU throughput, in: Proceedings of the 51th Annual IEEE/ACM International Symposium on Microarchitecture, 2018.

[26] M.K. Yoon, K. Kim, S. Lee, M. Annavaram, W.W. Ro, Virtual thread: maximizing thread-level parallelism beyond GPU scheduling limit, in: Proceedings of the 43th Annual International Symposium on Computer Architecture, 2016.

[27] Y. Oh, G. Koo, M. Annavaram, W.W. Ro, Linebacker: preserving victim cache lines in idle register files of GPUs, in: 46th Annual International Symposium on Computer Architecture, 2019.

[28] NVIDIA, NVIDIA's Next Generation CUDA Compute Architecture: Fermi, 2009.

[29] NVIDIA, NVIDIA's Next Generation CUDA Compute Architecture: Kepler GK110, 2012.

[30] NVIDIA, NVIDIA GeForce GTX 980: Featuring Maxwell, The Most Advanced GPU Ever Made, 2014.

[31] NVIDIA, NVIDIA Tesla P100: The Most Advanced Datacenter Accelerator Ever Built, 2016.

[32] NVIDIA, NVIDIA Turing GPU Architecture: Graphics Reinvented, 2018.

[33] NVIDIA. https://www.developer.nvidia.com/cudnn.

[34] NVIDIA, NVIDIA Deep Learning SDK Documentation, 2020.

[35] NVIDIA, NVIDIA Deep Learning TensorRT Documentation, 2020.

[36] NVIDIA. https://www.developer.nvidia.com/deep-learning-performance-training-inference#deeplearningperformance_inference.

[37] NVIDIA. https://www.developer.nvidia.com/blog/speeding-up-deep-learning-inference-using-tensorrt/.

[38] NVIDIA. https://www.developer.nvidia.com/blog/int8-inference-autonomous-vehicles-tensorrt/.

[39] NVIDIA. https://www.developer.nvidia.com/blog/fast-ai-data-preprocessing-with-nvidia-dali/.

[40] NVIDIA, NVIDIA Deep Learning DALI Documentation, 2020.

[41] NVIDIA, NVIDIA Deep Learning NCCL Documentation, 2020.

[42] NVIDIA, CUTLASS: Fast Linear Algebra in CUDA C++. https://devblogs.nvidia.com/cutlass-linear-algebra-cuda/.

[43] S. Chetlur, C. Woolley, P. Vandermersch, J. Cohen, J. Tran, B. Catanzaro, et al., cuDNN: Efficient Primitives for Deep Learning, 2014.

[44] J. Yu, A. Lukefahr, D. Palframan, G. Dasika, R. Das, S. Mahlke, Scalpel: customizing DNN pruning to the underlying hardware parallelism, in: 2017 ACM/IEEE 44th Annual International Symposium on Computer Architecture, 2017.

[45] P. Hill, A. Jain, M. Hill, B. Zamirai, C. Hsu, M.A. Laurenzano, et al., DeftNN: addressing bottlenecks for DNN execution on GPUs via synapse vector elimination and near-compute data fission, in: 2017 50th Annual IEEE/ACM International Symposium on Microarchitecture, 2017.

[46] M. Zhu, T. Zhang, Z. Gu, Y. Xie, Sparse tensor core: algorithm and hardware co-design for vector-wise sparse neural networks on modern GPUs, in: Proceedings of the 52nd Annual IEEE/ACM International Symposium on Microarchitecture (MICRO), ACM, pp. 359–371.

[47] C. Holmes, D. Mawhirter, Y. He, F. Yan, B. Wu, GRNN: low-latency and scalable RNN inference on GPUs, in: Proceedings of the Fourteenth EuroSys Conference 2019, 2019.

[48] S. Li, Y. Zhang, C. Xiang, L. Shi, Fast convolution operations on many-core architectures, in: 2015 IEEE 17th International Conference on High Performance Computing and Communications, 2015 IEEE 7th International Symposium on Cyberspace Safety and Security, and 2015 IEEE 12th International Conference on Embedded Software and Systems, 2015.

[49] H. Cui, H. Zhang, G.R. Ganger, P.B. Gibbons, E.P. Xing, GeePS: scalable deep learning on distributed GPUs with a GPU-specialized parameter server, in: Proceedings of the Eleventh European Conference on Computer Systems, 2016.

[50] H. Park, D. Kim, J. Ahn, S. Yoo, Zero and data reuse-aware fast convolution for deep neural networks on GPU, in: Proceedings of the Eleventh IEEE/ACM/IFIP International Conference on Hardware/Software Codesign and System Synthesis, 2016.

[51] H. Kim, S. Ahn, Y. Oh, B. Kim, W.W. Ro, W.J. Song, Duplo: lifting redundant memory accesses of deep neural networks for GPU tensor cores, in: Proceedings of the 2020 53rd Annual IEEE/ACM International Symposium on Microarchitecture, 2020.

[52] C. Li, Y. Yang, M. Feng, S. Chakradhar, H. Zhou, Optimizing memory efficiency for deep convolutional neural networks on GPUs, in: Proceedings of the International Conference for High Performance Computing, Networking, Storage and Analysis, 2016.

[53] X. Chen, D.Z. Chen, X.S. Hu, moDNN: memory optimal DNN training on GPUs, in: Proceedings of 2018 Design, Automation & Test in Europe Conference & Exhibition, 2018.

[54] X. Chen, J. Chen, D.Z. Chen, X.S. Hu, Optimizing memory efficiency for convolution kernels on kepler GPUs, in: Proceedings of the 54th Annual Design Automation Conference 2017, 2017.

[55] S. Dong, X. Gong, Y. Sun, T. Baruah, D.R. Kaeli, Characterizing the micro-architectural implications of a convolutional neural network (CNN) execution on GPUs, in: Proceedings of the 2018 ACM/SPEC International Conference on Performance Engineering, 2018.

[56] Q. Chang, M. Onishi, T. Maruyamam, Fast convolution kernels on pascal GPU with high memory efficiency, in: Proceedings of the High Performance Computing Symposium, 2018.

[57] M. Rhu, N. Gimelshein, J. Clemons, A. Zuliqar, S.W. Keckler, vDNN: virtualized deep neural networks for scalable, memory-efficient neural network design, in: Proceedings of the 49th Annual IEEE/ACM International Symposium on Microarchitecture, 2016.

[58] B. Zheng, N. Vijaykumar, G. Pekhimenko, Echo: compiler-based GPU memory foot-print reduction for LSTM RNN training, in: Proceedings of the ACM/IEEE 47th Annual International Symposium on Computer Architecture, 2020.

[59] T. Jin, S. Hong, Split-CNN: splitting window-based operations in convolutional neural networks for memory system optimization, in: Proceedings of the Twenty-Fourth International Conference on Architectural Support for Programming Languages and Operating Systems, 2019.

[60] C. Huang, G. Jin, J. Li, Swap, Advisor: pushing deep learning beyond the GPU memory limit via smart swapping, in: Proceedings of the Twenty-Fifth International Conference on Architectural Support for Programming Languages and Operating Systems, 2020.

[61] L. Wang, J. Ye, Y. Zhao, W. Wu, A. Li, S.L. Song, et al., Superneurons: dynamic GPU memory management for training deep neural networks, in: ACM SIGPLAN Notices, 2018.

[62] A.A. Awan, C. Chu, H. Subramoni, X. Lu, D.K. Panda, OC-DNN: exploiting advanced unified memory capabilities in CUDA 9 and Volta GPUs for out-of-core DNN training, in: Proceedings of 2018 IEEE 25th International Conference on High Performance Computing, 2018.

[63] A.A. Awan, K. Hamidouche, J.M. Hashmi, D.K. Panda, S-Caffe: co-designing MPI runtimes and caffe for scalable deep learning on modern GPU clusters, in: ACM SIGPLAN Notices, 2017.

[64] F.N. Iandola, M.W. Moskewicz, K. Ashraf, K. Keutzer, FireCaffe: near-linear acceleration of deep neural network training on compute clusters, in: Proceedings of 2016 IEEE Conference on Computer Vision and Pattern Recognition, 2016.

[65] Y. Yang, A. Buluç, J. Demmel, Scaling deep learning on GPU and knights landing clusters, in: Proceedings of the International Conference for High Performance Computing, Networking, Storage and Analysis, 2017.

[66] S. Han, J. Kang, H. Mao, Y. Hu, X. Li, Y. Li, et al., ESE: efficient speech recognition engine with sparse ISTM on FPGA, in: Proceedings of the 2017 ACM/SIGDA International Symposium on Field-Programmable Gate Arrays, 2017.

[67] S. Han, X. Liu, H. Mao, J. Pu, A. Pedram, M.A. Horowitz, et al., EIE: efficient inference engine on compressed deep neural network, in: Proceedings of the 43th Annual International Symposium on Computer Architecture, 2016.

[68] S. Han, H. Mao, W.J. Dally, Deep compression: compressing deep neural networks with pruning, in: trained quantization and Huffman coding, 4th International Conference on Learning Representations, 2016.

[69] A. Buluç, J.T. Fineman, M. Frigo, J.R. Gilbert, C.E. Leiserson, Parallel sparse matrix-vector and matrix-transpose-vector multiplication using compressed sparse blocks, in: Proceedings of the 21st Annual Symposium on Parallelism in Algorithms and Architectures, 2009.

[70] A. Krizhevsky. http://code.google.com/p/cuda-convnet2/.

[71] A. Krizhevsky, I. Sutskever, G.E. Hinton, ImageNet classification with deep convolutional neural networks, in: Advances in Neural Information Processing Systems, 2012, pp. 25:1106–1114.
[72] M.D. Zeiler, R. Fergus, Visualizing and understanding convolutional networks, in: European conference on computer vision, Springer, 2014, pp. 818–833.
[73] NVIDIA, NVIDIA CUDA 6.5 SDK Samples, 2014.

Further reading/References for advance

[74] D. Kirk, W.W. Hwu, Programming Massively Parallel Processors, third ed., Elsevier Inc., 2017.
[75] E. Stevens, L. Antiga, T. Viehmann, Deep Learning with PyTorch, Manning Publications Co., 2020.
[76] S. Mittal, S. Vaishay, A survey of techniques for optimizing deep learning on GPUs, J. Syst. Archit. 99 (2019), 101635.
[77] S.B. Shriram, A. Garg, P. Kulkarni, Dynamic memory management for GPU-based training of deep neural networks, in: IEEE International Parallel and Distributed Processing Symposium, IEEE, 2019, pp. 200–209.
[78] T.M. Aamodt, W.W.L. Fung, T.G. Rogers, General-Purpose Graphics Processor Architectures, Morgan & Claypool, 2018.
[79] T. Gale, M. Zahria, C. Young, E. Elson, Sparse GPU kernels for deep learning, arXiv, 2006, p. 10901.
[80] V. Sze, Y. Chen, T. Yang, J.S. Emer, Efficient processing of deep neural networks: a tutorial and survey, IEEE 105 (2017) 2295–2329.
[81] X. Li, Y. Liang, S. Yan, L. Jia, Y. Li, A coordinated tiling and batching framework for efficient GEMM on GPUs, in: Proceedings of the 24th Symposium on Principles and Practice of Parallel Programming, 2019.
[82] X. Peng, X. Shi, H. Dai, H. Jin, W. Ma, Q. Xiong, et al., Capuchin: tensor-based GPU memory management for deep learning, in: Proceedings of the 25th International Conference on Architectural Support for Programming Languages and Operating Systems, 2020.

About the authors

Won Jeon received the B.S. degree in electrical and electronic engineering from Yonsei University, Seoul, Korea, in 2014. He is currently working toward the Ph.D. degree in the Embedded Systems and Computer Architecture Laboratory, the School of Electrical and Electronic Engineering, Yonsei University. His current research interests are GPU memory systems, processing-in-memory architecture designs, and approximate computing for neural network applications. He is a student member of the IEEE.

Gun Ko received the B.S. degree in electrical engineering from the Pennsylvania State University, State College, Pennsylvania, United States, in 2017. He is currently working towards the Ph.D. degree in the Embedded Systems and Computer Architecture Laboratory, the School of Electrical and Electronic Engineering, Yonsei University. His current research interests are GPU microarchitecture and memory systems. He is a student member of the IEEE.

Jiwon Lee received the B.S. degree in electrical and electronic engineering from Yonsei University, Seoul, Korea, in 2018. He is currently working toward the Ph.D. degree in the Embedded Systems and Computer Architecture Laboratory, the School of Electrical and Electronic Engineering, Yonsei University. His current research interests are virtual memory, GPU memory systems, and heterogeneous computing. He is a student member of the IEEE.

Hyunwuk Lee received the B.S. degree in electrical and electronic engineering from Yonsei University, Seoul, Korea, in 2018. He is currently working toward the Ph.D. degree in the Embedded Systems and Computer Architecture Laboratory, the School of Electrical and Electronic Engineering, Yonsei University. His current research interests are a multi GPU system, GPU memory systems, and approximate computing for the neural network application. He is a student member of the IEEE.

Dongho Ha received the B.S. degree in electrical and electronic engineering from Yonsei University, Seoul, Korea, in 2019. He is currently working toward the Ph.D. degree in the Embedded Systems and Computer Architecture Laboratory, the School of Electrical and Electronic Engineering, Yonsei University. His current research interests are GPU programming language, microarchitecture and memory systems, and heterogeneous computing. He is a student member of the IEEE.

Won Woo Ro received the B.S. degree in electrical engineering from Yonsei University, Seoul, Korea, in 1996, and the MS and Ph.D. degrees in electrical engineering from the University of Southern California, in 1999 and 2004, respectively. He worked as a research scientist with the Electrical Engineering and Computer Science Department, University of California, Irvine. He currently works as a professor with the School of Electrical and Electronic Engineering, Yonsei University. Prior to joining Yonsei University, he worked as an assistant professor with the Department of Electrical and Computer Engineering, California State University, Northridge. His industry experience includes a college internship with Apple Computer, Inc. and a contract software engineer with ARM, Inc. His current research interests include high-performance microprocessor design, GPU microarchitectures, neural network accelerators, and memory hierarchy design. He is a senior member of the IEEE.

CHAPTER SEVEN

Architecture of neural processing unit for deep neural networks

Kyuho J. Lee

The School of Electrical and Computer Engineering, The Artificial Intelligence Graduate School, Ulsan National Institute of Science and Technology, Ulsan, South Korea

Contents

Abstract

Deep Neural Networks (DNNs) have become a promising solution to inject AI in our daily lives from self-driving cars, smartphones, games, drones, etc. In most cases, DNNs were accelerated by server equipped with numerous computing engines, e.g., GPU, but recent technology advance requires energy-efficient acceleration of DNNs as the modern applications moved down to mobile computing nodes. Therefore, Neural Processing Unit (NPU) architectures dedicated to energy-efficient DNN acceleration became essential. Despite the fact that training phase of DNN requires precise number representations, many researchers proved that utilizing smaller bit-precision is enough for inference with low-power consumption. This led hardware architects to investigate energy-efficient NPU architectures with diverse HW-SW co-optimization schemes for inference. This chapter provides a review of several design examples of latest NPU architecture for DNN, mainly about inference engines. It also provides a discussion on the new architectural researches of neuromorphic computers and processing-in-memory architecture, and provides perspectives on the future research directions.

Advances in Computers, Volume 122
ISSN 0065-2458
https://doi.org/10.1016/bs.adcom.2020.11.001

1. Introduction

With the rapid development of Artificial Intelligence (AI) technology, the 4th industrial revolution enabled us living in intelligence-assisted world. There is no doubt that the world is full of AI algorithms. In 2012, mobile wallet app for cryptocurrency (Bitcoin) used QR codes for payments. Tesla released autopilot system for vehicles and Amazon introduced smart speaker Echo that can receive voice commands in 2014. In 2016, Google's AlphaGo beat *Lee Sedol* in board game Go. As of 2020, smartphones provide automatic payments based on fingerprint recognition and face recognition, language translation, in-door navigation, etc. These dramatic change of our lives was possible thanks to the advanced AI technologies.

The AI technology includes logical systems, knowledge-based systems, machine learning, and deep learning. The logical systems and knowledge-based systems belong to classical AI approach. Machine learning covers searching algorithms, clustering algorithms, feature-based classification, and deep learning. Among the broad range of AI algorithms, the advances in Deep Neural Network (DNN) accelerated AI to blend in with our daily life to improve comfortability and safety of human being. What makes DNNs different from classical hand-crafted features is its ability to extract high-level features directly from raw data by training over enormous datasets. Then, the extracted high-level features effectively represent the input sensory data, and this fact boosted the performance of AI algorithm in a great amount. Also, internet became smart with the emergence of big data in the first stage of AI boom in early 2010s. The ability to treat big data gave DNNs outstanding performance over conventional machine learning algorithm. Google provided user-centric data based on what each user searches frequently by deep learning with big data. Facebook applied Convolutional Neural Network (CNN) to provide users with face recognition and object recognition. These early deployments of DNN derived benefit from servers equipped with powerful GPUs because hundreds of hidden layers of DNN required huge amount of computations and memory footprints. Therefore, most of the applications aimed at large systems that contain high-performance GPUs even though they consumed several kilo-watts of power.

As the needs of mobile DNN application is being increased, advanced DNN researches [1–7] provided successful mobile applications. However, DNN algorithms still suffer from harsh requirement of processing speed under limited hardware resources and power budget due to its massive computational cost. Other works that attempted to lighten the DNN architecture [8–16] to make it feasible for mobile devices failed to solve fundamental problem of heavy computational cost. For example, AlexNet [17] running on Intel Xeon CPU and NVIDIA Titan X consumes 130 and 250 W, respectively. Even hidden layer of AlexNet consumes too much power in the view of mobile devices, e.g., 73 W with Intel Core-i7 CPU, 159 W with Titan X GPU, and 5.1 W with Tegra K1 mobile GPU [18]. Therefore, energy-efficient processors dedicated to DNN acceleration gained lots of attention worldwide since power-hungry GPUs were not feasible for battery-driven mobile systems. Now, the use of DNN spread closer to our life; from advanced driver assistant system [33] and autonomous driving cars [34] to mobile devices such as smart-phones, head-mounted displays [35,36,37], always-on sensors [38], and broad applications of Internet-of-Things (IoT). The movement of AI from large systems to mobile systems was facilitated by the dedicated Neural Processing Units (NPUs) with high energy efficiency. This chapter introduces design considerations, hardware requirements, and hardware architectures of NPU with examples.

2. Background

The history of DNN started from neural networks (NNs) in 1940s. The very first neuron was capable of logic functions such as AND, OR, and NOT. Later its weights became trainable by learning algorithms and single neurons were combined to construct a multi-layer perceptron (MLP), which consists of neurons, layers, weights, and biases. The versatile NN topology resulted in various types of DNNs, including convolutional neural network (CNN), recurrent neural network (RNN), and MLP, which is sometimes called fully-connected network. Also the architecture became deeper with numerous neurons, hence called DNN, by the invention of hierarchical feature learning. The variants of DNN outperforms in different applications. CNN is widely used for image processing and RNN is used for natural language processing and time-series predictions. Therefore, a DNN architecture become different according to the target applications.

Fig. 1 shows an example of simple NN architecture. The gray neurons are hidden neurons and each column of the hidden neurons indicates a hidden layer. Input vector is multiplied by its corresponding weights and they are summed when going into a neuron in the next layer,

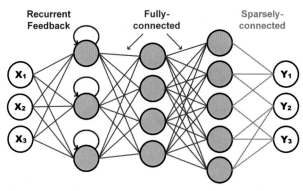

Fig. 1 Neural network architecture.

therefore Multiply-Accumulate (MAC) operation dominates the DNN operations. The output of a neuron is rectified by activation function. Sigmoid and hyperbolic tangent functions were commonly used in early DNNs as their activation function, where recent DNNs utilize rectified linear unit (ReLU) [60] or leaky ReLU [61], etc.

The network connections could be feedforward or recurrent feedback. Feedforward network passes the output of neuron to those in succeeding layer, and the final output at **Y** nodes represent confidence level, or probability that an input belongs to a certain class. The output of feedforward network is only dependent on current input, just like a combinational logic in digital circuits. On the other hand, the recurrent connection of RNN gives the ability to store internal states. The output becomes dependent on current input as well as previous state of the neurons.

Another characteristic of DNN architecture is that the connections between layers could be fully-connected (FC) or sparsely-connected. The fully-connected layers (FCLs) requires massive amount of memory for parameters and computations. However, recent researchers found that removing some of the connections does not affect the accuracy [62]. This technique reduces both storage and computations; but hardware designers should carefully utilize such data sparsity to increase efficiency and processing element (PE) utilization rate.

Fig. 2 shows an example of CNN architecture that is composed of up to thousands of convolution layers (CLs), pooling layers, and FCLs at the end for classification. The input to each layer is called an input feature maps (IFM) and the output corresponds to output feature map (OFM). Nonlinear activation usually takes places within CLs. Pooling and normalization layers are

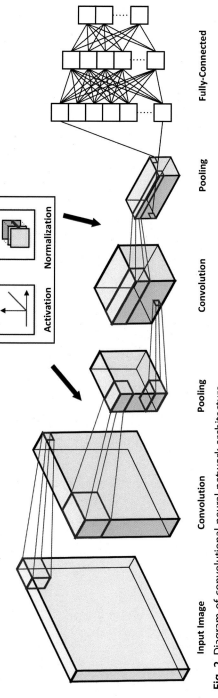

Fig. 2 Diagram of convolutional neural network architecture.

optionally inserted. Since image processing usually requires high dimension of color representations, e.g., 2-D image with three channels of RGB, sensory input data becomes 3-D matrix. Hence, the convolution filters (weights) are also 3-D. Convolution results between an IFM and 3-D filters are accumulated and activated to generate 1 channel of OFM. The filters in CL are shared among the IFM, mimicking local receptive field of visual system. This shared weights significantly reduces weight connections compared with FCL, but usually the channel depth becomes more than thousands as the DNN model gets deeper. This fact results in massive computations for CLs. In contrast, it is impossible (in most cases) to store the weights of FCLs on chip because they cannot be shared, resulting in large off-chip memory bandwidth.

Since 2-D convolution operation with shared filters (weights) dominates the CNN operation, both temporal and spatial locality of operands are very high. Therefore, CL is computation-centric as the operational intensity is very high. On the other hand, 2-D matrix-vector multiplication is dominant in RNN/FCL and their non-reusable weights result in memory-centric, which means the throughput of RNN/FCL is limited by external memory accesses for operands rather than operation itself.

3. Considerations in hardware design

Fig. 3 shows a typical hardware architecture of NPU, which consists of a massive array of PEs and hierarchical memory architecture. The entire data for large DNN models cannot be stored on–chip because DNN is composed of ~100 MB's of parameters, and the amount gets way larger when taking internal feature maps into account. Hence, the system essentially requires external memory. DRAMs are widely used for the external memory, however, the entire system requires 100s of GB/s memory bandwidth. In fact, it is well known that the highest energy expense is related to data movement,

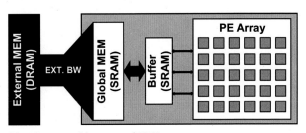

Fig. 3 Typical hardware architecture of NPU.

in particular DRAM accesses, rather than computation [20,21,22]. Therefore, reducing off-chip memory accesses (DRAM bandwidth) is crucial to DNN acceleration.

In addition to the off-chip memory accesses, on-chip memory bandwidth also becomes the bottleneck due to the enormous number of layers, parameters, and varying kernel sizes. Therefore, analyzing data movement pattern and optimizing memory access is very important. The energy efficiency could be enhanced with a hierarchical memory system and reuse of local data. CLs account for more than 90 % of the total operations and accelerating these calls for the efficient balancing of the computational vs. memory resources to achieve maximum throughput without hitting associated ceilings [44].

However, NPUs cannot solely solve the problem of DNN implementation; algorithm must be hardware-friendly to make DNN applications feasible in mobile and embedded platform. This emphasizes the importance of hardware-software co-optimization, and a lot of researches have been conducted to reduce the burden of software. Pruning [63] is a representative example of reducing model size without degrading DNN performance. This technique dramatically reduces the required number of computations and memory bandwidth in trade of resulting in sparsity in operands, which makes typical architecture NPU difficult to control efficiently. To efficiently utilize such sparsity, most NPUs deploy zero-skipping technique [41] to get rid of abundant but useless computations, such as multiplication or addition of 0 values.

To further reduce the computational and storage costs, many researches on precision reduction (number of bits) by quantization have been utilized [19]. Quantization can be applied to either/both weights or/and activations. Also, weight sharing can further reduce hardware costs [62]. In addition, NPUs should be capable of computing variable number precision to support versatile DNN model efficiently. And encoding/decoding techniques are frequently applied in NPUs to reduce communication overheads.

4. NPU architectures

4.1 NPU architectures for primitive neural networks

In fact, hardware architectures for relatively simple neural network acceleration were developed over 30 years. Mauduit et al. designed a single neuronal unit that integrates only memories and ALU for simple operations such as MUL, ADD, ACC, while non-linear activation is performed by off-chip component [23]. Viredaz et al. introduced 40×40 systolic array architecture for neural network acceleration [24]. Asanovic et al. designed a Very Large Instruction Word (VLIW) machine with SIMD array containing eight 32-bit fixed-point datapaths and register files [25]. In 1996, an image processor that integrates 12 DSPs working in SIMD mode, Vector-to-Scalar Unit, Scalar Unit, RISC processor, on-chip memories and serial interfaces is designed by Duranton et al. [26]. The communication paths among DSP slices adopted 1-D systolic ring to reduce data transaction overhead. The SIMD array architecture has been a typical hardware implementation for NPU for decades as shown in Fig. 3. Neural processing components are computed in each PE and the cluster of PEs makes the PE array. On-chip memories have hierarchy to hide external access time or reuse as many data as possible inside the chip. Many advanced DNNs with large network architecture require off-chip memory, usually DRAM, for data storage. Since DRAM access consumes lots of energy, reducing off-chip memory access and optimizing data movement is very important to efficient hardware design.

Recent designs of MLP and RNN continued to utilize SIMD architecture to predict workload of multi-core system for fast workload balancing [27,28]. Since the target MLP and RNN of each processor contains only one hidden layer, they implement 8-way SIMD neuron array where each neuron element is capable of multiplication (MUL), multiplication-and-accumulation (MAC), sigmoid function, and derivative of sigmoid function. For the target application, MUL and MAC require 16-bit precision to achieve high prediction accuracy as well as avoid overflow. Thus, the multiplier has reconfigurability (8-bit and 16-bit) for twice speed-up for 8-bit precision processing. Both of sigmoid and its derivative operations are generated by piecewise linear approximation, and accelerated by look-up table (LUT)-based architecture. This scheme removes bulky spaces for generation or storage of activation functions.

Apart from the primitive NPU architecture of Fig. 3, PE arrays in the cellular neural network processor introduced in [39] are shared among memory cells. The cellular neural network is composed of a 2-D cell array and synaptic connections among neighboring cells. The primitive hardware must have memory, inter-cell communication for synaptic connections, and PEs for neural computations. The hardware architecture is shown in Fig. 4. Each PE contains arithmetic logic unit (ALU) that can execute basic functions of cellular neural network and general-purpose image processing. The PEs are fully-pipelined with three stages (read, execute, write) for high

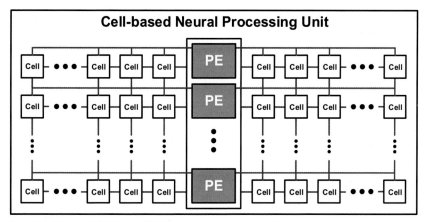

Fig. 4 Cell-based NPU for cellular neural network acceleration.

throughput. Cells are designed with register files for data storage and shift registers for efficient inter-cell communication, and all of the cells are connected by 2-D mesh data bus. The data in cells are transmitted to neighboring cells via shift register through the connection marked in red lines. Data bus for communication between PE and each cell is colored in blue lines. The target application of the processor is image processing with small image resolution (80 × 60), which are organized into two clusters of 60 PEs dealing with 40 × 60 cell arrays, i.e., each PE is shared by 40 cells. The architecture results in low data access overhead because each global shift operation along with the red lines takes only one cycle to complete. Also, it achieves 93% of PE utilization since all the data is delivered in one cycle.

A processor that is capable of MLP and CNN is designed in [29] supporting 16-bit fixed-point arithmetic. The accelerator divides neural processing into three stages: (1) multiplication of input and weights; (2) addition of the weighted vectors; and (3) non-linear activation function transfer. The 3-stage operation fits for FCL and CL, however, ALUs in the second stage also have shifter and max operators for pooling. The last stage must be performed after the weighted summation is done, therefore, pipeline of the accelerator is staggered. The processor utilizes 16-segment piecewise linear approximation for activation generation as in [28], where coefficients of each segment are stored in RAM. This processor deployed temporal reuse of IFM by implementing circular input buffer. The input buffer for IFM is also capable of local transpose for efficient pooling operation. However, the processor is inefficient for CNNs because the locality of 2-D data cannot be exploited by treating data as 1-D vectors.

The processor in [29] is expanded to execute complex network [31] in [30]. The PE array contains 2-D mesh architecture of the former neural functional unit. Each PE additionally contained local storage to enable inter-PE data propagation of input neurons. It can perform MAC, ADD, CMP (compare) operations every cycle, and its two outputs transfers data to memories and neighboring PEs for data reuse. The inter-PE communication greatly reduced memory bandwidth, i.e., approximately 75% in case of using 25 PEs benchmarked with LeNet-5 [31]. An additional ALU capable of 16-bit fixed-point arithmetic is integrated to compensate for neural functions that PE cannot compute, which are piecewise linear non-linear activation functions and division.

In [56], each PE array dedicated to 2-D sliding computation contains 5×5 MUL units for parallel computation of 5×5 convolution and an adder-tree to sum 25 multiplication results in one cycle. The 8 PE arrays share a host controller, memories, max-pooling ALU, and hierarchical adder-trees. Similar with [30], register files in MUL units are tightly connected to reduce memory bandwidth so that only the first row of MUL units must fetch data from on-chip memory where IFM and weights are stored. This inter-core connection also enables efficient 2-D sliding computation for image processing, resulting in 85% reduction of latency. However, the 2-D array architecture of processing units is not energy-efficient when handling 1-D data, because they have different data access pattern. To mitigate this issue, the processor deployed hierarchical adder-trees for dual-mode configuration (CL and FCL). Unlike the PE arrays share max-pooling ALU for CL computation, two PE arrays share level-1 adder-tree in MLP configuration and level-2 adder-tree sums the results of two level-1 adder-trees in hierarchical manner to output 100 MAC results. This architecture lets the processor compute 4 hidden neuron values in parallel with the 8 PE arrays, efficiently utilizing the 2-D arrays of unit for 1-D vector computation. Although its target CNN architecture is quite simple, having only two consecutive CONV and POOL layers with 50-100-1 MLP at the end as a classifier, the processor effectively reduces region-of-interests for object recognition. The 80% reduction of region-of-interests increases energy efficiency of the entire object recognition pipeline.

Another accelerator for RNN and fuzzy inference system is introduced in [33] to provide high-level intelligence in autonomous application. The purpose of RNN is to estimate future state of objects, represented in its distances and velocity in longitudinal and lateral directions; then the fuzzy inference system classifies if the future motion of an object has potential risk to collide with the driving vehicle. The hardware architecture is shown in Fig. 5. MAC operations of RNN are computed by the massively-parallel matrix processing unit that contains 8 neuron units. Each neuron unit is designed with SIMD extension PE, which selectively performs 32 *8-bit*, 16 *16-bit*, or 8 *32-bit* operations in parallel. This enables accurate computation as the precision becomes higher in deeper layers of RNN. The fuzzy inference system is also accelerated by 64 parallel processing units. All of the complex calculations for membership function generation are replaced with LUT-based in order to save area and power.

4.2 NPU architectures for DNN

The major difference of Sections 4.1 and 4.2 is the targeted DNN architecture. The processors reviewed in this section aims at accelerating much complex and large DNN architectures. Fig. 6 shows two different approaches to NPU designs. Many processors utilized the global SRAM approach for

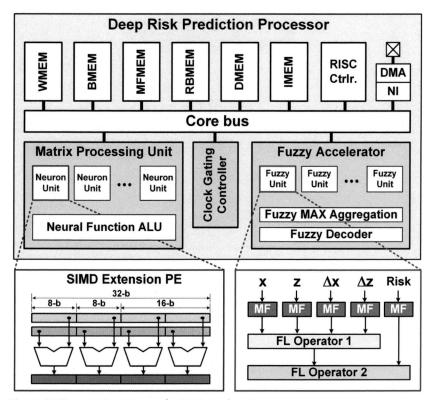

Fig. 5 SIMD extension PE array for RNN acceleration.

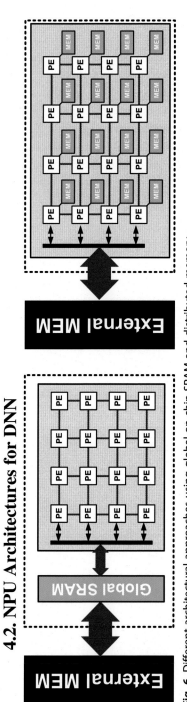

Fig. 6 Different architectural approaches using global on-chip SRAM and distributed memory.

acceleration of general CNN. But the problem of global SRAM implementation rises on-chip communication between memory and processing parts and complexity of global routings. The other non von Neumann approach distributes the global memory into the PE arrays are proposed to reduce communication latency and complexity. But it is not sufficient for different CNN architecture if the model becomes very large. The processors introduced in this Section are on the basis of the two approaches of Fig. 6.

A CNN processor with four homogeneous cores is presented for IoT device [42]. Each core has buffers and two neuron processing arrays each of which deals different IFM and OFM while sharing kernel buffers to reduce memory accesses. The array contains 32 dual-range MAC, 32 ReLU units, and 8 max-pooling units where the numbers fit the single recognition process. ReLU units and max-pooling units are turned off until convolution is done. The dual-range MAC reduces power by selectively processing 16-bit/24-bit precision by masking the upper 8 bits of inputs of a 24-bit MAC unit. This eliminates the switching activity of the upper 8 bits, hence reducing 56% of switching power. The processor applies tile-based operation that reuses IFM as many kernels as possible to reduce off-chip memory accesses. The on-chip input/output buffers switch their role for the subsequent CNN layer processing. To reduce off-chip accesses for fetching kernel data, the processor generates original kernel by weighted sums of basic kernels which are extracted from off-chip principle component analysis. The kernels are generated from dual-range MAC units. This on-chip kernel generation scheme reduces off-chip access for kernels by 92% at maximum.

Another famous CNN processor (Eyeriss) was proposed to optimize for the energy efficiency of the accelerator chip and off-chip DRAM in [41]. Eyeriss consists of 12×14 PE arrays designed for Row Stationary CNN dataflow, a Network-on-Chip for multicasting and point-to-point data transmission, a global buffer with memory hierarchy, a ReLU unit, and a pair of run-length compression encoder and decoder. Thanks to the inter-PE communication fabric and low-cost scratch pad memories in each PE, data is maximally reused to minimize the high-cost DRAM and global buffer accesses. The local scratch pads enable reuse of convolutional data and accumulation of partial sums. Since all the PEs comprise spatial architecture, the IFMs with large width can be divided into separate segments, but the case of IFMs taller than the PE arrays were not supported limiting the vertical filter size to 12. Each PE contains separate scratch pads for IFM, filter, and partial sum to maintain high bandwidth. 16-bit multipliers are pipelined into two stages, followed by an adder, and the truncation of resulting 32-bit to

16-bit is configurable based on the experiments. In addition, the PE exploits zero-skipping logic, which detects zero values in IFM and skips MAC operation, to reduce power consumption.

An efficient inference engine is proposed for DNN acceleration by supporting sparse matrix-vector multiplication and weight sharing [32] that are based on the authors' prior publication on network compression method [62]. A scalable PE array of the processor holds 131 K weights of compressed model from 1.2 M weights of original network model. Since weight matrices are not reusable during matrix-vector multiplication of FCL, conventional CPUs and GPUs utilized augmented matrices of vectors which directly increases latency. Thus, DNN network is compressed and parallelized in this work. The sparse weight matrix is encoded in compressed sparse column format, which takes non-zero values and number of zero values as pointers. The hardware architecture of each PE, then, receives activation queue and starts with the pointer read operation that specifies start and end pointers for each column of matrix. The pointers are used to access sparse matrix SRAM, and the arithmetic unit performs MAC operation. Activations are stored in two register files according to their source and destination, and the role of them are exchanged as a ping-pong manner for next layer. The architecture consumed $2700 \times$ less energy compared to mobile GPU.

Desoli et al. introduced a CNN SoC that integrates ARM microprocessor, DSP clusters, high-speed interfaces, and 8 convolutional accelerators [44]. 4 MB ($4 \times 16 \times 64$ KB) of SRAM is shared to sustain peak CNN throughput, where each 64 KB block is individually power-gated. Eight convolutional cores are chained through configurable stream switch to support different IFM size and multiple kernels in parallel. Kernels are divided in channel direction to store intermediate data on chip. The feature line buffer within the convolution core can fetch 12 feature map in parallel. There are 36 16-bit fixed-point MAC units and their results for each kernel column are accumulated by an adder-tree, allowing optimal reuse of FM for multiple MAC operations.

Whatmough et al. designed a programmable hardware for FC-DNN [43]. The DNN engine is composed of 5-stage SIMD accelerator that supports sparse matrix-vector multiplications with 8-way MAC units under 16-bit fixed-point precision. The key idea of this architecture is exploiting dynamic sparsity in IFM and activations effectively. Unlike [41] consumes clock cycles for pipeline hazards even though it clock-gated zero operands, its dynamic elision of zero operands at write back stage eliminates the pipeline hazards. In addition, output activations are compared with layer-wise thresholds and skips small non–zero data to improve energy and throughput. The MAC datapath utilizes time–borrowing technique with Razor flip–flops to enhance error tolerance. The large on–chip memory (4 × 256 kB) provides high bandwidth to the processing cores.

Another CNN processor with subword–parallel dynamic–voltage–accuracy–frequency scaling is introduced in [40] to provide 40× energy–precision scalability. Previous to this architecture, Moons et al. showed that dynamic–precision–scaling reduces switching activity by approximating LSBs at the inputs of MAC units, and dynamic–precision–and–voltage–scaling combines voltage to exploit shorter critical paths to increase energy in [40]. Frequency control, which lowers all run–time adaptable parameters at the same time, is added in the new architecture. The NPU integrates 16 × 16 2-D SIMD MAC arrays for convolution and 1-D SIMD arrays for ReLU and max–pooling, and a scalar processing unit. Also it deploys Huffman encoder/decoder to compress I/O bandwidth to 5.8×. The 2-D SIMD MAC arrays consists of 6-stage pipelined processor, where each MAC is composed of subword–parallel multiplier with reconfigurable accumulation adder and register. Intermediate data generated by the 16 × 16 array every clock cycle are reused. In addition, the FMs are shifted along the x-axis in one cycle, reducing memory bandwidth. SRAM access is reduced by storing intermediate output values in accumulation register. The sparsity that requires zero–skipping, i.e., large amount (up to 89%) of zero–valued data due to ReLU or use of low–precision words, is supported by guarding memory fetches and MAC operations. The sparsity of a new layer is stored in flag memory, then only the flags are fetched to prevent MAC operations and SRAM accesses. This scheme reduces energy consumption together with dynamic–voltage–accuracy–frequency scaling.

A heterogeneous multi–core NPU architecture, named DNN processing unit (DNPU), is porposed to accelerate CNN and RNN in mobile environment [45]. The high spatio–temporal locality of operands in 2-D convolution makes CNN computation–bounded. However, matrix multiplication with large amount of non–reusable weights characterizes RNN memory–bounded; its throughput is limited by external memory accesses. The DNPU exploits heterogeneous architecture to satisfy the conflicting requirements; CNN processor exploits reusability of convolution with large number of PEs, while RNN processor is optimized for reducing off–chip accesses. The CNN processor consists of an aggregation core and 16 convolution cores, which has 3 × 4 × 4 PE arrays. Four convolution cores are serially connected to transfer partial sums to the consecutive core. The RNN processor deploys quantization tables and 16-bit fixed-point multipliers to reduce external memory bandwidth. The baseline of quantization tables comes from the observation that weight quantization of RNN greatly reduces external memory accesses by fetching weight indices only while its accuracy loss is very small. Also, partial-sums are quantized by multiplying inputs with quantized weights. Once the quantization table is constructed, only quantized indices are used for multiplication, resulting in 75% reduction of off-chip accesses. Moreover, matrix multiplication is computed by accessing the table values. This scheme reduces 99% of multiplications by removing the needs of multipliers. In addition to the dedicated hardware architecture, DNPU proposed a new computation methodology to efficiently utilize limited capacity of on-chip memory. Since IFM, OFM, and kernel weights reaches up to several hundreds of MB, image and channel are divided for efficient CNN processing. When the NPU processes IFMs divided into 2-D direction, same values of kernels should be fetched repeatedly for the divided images. This approach is useful when the entire image map cannot be stored on-chip. On the other hand, multiple off-chip accesses are required if the NPU processes along with divided channels because OFM cannot be calculated with a single image. It takes advantages when the whole weights cannot be stored on-chip, or the channel is deep. The DNPU combines these two methods so that the processor can utilize efficient processing method for large CNN model with limited on-chip memory. But its heterogeneous architecture suffers from limited hardware utilization rate. For example, the DNPU wastes hardware resources if the DNN application requires only one of CNN or RNN.

Following the previous NPU architecture, Lee et al. fabricated the unified neural network processing unit (UNPU) to resolve the limitations of the DNPU [48]. It focused on finding energy-optimal point and obtain high utilization. Unlike the DNPU utilized reuse of weights, the UNPU reuses feature maps according to their observation that the feature maps become dominant if the weight bit precision is reduced on the accuracy-energy optimal point. Its unified datapath facilitates efficient processing of both CNN and RNN, achieving $1.15 \times$ and $13.8 \times$ higher peak-performance with same area. Each LUT-based PE performs 1-bit operation and they are bit-serially connected to support fully variable precision. The entire PE arrays can perform 18,432 1-bit MAC operations in one cycle, or 13,824 16-bit MAC operations per 16 cycles. Each LUT-based PE contains four LUT Bundles with four 8×16-bit LUT and multiplexer. For a feature vector, weights are put into the PEs bit-serially and multiple PEs produce partial-products in every cycle. The architecture supports weight bit precision from 1-bit to 16-bit, sacrificing the latency. LUTs in each PE store possible bit-serial partial-products, so that it can output MAC results without bulky MAC units. It reduces energy consumption by 23.1% (16-bit), 27.2% (8-bit), 41.0% (4-bit), and 53.6% (1-bit) compared with the fixed-point MAC PEs. The UNPU integrated *aligned feature map loaders* which transfer 1-D vectorized IFM to LUT bundles in order for the vectors to be computed based on LUT values. The impact of ability to support wide range of bit precision is important. The multi-bit precision that the DNPU supports (4-bit, 8-bit, 16-bit) cannot find the optimal bit precision, which differs by different layers and CNN networks [50,51]. Advanced DNN applications [52,53,54] uses small bit-precision. Therefore, the UNPU can accelerate more CNN networks by providing wide range of precision. The sacrifice in latency arouse from the bit-serial PE architecture could be resolved by using small bit precision CNNs.

Another CNN processor with always-on sensor system is proposed in [38]. The system integrates CMOS image sensor dedicated to face detection and the CNN processor for face recognition with ultra-low-power consumption. The distributed-memory NPU architecture consists of 4×4 PE array in which make use of local memory. Each PE consists of local SRAM, a weight buffer, a register file, four convolution units, and two ALU units. The distributed memory architecture reduces global routing complexity by storing CNN parameters on chip. The convolution unit is composed of 16-way SIMD MAC datapath and registers. It deals with 16-bit fixed-point IFM and 8-bit floating-point weights. The MAC operation basis on shifter and adders instead of multipliers, resulting in energy reduction by 43%. In software-level, the processor utilized separable filter approximation that approximates 2-D convolution with cascaded 1-D vector convolutions. For example, 2-D $m \times m$ convolution is approximately obtained with 1-D $m \times 1$ vertical filter and $1 \times m$ horizontal filter. However, conventional SRAM is not efficient for reading vertical data because it requires multiple wordline access. Therefore, the processor proposed an SRAM architecture which can access the data shared over the same bitline so that both horizontal and vertical data can be fetched by single SRAM access. This technique reduces the more than 60% of computation with less than 1% accuracy loss for face recognition.

However, the distributed memory architecture of the NPU [38] has limited hardware utilization especially when the size of IFM is small. Therefore, it is not efficient for other CNN algorithm, e.g., stereo matching CNN [55] of which the FM becomes small in deeper layers, and NPU architecture that maximizes the hardware utilization is presented in [46]. It computes channel-wise 1–D MAC operations in parallel to avoid redundant operations that is caused by the 2-D workload assignment. The NPU core consists of a pipelined CNN PE, memories, and local DMA for inter-core communication. The 2-D convolutions are obtained by repetitive use of 3-stage shift MAC units. The CNN PE has 7-stage pipelined architecture and it processes 48 MACs in one cycle with 96% of core utilization. The pipeline hides both of accumulation latency (among channels) and data fetch latency, achieving 20% higher energy efficiency. In order to improve the overall performance, the NPU utilizes inter-core communication in local DMA unit to balance workload between two CNN cores. This reduces the overall processing time by 23.9%.

Song et al. proposed a sparsity-aware NPU architecture that operates stationary weights and streaming features based on the layers [49]. The NPU with 1024 MAC units is divided into two cores to resolve logic costs and wiring congestion. Two data-staging units are in charge of buffering features, storing weights, and decompressing weight streams in each core. Then, non-zero weights of CL and features of FCL are selected in the data-staging units, then are fetched into the dual 8-bit MAC arrays. The dual MACs holds 16 partial sums, which is delivered to a data–returning unit for non–linear activation. For CL operation, the processor utilizes channel-division method for both IFM and kernels. 16 weight kernels are reused and computed in parallel until OFM is calculated, then 4×4 OFM is computed in parallel. For FCL, IFM is divided into four blocks and assigned to the data-staging units, then the NPU computes 256 rows of a matrix in parallel and 16 OFM are computed in parallel. The zero-skipping is supported by feature-map selection circuit that selects 2-D feature array corresponds to a non-zero weights. The NPU architecture improves energy efficiency and performance of [48] by $2.06 \times$ and $10 \times$, respectively.

A primitive NPU architecture for training was introduced in [57]. The processor used the same arithmetic precision (16-bit fixed-point) but has separate cores for inference and training, and a random number generator. More advanced version of the processor was designed for user-adaptive UI/UX based on on-line training with user-centric data in head-mounted display application [35]. The processor consists of a hand-segmentation core and a speech segmentation core for image preprocessing, and a cluster of four dropout deep-learning engines and a PVT-robust true-random-number generator (TRNG) for DNN training. In the training mode, the dropout

DL engine receives training data from off-chip and dropout learning [59], which neglects randomly selected neurons during training to reduce error rate, is accelerated according to the random numbers generated from the TRNG. After the TRNG generates random bits, a drop-connect decider selects which neurons and corresponding weights should be gated. The high-V_T PMOS capacitor bank connected to an inverter pair architecture ensures controlling small values of capacitance. After PVT-variation causes random shift of the TRNG, it is monitored to control the sizes of inverter pairs and output voltages to increase randomness. The NPU core with TRNG improved gesture and speech recognition accuracy by 2.1% and 1.9%, respectively.

5. Discussion

There exist two phases of DNN operation: training and inference. The values of synaptic weights are updated during the training phase in the way that minimizes its output error by using back-propagation algorithm until the whole DNN network reaches some reasonable accuracy. Big data enabled highly-accurate DNN performance since DNN is trained with enormous dataset, i.e., tens of thousands of training samples. The changes of the weight values become very tiny as the training phase goes deeper. Therefore, the training phase generally requires floating-point arithmetic (FP64) to represent small numbers for higher accuracy. Large amount of memory is also required for the training phase to deal with the whole dataset, thus, the training phase is computationally expensive. On the other hand, the inference phase takes only one input and computes with the pre-trained network, which results in less computation. The DNNs even can achieve marginal accuracy without using such wide bit-widths. The optimal bit-width varies depending on the architecture and layer. Therefore, recent NPU architectures are designed to support variable bit-width for energy optimization, usually from 1-bit to 16-bit.

The processors dealt in this article attempted to maximize on-chip data reuse for high energy-efficiency. Most of them utilized 2-D MAC arrays with dedicated resource management and scheduler, by implementing hierarchical memory architecture similar to cache to reduce off-chip bandwidth. Even though the DNN processors resolved massive off-chip data transaction for high energy efficiency, their approaches require large off-chip memory bandwidth and complex global routing to the large PE arrays, therefore, still suffer from memory bottleneck in many applications. In one aspect, researchers are investigating new paradigm of computer architecture to solve this problem. Bong et al. [38] proposed distributed memory architecture that eliminates global memory hence global routing to reduce complexity and bottleneck of memory bandwidth. Such distributed memory architecture expands to Processing-in-Memory (PIM), which is one of the appealing approach to the non-von Neumann architecture that deploys data locality to maximize data reuse. Inspired by the concept of mixed-mode computing in image sensors, Kang et al. [70] utilized a computation scheme in the periphery of memory cell to minimize the costs of data access. The data, which is stored in SRAM cell, is computed in advance of analog-digital conversion. As a result, the processor saved $4.9 \times$ energy and achieved $2.4 \times$ improvement in throughput than von-Neumann architecture. Lee et al. [71] designed cost generation peripheral circuit, which is sandwiched between the SRAM cell arrays. The census transform data from left and right images are temporarily stored in each stereo-SRAM. Then, cost values are directly generated by XOR function of the data stored in the same location but in separate SRAM cells when the data are to be read out. Yang et al. [72] accelerated binary-weight network by combining pulse width modulation and shift memory in to the bitcell array. Binary-weight network is accelerated using analog computing in [72]. They implemented 10-T SARM cell array with 1-bit multiplication-and-average operation. Also, Yin et al. [47] designed SRAM macro capable of ternary-XNOR-and-accumulate operations. Despite the fact that such approaches are energy-efficient, their architectures are yet primitive so they are limited to only light-weighted NNs and simple tasks. In addition to the new computing paradigm, PIM, 3-D stacking technique with NPU and memory is also investigated to overcome massive external memory bandwidth.

Another trend, neuromorphic processor, is to mimic biological neurons with semiconductors for ultra-low-power consumption. Neurogrid [64] is a mixed-mode multichip system of which the neuron array is implemented with analog circuits and spike TRx as well as memories are digital implementation. IBM revealed a large-scale chip (4.3cm^2 @ 28nm) TrueNorth [65] composes 4096 neurosynaptic cores of spiking neural networks. In each core, 256 digital axons and neurons are connected by 256×256 crossbar. It consumed 63 mW when processing multiobejct detection with 400×240 video running at 30 fps. Later, Intel introduced Loihi [66] that consists of 128 cores of 1024 spiking neural network units together with three x86 cores. The entire SoC is implemented in 14nm FinFET technology and occupies 60 mm^2. Implemented with fully-digital circuits, both TrueNorth and

Loihi consumed only tens of milliwatt. On the other hand, some of the recent neuromorphic processors take advantage of analog computing arrays. Miyashita et al. [67] proposed time-domain neural network with analog MAC units, resulting in high energy efficiency of 12.9 fJ per synaptic operation. Another approach is to make use of nonvolatile memories (NVM) in their analog arrays to replace floating-point MAC operation with parallel analog processing array [64,68,69]. Fig. 7 shows the conceptual diagram of analog computing. NVMs are place as crossbar switches and their transconductance represents corresponding weight values. When a pattern of signal is input, then current flowing through each column becomes the sum-of-products of the input vector and weight. As an example, Bayat et al. [58] utilized the concept with ReRAM crossbar arrays, where the target MLP has only 10 hidden neurons due to the physical array size. Sometimes, analog accelerators are said to achieve significant performance improvement than digital processors. According to [68], analog processor achieved $270 \times$ energy, $1040 \times$ latency, and $1.8 \times$ area than digital ReRAM accelerators; $430 \times$, $34 \times$, and $11 \times$ compared to SRAM-based digital accelerators, respectively. But implementation details and material properties of the NVM must be aligned with the requirements of NN algorithms [69]. Therefore, neuromorphic processors yet have limited functions and application tasks.

Since much more transistors will be integrated while power consumption becomes low as the process technology advances, the future researches on NPU for high performance applications will be digital implementations

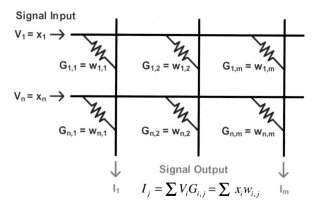

Fig. 7 Conceptual diagram of analog vector MAC array.

to ensure high system throughput and precise computation. This approach will facilitate integration of higher-level intelligence embedding complicate DNN with high energy efficiency. In another respect, neuromorphic processors will be researched for ultra-low-power applications (e.g., IoT) but the application tasks will be simpler than digital implementations. Analog computations will gain lots of attention for energy-efficient neuromorphic computing. However, majority of the neuromorphic processors highly rely on the development of new technologies, such as ReRAM and STT-MRAM, while those technologies are still in their early stages. In the meantime the technologies become public and mass production, it is believed that mixed-mode circuit implementations as well as PIM architectures based on CMOS technology is promising for ultra-low-power DNN applications until the neuromorphic devices become mature and mass production.

6. Summary

Neural Processing Units (NPUs) are built to efficiently accelerate various types of deep neural networks (DNNs). The main concerns of NPU designs are high throughput and energy efficiency. Since DNNs require large amount of data for both training and inference, memory bandwidth becomes crucial in NPU designs. Most of the NPUs utilizes data-reuse and skipping useless computation techniques to resolve large off-/on-chip memory bandwidth. Also, NPUs equip with PE arrays in massively-parallel SIMD architecture because most of the DNN computations are linear algebraic operations. What makes NPU design very hard is the fact that optimized precision for DNN processing varies by layers and DNN architectures; researchers are deploying hardware architectures that are capable of variable bit-width operations to enable much energy-efficient computing.

In addition to the NPUs, new computing paradigms such as Processing-in-Memory (PIM) architectures and neuromorphic processors are briefly introduced and discussed for future research directions. PIM architectures attempt to push computing cores closer to on-chip memories to reduce memory bandwidth. Current stage of PIM is investigated using CMOS and memory technologies, however, advanced process technologies are also being developed for PIM. On the other hand, neuromorphic processors are yet highly dependent on the development of new technologies, such as ReRAM and STT-MRAM. Therefore, traditional approach to NPUs and PIM using CMOS technology would be dominant research topic until the memory technologies become mature and public to designers.

Acknowledgments

This research was support in part by the Korea Institute for Advancement of Technology (KIAT) grant funded by the Korean Government (MOTIE) (N0001883, The Competency Development Program for Industry Specialist), in part by Samsung Electronics, and in part by the National Research Foundation of Korea (NRF) grant funded by the Korean Government (MSIT) under Grant 2019R1C1C1009857.

References

[1] C. Szegedy, W. Liu, Y. Jia, P. Sermanet, S. Reed, D. Anguelov, D. Erhan, V. Vanhoucke, A. Rabinovich, Going deeper with convolutions, in: Proceedings of IEEE Conference on Computer Vision and Pattern Recognition (CVPR), Boston, MA, USA, 2015, pp. 1–9.

[2] K. He, X. Zhang, S. Ren, J. Sun, Deep residual learning for image recognition, in: Proceedings of IEEE Conference on Computer Vision and Pattern Recognition (CVPR), Las Vegas, NV, USA, 2016, pp. 770–778.

[3] S. Xie, R. Girshick, P. Dollar, T. Zhuowen, K. He, Aggregated residual transformations for deep neural networks, in: Proceedings of IEEE Conference on Computer Vision and Pattern Recognition (CVPR), Honolulu, HI, USA, 2017, pp. 5987–5995.

[4] J. Redmon, S. Divvala, R. Girshick, A. Farhadi, You only look once: unified, real-time object detection, in: Proceedings of IEEE Conference on Computer Vision and Pattern Recognition (CVPR), Las Vegas, NV, USA, 2016, pp. 779–788.

[5] K. Ota, M.S. Dao, V. Mezaris, F.G.B. De Natale, Deep learning for mobile multimedia: a survey, in: ACM Transactions on Multimedia Computing, Communications, and Applications (TOMM), vol. 13, 2017, pp. 1–22. no. 3s.

[6] S. Yao, S. Hu, Y. Zhao, A. Zhang, T. Abdelzaher, DeepSense: a unified deep learning framework for time-series mobile sensing data processing, in: Proceedings of the 26th International Conference on World Wide Web (WWW), Perth, Australia, 2017, pp. 351–360.

[7] A.G. Howard, M. Zhu, B. Chen, D. Kalenichenko, W. Wang, T. Weyand, M. Andreetto, H. Adam, MobileNets: Efficient Convolutional Neural Networks for Mobile Vision Applications, 2017, [Online]. Available https://arxiv.org/abs/1704.04861.

[8] D. Li, X. Wang, D. Kong, DeepRebirth: accelerating deep neural network execution on mobile devices, in: Proceedings of 32nd AAAI Conference on Artificial Intelligence (AAAI), New Orleans, USA, 2018, pp. 2322–2330.

[9] A. Mathur, N.D. Lane, S. Bhattacharya, A. Boran, C. Forlivesi, F. Kawsar, Deepeye: resource efficient local execution of multiple deep vision models using wearable commodity hardware, in: Proceedings of the 15th Annual International Conference on Mobile Systems (MobiSys'17), Niagara Falls, New York, USA, 2017, pp. 68–81.

[10] Q. Cao, N. Balasubramanian, A. Balasubramanian, MobiRNN: efficient recurrent neural network execution on mobile GPU, in: Proceedings of the 1st International Workshop on Deep Learning for Mobile Systems and Applications (EMDL'17), Niagara Falls, New York, USA, 2017, pp. 1–6.

[11] N.D. Lane, S. Bhattacharya, P. Georgiev, C. Forlivesi, L. Jiao, L. Qendro, F. Kawsar, DeepX: a software accelerator for low-power deep learning inference on mobile devices, in: Proceedings of the 15th International Conference on Information Processing in Sensor Networks (IPSN), Vienna, Austria, 2016, pp. 23:1–23:12.

[12] S.I. Venieris, C.-S. Bouganis, fpgaConvNet: a toolflow for mapping diverse convolutional neural networks on embedded FPGAs, in: Proceedings of 31st Conference on Neural Information Processing Systems (NeurIPS), Long Beach, CA, USA, 2017, pp. 1–5.

[13] J. Wu, C. Leng, Y. Wang, Q. Hu, J. Cheng, Quantized convolutional neural networks for mobile devices, in: 2016 IEEE Conference on Computer Vision and Pattern Recognition (CVPR), Las Vegas, NV, USA, 2016, pp. 4820–4828.

[14] P. Molchanov, S. Tyree, T. Karras, T. Aila, J. Kautz, Pruning convolutional neural networks for resource efficient inference, in: Proceedings of 5th International Conference on Learning Representations (ICLR), Toulon, France, April 2017, 2017.

[15] A. Tulloch, Y. Jia, High performance ultra-low-precision convolutions on mobile devices, in: Proceedings of 8th International Conference on Learning Representations (ICLR), New Orleans, USA, 2019.

[16] Y.-D. Kim, E. Park, S. Yoo, T. Choi, Y. Lu, D. Shin, Compression of deep convolutional neural networks for fast and low power mobile applications, in: Proceedings of 4th International Conference on Learning Representations (ICLR), San Juan, Puerto Rico, 2016.

[17] A. Krizhevsky, I. Sutskever, G.E. Hinton, ImageNet classification with deep convolutional neural networks, in: Proceedings of 25th Neural Information Processing Systems (NeurIPS) 25, Lake Tahoe, USA, 2012.

[18] K.J. Lee, J. Lee, S. Choi, H.-J. Yoo, The development of silicon for AI: different design approaches, in: IEEE Transactions on Circuits and Systems I—Regular Papers (Early Access), 2020.

[19] V. Sze, Y.-H. Chen, T.-J. Yang, J.S. Emer, Efficient processing of deep neural networks: a tutorial and survey, in: Proceedings of the IEEE, vol. 105, 2017, pp. 2295–2329. no. 12.

[20] R. Hameed, W. Qadeer, M. Waches, O. Azizi, A. Solomatnikov, B.C. Lee, S. Richardson, C. Kozyrakis, M. Horowitz, Understanding sources of inefficiency in general-purpose chips, in: ACM Proceedings of Annual International Symposium on Computer Architecture (ISCA), New York, USA, 2010, pp. 37–47.

[21] S.W. Keckler, W.J. Dally, B. Khailany, M. Gariand, D. Glasco, GPUs and the future of parallel computing, in: IEEE Micro, vol. 31, 2011, pp. 7–17. no. 5. September/October.

[22] M. Horowitz, 1.1 Computing's energy problem (and what we can do about it), in: IEEE International Solid-State Circuits Conference (ISSCC) Digest of Technical Papers, San Francisco, CA, USA, 2014, pp. 10–14.

[23] N. Mauduit, M. Duranton, J. Gobert, Lneuro 1.0: a piece of hardware LEGO for building neural network systems, in: IEEE Transactions on Neural Networks (TNN), vol. 3, 1992, pp. 414–422. no. 3.

[24] M.A. Viredaz, P. Ienne, MANTRA I: a systolic neuro-computer, in: Proceedings of IEEE International Joint Conference on Neural Networks (IJCNN), Nagoya, Japan, 1993, pp. 3054–3057.

[25] K. Asanovic, J. Beck, B.E.D. Kingsbury, P. Kohn, N. Morgan, J. Wawrzynek, SPERT: a VLIW/SIMD neuro-microprocessor, in: Proceedings of IEEE International Joint Conference on Neural Networks (IJCNN), Baltimore, MD, USA, 1992, pp. 577–582.

[26] M. Duranton, Image processing by neural networks, in: IEEE Micro, vol. 16, 1996, pp. 12–19. no. 5.

[27] K. Lee, J. Park, I. Hong, H.-J. Yoo, Intelligent task scheduler with high throughput NoC for real-time mobile object recognition SoC, in: Proceedings of IEEE European Solid-State Circuits Conference (ESSCIRC), Graz, Austria, 2015, pp. 100–103.

[28] Y. Kim, G. Kim, I. Hong, D. Kim, H.-J. Yoo, A 4.9 mW neural network task scheduler for congestion-minimized network-on-chip in multi-core systems, in: Proceedings of IEEE Asian Solid-State Circuits Conference (A-SSCC), KaoHsiung, Taiwan, 2014, pp. 213–216.

[29] T. Chen, D. Zidong, N. Sun, J. Wang, C. Wu, Y. Chen, O. Temam, DianNao: a small-footprint high-throughput accelerator for ubiquitous machine learning, in: ACM

Proceedings of the 19th International Conference on Architectural Support for Programming Languages and Operating Systems (ASPLOS), Salt Lake City, UT, USA, 2014, pp. 269–284.

[30] D. Zidong, R. Fasthuber, T. Chen, P. Ienne, L. Li, T. Luo, X. Feng, Y. Chen, O. Temam, ShiDianNao: shifting vision processing closer to the sensor, in: ACM Proceedings of Annual International Symposium on Computer Architecture (ISCA), Portland, OR, USA, 2015, pp. 92–104.

[31] Y. Lecun, L. Bottou, Y. Bengio, P. Haffner, Gradient-based learning applied to document recognition, in: Proceedings of the IEEE, vol. 86, 1998, pp. 2278–2324. no. 11.

[32] S. Han, X. Liu, H. Mao, P. Jing, A. Pedram, M.A. Horowitz, W.J. Dally, EIE: efficient inference engine on compressed deep neural network, in: ACM/IEEE Annual International Symposium on Computer Architecture (ISCA), Seoul, South Korea, 2016, pp. 243–254.

[33] K.J. Lee, K. Bong, C. Kim, J. Jang, K.-R. Lee, J. Lee, G. Kim, H.-J. Yoo, A 502-GOPS and 0.984-mW dual-mode intelligent ADAS SoC with real-time semiglobal matching and intention prediction for smart automotive black box system, in: IEEE Journal of Solid-State Circuits (JSSC), vol. 52, 2017, pp. 139–150. no. 1.

[34] E. Talpes, D.D. Sarma, G. Venkataramanan, P. Bannon, B. McGee, B. Floering, A. Jalote, C. Hsiong, S. Arora, A. Gorti, G.S. Sachdev, Compute solution for tesla's full self-driving computer, in: IEEE Micro, vol. 40, 2020, pp. 25–35. no. 2.

[35] S. Park, S. Choi, J. Lee, M. Kim, J. Park, H.-J. Yoo, 14.1 A 126.1mW real-time natural UI/UX processor with embedded deep-learning core for low-power smart glasses, in: IEEE International Solid-State Circuits Conference (ISSCC) Digest of Technical Papers, San Francisco, CA, USA, 2016, pp. 254–255.

[36] G. Kim, K. Lee, Y. Kim, S. Park, I. Hong, K. Bong, H.-J. Yoo, A 1.22 TOPS and 1.52 mW/MHz augmented reality multicore processor with neural network NoC for HMD applications, in: IEEE Journal of Solid-State Circuits (JSSC), vol. 50, 2015. no. 1.

[37] I. Hong, K. Bong, D. Shin, S. Park, K.J. Lee, Y. Kim, H.-J. Yoo, A 2.71 nJ/pixel gaze-activated object recognition system for low-power mobile smart glasses, in: IEEE Journal of Solid-State Circuits (JSSC), vol. 51, 2016, pp. 45–55. no. 1.

[38] K. Bong, S. Choi, C. Kim, D. Han, H.-J. Yoo, A low-power convolutional neural network face recognition processor and a CIS integrated with always-on face detector, in: IEEE Journal of Solid-State Circuits (JSSC), vol. 53, 2018, pp. 115–123. no. 1.

[39] S. Lee, M. Kim, K. Kim, J.-Y. Kim, H.-J. Yoo, 24-GOPS 4.5-mm2 digital cellular neural network for rapid visual attention in an object-recognition SoC, in: IEEE Transactions on Neural Networks (TNN), vol. 22, 2011, pp. 64–73. no. 1.

[40] B. Moons, M. Verhelst, A 0.3–2.6 TOPS/W precision-scalable processor for real-time large-scale ConvNets, in: Proceedings of IEEE Symposium on VLSI Circuits (VLSIC), Honolulu, HI, USA, 2016.

[41] Y.-H. Chen, T. Krishna, J.S. Emer, V. Sze, Eyeriss: an energy-efficient reconfigurable accelerator for deep convolutional neural networks, in: IEEE Journal of Solid-State Circuits (JSSC), vol. 52, 2017, pp. 127–138. no. 1.

[42] J. Sim, J.-S. Park, M. Kim, D. Bae, Y. Choi, L.-S. Kim, 14.6 A 1.42TOPS/W deep convolutional neural network recognition processor for intelligent IoE systems, in: IEEE International Solid-State Circuits Conference (ISSCC) Digest of Technical Papers, San Francisco, CA, USA, 2016, pp. 264–265.

[43] P.N. Whatmough, S.K. Lee, H. Lee, S. Rama, D. Brooks, G.-Y. Wei, 14.3 A 28nm SoC with a 1.2GHz 568nJ/prediction sparse deep-neural-network engine with >0.1 timing error rate tolerance for IoT applications, in: IEEE International Solid-State Circuits Conference (ISSCC) Digest of Technical Papers, San Francisco, CA, USA, 2017, pp. 242–243.

[44] G. Desoli, N. Chawla, T. Boesch, S.-p. Singh, E. Guidetti, F. De Ambroggi, T. Majo, P. Zambotti, M. Ayodhyawasi, H. Singh, N. Aggarwal, 14.1 A 2.9 TOPS/W deep convolutional neural network SoC in FD-SOI 28 nm for intelligent embedded systems, in: IEEE International Solid-State Circuits Conference (ISSCC) Digest of Technical Papers, San Francisco, CA, USA, 2017, pp. 238–239.

[45] D. Shin, J. Lee, J. Lee, J. Lee, H.-J. Yoo, DNPU: an energy-efficient deep-learning processor with heterogeneous multi-core architecture, in: IEEE Micro, vol. 38, 2018, pp. 85–93. no. 5. September/October.

[46] S. Choi, J. Lee, K. Lee, H.-J. Yoo, A 9.02mW CNN-stereo-based real-time 3D hand-gesture recognition processor for smart mobile devices, in: 2018 IEEE International Solid-State Circuits Conference (ISSCC) Digest of Technical Papers, San Francisco, CA, USA, 2018, pp. 220–222.

[47] S. Yin, P. Ouyang, S. Tang, T. Fengbin, X. Li, L. Liu, S. Wei, A 1.06-to-5.09 TOPS/W reconfigurable hybrid-neural network processor for deep learning applications, in: Proceedings of IEEE Symposium VLSI Circuits (VLSIC), Kyoto, Japan, 2017, pp. C26–C27.

[48] J. Lee, C. Kim, S. Kang, D. Shin, S. Kim, H.-J. Yoo, UNPU: an energy-efficient deep neural network accelerator with fully variable weight bit precision, in: IEEE Journal of Solid-State Circuits (JSSC), vol. 54, 2019, pp. 173–185. no. 1.

[49] J. Song, Y. Cho, J.-S. Park, J.-W. Jang, S. Lee, J.-H. Song, J.-G. Lee, I. Kang, 7.1 An 11.5TOPS/W 1024-MAC butterfly structure dual-core sparsity-aware neural processing unit in 8nm flagship mobile SoC, in: IEEE International Solid-State Circuits Conference (ISSCC) Digest of Technical Papers, San Francisco, CA, USA, 2019, pp. 130–131.

[50] L. Lai, N. Suda, V. Chandra, Deep Convolutional Neural Network Inference with Floating-point Weights and Fixed-point Activations, 2017, [Online]. Available: https://arxiv.org/abs/1703.03073.

[51] P. Judd, J. Albericio, T. Hetherington, T.M. Aamodt, A. Moshovos, Stripes: bit-serial deep neural network computing, in: Proceedings of 49th IEEE/ACM International Symposium on Microarchitecture (MICRO), Taipei, Taiwan, 2016.

[52] C. Zhu, S. Han, H. Mao, W.J. Dally, Trained ternary quantization, in: Proceedings of 5th International Conference on Learning Representations (ICLR), Toulon, France, 2017. April 2017.

[53] M. Rastegari, V. Ordonez, J. Redmon, A. Farhadi, XNOR-Net: ImageNet classification using binary convolutional neural networks, in: Proceedings of European Conference on Computer Vision (ECCV), Amsterdam, the Netherlands, 2016, pp. 525–542.

[54] Q. He, W. He, S. Zhou, Y. Wu, C. Yao, X. Zhou, Y. Zou, Effective Quantization Methods for Recurrent Neural Networks, 2016, [Online]. Available: https://arxiv.org/abs/1611.10176.

[55] W. Luo, A.G. Schiwing, R. Urtasun, Efficient deep learning for stereo matching, in: Proceedings of IEEE Conference on Computer Vision and Pattern Recognition (CVPR), Las Vegas, NV, USA, 2016, pp. 5695–5703.

[56] S. Park, I. Hong, J. Park, H.-J. Yoo, An energy-efficient embedded deep neural network processor for high speed visual attention in mobile vision recognition SoC, in: IEEE Journal of Solid-State Circuits (JSSC), vol. 51, 2016, pp. 2380–2388. no. 10.

[57] S. Park, J. Park, K. Bong, D. Shin, J. Lee, S. Choi, H.-J. Yoo, An energy-efficient and scalable deep learning/inference processor with tetra-parallel MIMD architecturre for big data applications, in: IEEE Transactions on Biomedical Circuits and Systems, vol. 9, 2015, pp. 838–848. no. 6.

[58] F. Merrikh Bayat, M. Prezioso, B. Chakrabarti, H. Nili, I. Kataeva, D. Strukov, Implementation of multilayer perceptron network with highly uniform passive memristive crossbar circuits, in: Nature Communications, vol. 9, Nature Publishing Group (NPG), 2018, p. 2331.

[59] G.E. Hinton, N. Srivastava, A. Krizhevsky, I. Sutskever, R.R. Salakhutdinov, Improving Neural Networks by Preventing Co-Adaptation of Feature Detectors, [Online]. Available: https://arxiv.org/abs/1207.0580.

[60] V. Nair, G.E. Hinton, Rectified linear units improve restricted Boltzmann machines, in: Proceedings of the 27th International Conference on Machine Learning (ICML), Haifa, Israel, 2010, pp. 807–814.

[61] A.L. Maas, A.Y. Hannun, A.Y. Ng, Rectifier nonlinearities improve neural network acoustic models, in: Proceedings of the 30th International Conference on Machine Learning (ICML), Atlanta, Georgia, USA, 2013, pp. 1–6.

[62] S. Han, H. Mao, W.J. Dally, Deep compression: compressing deep neural networks with pruning, trained quantization and Huffman coding, in: Proceedings of 4th International Conference on Learning Representations (ICLR), San Juan, Puerto Rico, 2016.

[63] S. Han, J. Pool, J. Tran, W.J. Dally, Learning both weights and connections for efficient neural networks, in: Proceedings of 28th Neural Information Processing Systems (NeurIPS) 28, Montreal, Quebec, Canada, 2015.

References for advance

[64] B.V. Benjamin, P. Gao, E. McQuinn, S. Choudhary, A.R. Chandrasekaran, J.-M. Bussat, R. Alvarez-Icaza, J.V. Arthur, P.A. Merolla, K. Boahen, Neurogird: mixed-analog-digital multichip system for large-scale neural simulations, in: Proceedings of the IEEE, vol. 102, 2014, pp. 699–716. no. 5.

[65] P.A. Merolla, et al., A million spiking-neuron integrated circuit with a scalable communication network and interface, Science 345 (2014) 668–673.

[66] M. Davies, et al., Loihi: a neuromorphic manycore processor with on-chip learning, in: IEEE Micro, vol. 38, IEEE, 2018, pp. 82–99. no. 1.

[67] D. Miyashita, et al., A neuromorphic chip optimizec for deep learning and CMOS technology with time-domain analog and digital mixed-signal processing, in: IEEE Journal of Solid-State Circuits (JSSC), vol. 52, 2017, pp. 2679–2689. no. 10.

[68] M.J. Marinella, et al., Multiscale co-deisng analysis of energy, latency, area, and accuracy of a ReRAM analog neural training accelerator, in: IEEE Journal on Emerging and Selected Topics in Circuits and Systems (JETCAS), vol. 8, 2018, pp. 86–101. no. 1.

[69] W. Haensch, T. Gokmen, R. Puri, The next generation of deep learning hardware: analog computing, in: Proceedings of the IEEE, vol. 107, 2019, pp. 108–122. no. 1.

[70] M. Kang, S. Lim, S. Gonugondla, N.R. Shanbhag, An in-memory VLSI architecture for convolutional neural network, in: IEEE Journal on Emerging and Selected Topics in Circuits and Systems (JETCAS), vol. 8, 2018, pp. 494–505. no. 3.

[71] J. Lee, D. Shin, K. Lee, H.-J. Yoo, A 31.2pJ/disparity·pixel stereo matching processor with stereo SRAM for mobile UI application, in: Procedeeings of IEEE Symposium on VLSI Circuits (VLSIC), Kyoto, Japan, 2017, pp. 158–159.

[72] J. Yang, Y. Kong, Z. Wang, Y. Liu, B. Wang, S. Yin, L. Shi, Sandwich-RAM: an energy-efficient in-memory BWN architecture with pulse-width modulation, in: IEEE International Solid-State Circuits Conference (ISSCC) Digest of Technical Papers, San Francisco, CA, USA, 2019, pp. 394–395.

Further reading

[73] S. Zhang, D. Zidong, L. Zhang, H. Lan, S. Liu, L. Li, Q. Guo, T. Chen, Y. Chen, Cambricon-X: an accelerator for sparse neural network, in: Proceedings of 49th IEEE/ACM International Symposium on Microarchitecture (MICRO), Taipei, Taiwan, 2016.

[74] N.P. Jouppi, et al., In-datacenter performance analysis of a tensor processing unit, in: ACM SIGRACH Computer Architecture News (ISCA), vol. 45, 2017, pp. 1–12. no. 2.

[75] B. Moons, R. Uytterhoeven, W. Dehaene, M. Verhelst, 14.5 Envision: a 0.26-to-10 TOPS/W subword-parallel dynamicvoltage-accuracy-frequency-scalable convolutional neural network processor in 28 nm FDSOI, in: IEEE International Solid-State Circuits Conference (ISSCC) Digest of Technical Papers, San Francisco, CA, USA, 2017, pp. 246–257.

[76] K. Bong, S. Choi, C. Kim, H.-J. Yoo, Low-power convolutional neural network processor for a face-recognition system, in: IEEE Micro, vol. 37, 2017, pp. 30–38. no. 6.

[77] Z. Yuan, Y. Yang, J. Yue, R. Liu, X. Feng, Z. Lin, X. Wu, X. Li, H. Yang, Y. Liu, A 65nm 24.7μJ/frame 12.3mW activation-similarity-aware convolutional neural network video processor using hybrid precision, inter-frame data reuse and mixed-bit-width difference-frame data codec, in: IEEE International Solid-State Circuits Conference (ISSCC) Digest of Technical Papers, San Francisco, CA, USA, 2020, pp. 232–233.

[78] J. Lee, D. Shin, H.-J. Yoo, A 21mW low-power recurrent neural network accelerator with quantization tables for embedded deep learning applications, in: Proceedings of IEEE Asian Solid-State Circuits Conference (A-SSCC), Seoul, South Korea, 2017, pp. 237–240.

[79] J. Lee, J. Lee, D. Han, J. Lee, G. Park, H.-J. Yoo, An energy-efficient sparse deep neural network learning accelerator with fine-grained mixed precision of FP8-FP16, in: IEEE Solid-State Letters (SSCL), vol. 2, 2019, pp. 232–235. no. 11.

[80] C.-H. Lin, C.-C. Cheng, Y.-M. Tsai, S.-J. Hung, Y.-T. Kuo, P.H. Wang, P.-K. Tsung, J.-Y. Hsu, W.-C. Lai, C.-H. Liu, S.-Y. Wang, C.-H. Kuo, C.-Y. Chang, M.-H. Lee, T.-Y. Lin, C.-C. Chen, A 3.4-to-13.3TOPS/W 3.6TOPS dual-core deep-learning accelerator for versatile AI applications in 7nm 5G smartphone SoC, in: IEEE International Solid-State Circuits Conference (ISSCC) Digest of Technical Papers, San Francisco, CA, USA, 2020, pp. 134–135.

[81] K. Bong, I. Hong, G. Kim, H.-J. Yoo, A 0.5° error 10mW CMOS image sensor-based gaze estimation processor, in: IEEE Journal of Solid-State Circuits (JSSC), vol. 51, 2016, pp. 1032–1040. no. 4.

[82] A. Biswas, A.P. Chandrakasan, CONV-SRAM: an energy-efficient SRAM with in-memory dot-product computation for low-power convolutional neural networks, in: IEEE Journal of Solid-State Circuits (JSSC), vol. 54, 2019, pp. 217–230. no. 1.

[83] S. Yin, Z. Jiang, J.-S. Seo, M. Seok, XNOR-SRAM: in-memory computing SRAM macro for binary/ternary deep neural networks, in: IEEE Journal of Solid-State Circuits (JSSC), vol. 55, 2020, pp. 1733–1743. no. 6.

[84] J. Yue, Z. Yuan, X. Feng, Y. He, Z. Zhang, X. Si, R. Liu, M.-F. Chang, X. Li, H. Yang, Y. Liu, A 65nm computing-in-memory-based CNN processor with 2.9-to-35.8TOPS/W system energy efficiency using dynamic-sparsity performance-scaling architecture and energy-efficient inter/intra-macro data reuse, in: IEEE International Solid-State Circuits Conference (ISSCC) Digest of Technical Papers, San Francisco, CA, USA, 2020, pp. 234–235.

About the author

Prof. Kyuho J. Lee is an Assistant Professor of the School of Electrical and Computer Engineering, and an Adjunct Assistant Professor of the Artificial Intelligence Graduate School of Ulsan National Institute of Science and Technology (UNIST), Ulsan, South Korea, since 2018. He received B.S. (2012), M.S. (2014), and Ph.D. (2017) degrees from the School of Electrical Engineering, Korea Advanced Institute of Science and Technology (KAIST), Daejeon, South Korea. Before joining UNIST, he had worked for Samsung Research America as a Hardware Designer in 2016. From 2017 to 2018, he was a Post-Doctoral Researcher in the Information Engineering and Electronics Research Institute, KAIST. He has led or participated in more than 10 projects in the field of Artificial Intelligence System-on-Chip designs.

His main research interests include Neuromorphic System-on-Chip, Deep Learning Processor, In-Memory Processing Architecture, Network-on-Chip architectures, and Intelligent AI Processors for mobile devices and autonomous vehicles. He is a Senior Member of the IEEE, and he has been serving as a TPC member for the IEEE Asian Solid-State Circuits Conference (A-SSCC) and ACM/IEEE Design, Automation and Test in Europe (DATE). He is the coauthor of over 30 papers in the field of intelligent computer vision and deep learning processor designs.

CHAPTER EIGHT

Energy-efficient deep learning inference on edge devices

Francesco Daghero, Daniele Jahier Pagliari, and Massimo Poncino
Department of Control and Computer Engineering, Politecnico di Torino, Turin, Italy

Contents

Advances in Computers, Volume 122
ISSN 0065-2458
https://doi.org/10.1016/bs.adcom.2020.07.002

Abstract

The success of deep learning comes at the cost of very high computational complexity. Consequently, Internet of Things (IoT) edge nodes typically offload deep learning tasks to powerful cloud servers, an inherently inefficient solution. In fact, transmitting raw data to the cloud through wireless links incurs long latencies and high energy consumption. Moreover, pure cloud offloading is not scalable due to network pressure and poses security concerns related to the transmission of user data.

The straightforward solution to these issues is to perform deep learning inference at the edge. However, cost and power-constrained embedded processors with limited processing and memory capabilities cannot handle complex deep learning models. Even resorting to hardware acceleration, a common approach to handle such complexity, embedded devices are still not able to directly manage models designed for cloud servers. It becomes then necessary to employ proper optimization strategies to enable deep learning processing at the edge.

In this chapter, we survey the most relevant optimizations to support embedded deep learning inference. We focus in particular on optimizations that favor hardware acceleration (such as quantization and big-little architectures). We divide our analysis in two parts. First, we review classic approaches based on static (design time) optimizations. We then show how these solutions are often suboptimal, as they produce models that are either over-optimized for complex inputs (yielding accuracy losses) or under-optimized for simple inputs (losing energy saving opportunities). Finally, we review the more recent trend of dynamic (input-dependent) optimizations, which solve this problem by adapting the optimization to the processed input.

1. Introduction

In recent years, machine learning techniques have become pervasive in our society, as the backbone of an increasing number of applications in the mobile and IoT domains. This spread has been mainly fueled by the advent of deep learning. In fact, one of the limitations of classical (i.e., pre-deep-learning) ML models is their reliance on carefully hand-engineered feature extractors, which makes model design a long, complex and costly process. Furthermore, the resulting models are often not reusable if the specifications of the problem change even slightly [1]. Deep learning, in contrast, overcomes the need for hand-crafted features [1], by using representation learning to extract meaningful features directly from raw data. This approach has been applied successfully to a number of applications, from smart manufacturing [2], to medical analysis [3, 4] and agriculture [5]. In particular, for tasks such as computer vision [6], natural language processing [7] and speech recognition [8], deep learning models have achieved outstanding results, sometimes even outperforming humans [1].

In practice, although most of the theoretical concepts at the basis of deep learning have been around for decades, this approach has only become mainstream in the last decade, mostly due to the availability of parallel hardware with increasingly large storage and computing power [9]. Indeed, in order to generate complex representations from raw data, deep learning models require both large datasets and huge amounts of computations. For example, a relatively small (by modern standards) computer vision model such as AlexNet [6] requires 61 M weights and 724 M multiply and accumulate operations (MACs) to process a single 227×227 image [9].

Besides being data hungry, deep learning workloads are also "embarrassingly parallel," which makes them perfectly suited for Graphical Processing Units (GPUs) or cloud clusters [9]. In contrast, executing these models on cost and power-constrained edge devices requires a synergy of hardware specialization (i.e., accelerators) and model optimization. In this chapter, we describe some of the most relevant research efforts in this sense, focusing in particular on model optimizations for energy efficiency.

The rest of the document is organized as follows. Section 2 provides the required theoretical background on deep learning models from a computational perspective. Section 3 briefly describes the available frameworks for the development and deployment of deep learning models. Sections 4 and 5 provide the motivation for performing deep learning at the edge, while Section 6 describes the main hardware platforms available for such task. Finally, Sections 7 and 8 describe the most relevant model optimizations for efficient deep learning at the edge. In particular, Section 7 presents static (i.e., input-independent) optimizations, while Section 8 focuses on dynamic (i.e., input-dependent) methods.

2. Theoretical background

In this section, we provide a (noncomprehensive) background on deep learning models. We focus mostly on computational aspects, which are relevant for the optimizations presented in the rest of the chapter, providing examples based on "standard" architectures, while intentionally skipping some theoretical details related to the most advanced and exotic models. Readers interested in those aspects can refer to [10].

2.1 Neurons and layers

The atomic blocks of a deep learning model are neurons, generic computational units performing a weighted and biased sum of their inputs.

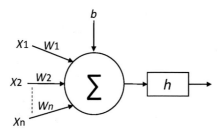

Fig. 1 The conceptual view of an artificial neuron.

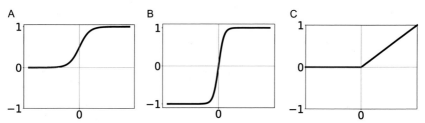

Fig. 2 Some of the most common activation functions in deep neural networks. (A) Sigmoid; (B) Tanh; (C) ReLU.

A nonlinear *activation function* is then applied to the output of each neuron. Fig. 1 shows a graphical representation of a neuron whose computation can be summarized as:

$$z = \sum_{i=1}^{n} w_i x_i + b$$

$$y = h(z) \tag{1}$$

where w_i are the weights applied to the inputs x_i, b is the bias and h is the activation function. During the training phase, weights and biases are iteratively updated to approximate the target function, as explained below. Fig. 2 shows some of the most commonly used activation functions, in particular:

- The *sigmoid (sigm)* function:

$$h(z) = \frac{1}{1 + e^{-z}} \tag{2}$$

  which squeezes its input onto the range (0, 1), as shown in Fig. 2A.
- The *hyperbolic tangent (tanh)*:

$$h(z) = \tanh(z) = \frac{e^z - e^{-z}}{e^z + e^{-z}} \tag{3}$$

which is similar to sigmoid but maps its input to the interval $(-1, 1)$, as shown in Fig. 2B.

- the *Rectified Linear Unit* (ReLU) [11]:

$$h(z) = \max(0, z) = \begin{cases} 0 & z \leq 0 \\ z & z > 0 \end{cases} \tag{4}$$

which is often used in modern deep learning models to solve the vanishing gradient problem of sigmoid and tanh, i.e., the fact that for very large input magnitude, the gradients of those two functions become very small, complicating the back-propagation of errors during training. In contrast, the gradient of ReLU is piece-wise constant, as clear from Fig. 2C.

Besides reducing the impact of vanishing gradients, ReLU is also advantageous from a computational perspective. Indeed, its evaluation simply consists of a hardware-friendly *max*() function. In contrast, sigmoid and tanh must be approximated either via a software routine or using table lookup, depending on the hardware platform [12]. Furthermore, ReLU maps all negative inputs to zero, leading to sparse activation outputs, which favor optimization techniques such as pruning (see Section 7.2) [9].

Most deep learning models are neural networks (NNs), i.e., combinations of neurons organized in a sequence of representation *layers*. Layers can be divided in three main categories with respect to their position in the network. The *input layer* processes raw input data and contains one neuron per input variable. Its outputs are then fed to one or more *hidden layers*, which are at the core of NN processing. Each hidden layer builds an increasingly complex representation of the input, projecting it into a high-dimensional *feature* space. Finally, the *output layer* takes the last hidden layer output and produces the final result of the NN computation. Clearly, the structure of this layer and its activation function depend on the task for which the NN is used. For classification, the output layer typically includes a number of neurons equal to the number of classes, each producing an estimate of the probability that the input belongs to the corresponding class. In this case, a common activation function is the *softmax*:

$$q_i = \frac{\exp(z_i/T)}{\sum_j \exp(z_j/T)} \tag{5}$$

which converts the preactivation output z for each class i into a probability q. T is the so-called temperature and is normally set to 1.

Input layer   Hidden layer   Output layer

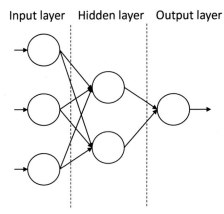

Fig. 3 A neural network with three layers.

The *depth* of a NN is defined as its number of layers. NNs are generally called deep when they include more than one hidden layer. Fig. 3 shows an overview of a network with fully connected layers (see Section 2.3). Deep NNs represent the majority of the models, with the average number of hidden layers for state-of-the-art NNs steadily increasing from a few to some thousands in recent years [9].

2.2 Training and inference

Neural networks have to undergo a *learning* or *training* phase before being able to accurately approximate a given function. Usually, training is performed iteratively feeding the NN with chunks of data, called *mini-batches*, in order to exploit the available hardware parallelism while still using a small amount of memory compared to the entire training dataset. In the so-called *forward pass*, the NN processes the mini-batch and produces a predicted output $\hat{y}$. A *cost* or *loss* function $\mathcal{L}$ is then applied to such output. In *supervised* learning, which is the most common approach for many deep learning applications, the loss measures the difference between $\hat{y}$ and the expected output y. Finally, in the *backward pass*, the network's weights are updated based on the value of the loss using *gradient descent*. Gradients of the loss with respect to neuron weights (w_i) and biases (b_i) are obtained through a process known as *back-propagation* [10], This sequence of forward and backward passes is repeated for a large number of iterations, spanning the entire training dataset multiple times.

Once the learning phase has concluded, the network can be used to perform predictions on unknown data. This process, called *inference*, consists only of the forward pass and uses fixed weight values. While it is possible to use mini-batches also in this phase, most inference tasks, especially for embedded and IoT applications, have tight latency constraints, i.e., outputs have to be produced as soon as possible after inputs become available. In those scenarios, batching is not an option, and inference has to be performed on *single* inputs, in a streaming fashion. Clearly, this negatively affects the exploitable parallelism and data sharing opportunities.

2.3 Feed-forward models

Feed-forward NNs are arguably the most popular type of deep learning model and are characterized by the absence of feedback connections. In other words, neuron outputs in a given layer are only fed as inputs to subsequent layers. The two most popular types of feed-forward NNs are *fully connected neural networks* and *convolutional neural networks*.

2.3.1 Fully connected neural networks

Fully connected neural networks have been among the first types of NN to be developed. They are constructed as a sequence of fully connected layers, i.e., layers in which each neuron receives as input *all* outputs from the previous layer [10]. This type of connectivity implies the presence of a separate parameter $w_{i,j}$ for each pair of neurons in adjacent layers. Therefore, the forward pass in fully connected networks can be described compactly as a large matrix multiplication, as depicted in Fig. 4. This is realized in hardware as a sequence of multiply-and-accumulate (MAC) operations. As shown in the figure, the input x and preactivation output z can be either vectors or matrices, depending on the use of batching during inference. Given the large number of neurons in each layer of modern NNs, the weight matrix in Fig. 4 can easily contain thousands or millions of elements. Therefore, fully connected forward passes tend to be strongly *memory bound*, i.e., to require a large number of data transfers to/from memory for each

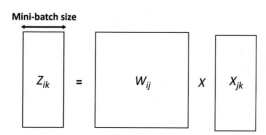

Fig. 4 Matrix multiplication performed for each fully connected layer.

MAC operation. This is only partially mitigated by the weight reuse made available by batching, when it is an option [13].

2.3.2 Convolutional neural networks

Convolutional neural networks (CNNs) have achieved outstanding results for computer vision tasks [1]. These networks typically process 3D tensors as inputs rather than flat 1D arrays, where the three dimensions may correspond to the height, width and number of channels of an image. The fundamental layers used in these networks are Convolutional (or Conv) layers, which apply a number of sliding window filters to the input tensors, followed by an element-wise activation function, most commonly a ReLU [10]. A representation of a Conv layer is shown in Fig. 5. Mathematically, a Conv layer output is computed as:

$$y_{i,j,c_o} = h\left(\sum_{l=0}^{K}\sum_{m=0}^{K}\sum_{c_i=0}^{C_i} w_{c_o,l,m,c_i} \cdot x_{i-\frac{k}{2}+l,\, j-\frac{k}{2}+m,\, c_i} + b\right),$$

$$\forall i \in [0,H),\ \forall j \in [0,W),\ \forall c_o \in [0,C_o) \tag{6}$$

where K corresponds to the height and width of the filter's *weight kernel* (assumed square), H and W are the height and width of the input tensor, and C_i and C_o are the number of input and output channels respectively. There exist several variants of this basic equation, for example to implement *strided* convolution, or for different *padding* strategies for boundary pixels. Moreover, many recent architectures resort to *depth-wise* or *group-wise* convolutions, where the input tensor slice only spans one or few channels in C_i.

Besides Conv layers, CNNs often include other types of layers, such as *pooling* and *batch normalization* (or *batch norm*). The former are used to reduce

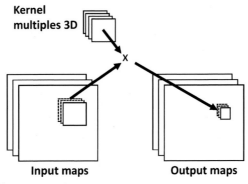

Fig. 5 The convolution operation.

the dimensionality of the input and to provide some degree of translation invariance, and typically compute either a *max* or an *average* over a spatial tensor window [1]. The latter, instead, apply a linear transformation to all elements in each channel, using the following equation:

$$y = \frac{x - \mu_B}{\sqrt{\sigma_B^2 + \epsilon}} \gamma + \beta \tag{7}$$

where γ and β are parameters learned during the training, while μ_B and σ_B are the mini-batch mean and standard deviation; ϵ is a small constant added for numerical stability [15]. During inference, μ_B and σ_B are replaced with the entire training set mean and standard deviations [16]. Batch norm has been shown to improve training speed by making gradients more stable, besides favoring the application of quantization (see Section 7.1) [17]. Features extracted by CNNs are often flattened and fed to one or more fully connected (FC) layers for the final classification. An example of a classical CNN architecture is shown in Fig. 6.

Conv layers dominate the inference phase of a CNN from a computational standpoint. This is due to fact that they process larger tensors with respect to the final FC layers, which operate on a *compressed* feature space. Moreover, the number of Conv layers is typically much larger than the number of FC layers in a modern CNN. The naive realization of a set of (standard) convolution filters on a 3D tensor consists of 6 indented loops [13]. More specifically, the 3 innermost loops perform the weighted sum of a 3D slice of the input tensor with a 3D filter kernel, i.e., the three summations of (6). Two additional loops move the slice across the width and height of the input tensor, and the last loop repeats the entire procedure with multiple filters, thus generating the various channels of the output tensor. Differently from fully connected layers, Conv operations are typically compute bound, even without batching [13].

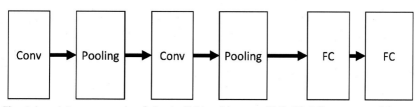

Fig. 6 Lenet-5, an example of classic CNN architecture [14]. FC, fully connected. Layers activations are not shown for simplicity.

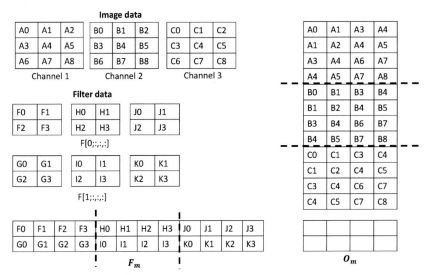

Fig. 7 An example of the im2col procedure on a 2 × 2 convolution. The image data is reorganized to obtain a single matrix where columns are the elements in a 2 × 2 window. F_m and o_m are the filter and output matrices.

Alternatively, convolution can be transformed into a single large matrix multiplication, using a procedure known as *im2col*. This operation is used in particular on CPUs and GPUs, and requires a reorganization of the input tensors and of the filter kernels, as shown in Fig. 7. While this reorganization *duplicates* some data, thus increasing the memory space required for the operation, it allows to exploit the extremely optimized CPU/GPU implementations of GEneral Matrix Multiply (GEMM) available in many mathematical libraries [18].

2.4 Sequential models

Feed-forward models are unable to model time-correlations, nor to process inputs of variable-size. Therefore, different types of neural networks have been designed to explicitly work with variable-length temporal sequences of data. In order to be able to *remember* information about previous input in the sequence, these networks typically include feedback connections.

Recurrent neural networks (RNNs) are among the most popular deep sequence models, and have achieved outstanding results in tasks such as neural machine translation, speech recognition, and summarization [1, 7, 19]. These networks process each input from a temporal sequence using one or more

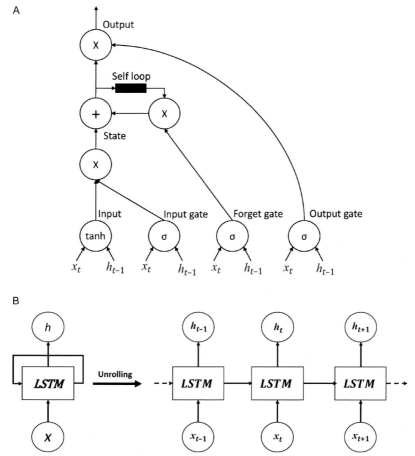

Fig. 8 An overview of a LSTM cell (A) and its unrolling during inference (B). In the cell diagram, *circles* labeled with σ represent a multiplication with a weight matrix followed by an element-wise sigmoid or tanh operation, while *circles* labeled with x or + represent element-wise products and sums.

layers of so-called *cells*, i.e., complex "neurons" with memory. Common cell architectures include the long-term short memory (LSTM) and the gated recurrent unit (GRU). An example of LSTM cell is shown in Fig. 8; for the details of the various operations involved, the reader can refer to [10]. The figure also shows how cells are used to process an entire sequence, by *unrolling* them a number of times equal to the sequence length. Notice that each cell "replica" shares the same learned weights.

The sequential nature of RNNs comes at the price of an even higher complexity compared to feed-forward models. Indeed, *for each time-step*, a LSTM cell has to compute the following operations:

$$i_t = \sigma(W_{ii}x_t + b_{ii} + W_{hi}h_{t-1} + b_{hi}) \tag{8}$$

$$f_t = \sigma(W_{if}x_t + b_{if} + W_{hf}h_{t-1} + b_{hf}) \tag{9}$$

$$g_t = \tanh(W_{ig}x_t + b_{ig} + W_{hg}h_{t-1} + b_{hg}) \tag{10}$$

$$o_t = \sigma(W_{io}x_t + b_{io} + W_{ho}h_{t-1} + b_{ho}) \tag{11}$$

$$c_t = f_t \circ c_{t-1} + i_t \circ g_t \tag{12}$$

$$h_t = o_t \circ tanh(c_t) \tag{13}$$

where h_t, c_t, and x_t are, respectively, the hidden state, cell state, and input vectors at time t and are multiplied with eight weight matrices W_{xx}. Vectors i_t, f_t, g_t, and o_t are called the input, forget, cell, and output gates, respectively; σ is the sigmoid function and $\circ$ is the element-wise product.

Matrix-vector products in (8)–(13) are the dominant operations from a computational perspective. In particular, the eight weight matrices can be combined into a single bigger matrix, reconducting the entire time-step computation to a large matrix multiplication, as in FC layers [10]. Differently from the feed-forward case, however, this multiplication is repeated for each time step. Moreover, parallelism between time-steps is limited, as each evaluation of (8)–(13) requires to have available the outputs c_{t-1} and h_{t-1} from the previous step. RNNs can be also organized in more complex architectures, such as the *encoder–decoder* one, which relies on two RNNs to perform sequence-to-sequence mapping (e.g., for translation). However, the basic computations involved remain identical.

Finally, it is worth noticing that deep sequence models have seen a very quick development in recent years, with the advent of *attention-based* architectures and *transformers* [20]. While extremely effective, however, these networks are still relatively unexplored from the point of view of model optimization for efficient edge processing and hardware acceleration.

3. Deep learning frameworks and libraries

A reason for the increasing popularity of deep learning is the availability of open-source frameworks, which simplify the development of new models, eliminating the need of rewriting state-of-the-art building blocks (activation functions, layers, etc.) from scratch. For high-performance systems, PyTorch [21] and TensorFlow (TF) [22] are currently the two most popular frameworks. Both provide high-level Python APIs for model development, training, and deployment, while leveraging optimized C/C++ libraries for CPU or GPU acceleration under the hood. Cross-framework model description formats such as the one provided by ONNX [23] are also increasingly popular.

These frameworks are extremely powerful and flexible, but they are not suited for low-power edge devices, mostly due to their significant requirements in terms of runtime memory occupation. Therefore, recent years have seen the development of many edge-oriented libraries and frameworks for deep learning. Both PyTorch and TensorFlow now offer lightweight inference engines (called PyTorch Mobile and TF Lite, respectively) targeting resource efficiency. With respect to their full-fledged counterparts, these stripped-down versions only permit the execution of the inference phase (i.e., no model development nor training), thus greatly reducing the size of the corresponding runtime. Conversion tools allow to export a standard PyTorch/TF model for these engines, while simultaneously applying efficiency-oriented optimizations to the network (mostly quantization, see Section 7.1).

While PyTorch Mobile and TF Lite target powerful edge devices such as smartphones or tablets, other projects have addressed the implementation of deep learning on even more constrained targets, such as Microcontrollers (MCUs). Tensorflow Lite for Microcontrollers is a recent effort in this sense from TF developers, targeting devices such as ARM Cortex-M processors; it runs without the need of operating system support, and occupies only a few kB of memory. Moreover, several companies and academic researchers are developing MCU-oriented libraries for deep learning inference, such as ARM's CMSIS-NN [18], STMicroelectronics CUBE AI [24], the PULP Platform library PULP-NN [25], and many others. More details on these libraries are provided in Section 6.3.

4. Advantages of deep learning on the edge

Nowadays, deep learning is one of the core components of many mobile and IoT applications. However, due to the heavy processing and memory requirements of DNNs, the standard approach to implement deep learning for mobile and IoT is pure cloud offloading, in which raw data are transmitted to a remote server, which runs the inference using high-performance hardware (e.g., a GPU) and returns the final result. While this approach is acceptable for training, which is an "offline" task, pure cloud offloading of DNN inference can be highly inefficient [9, 26]. In contrast, near-sensor *edge computing* might provide several benefits, as long as designers have available a set of optimization strategies that permit the efficient execution of DNNs on cost-constrained edge devices.

First, cloud offloading might incur long and unpredictable latencies when devices have a slow or intermittent internet connection. This might be critical for deep learning applications with tight real-time constraints, such as autonomous driving [26]. Local computation, instead, can be made predictable in terms of latency much more easily. Moreover, the cloud approach also has scalability issues, since having a large number of devices connected to the cloud increases the network pressure, deteriorating the quality of service. This is especially true for bandwidth-intensive inputs such as videos. Performing the entire inference, or at least a part of it, at the edge would reduce this bottleneck.

Besides latency and bandwidth problems, the transmission of high-dimensional data over the network is also highly energy inefficient, especially for wirelessly connected edge devices. In comparison, an optimized local processing can be orders of magnitude more efficient [13]. This is particularly critical given that mobile and IoT devices are mostly battery operated and run on tight energy constraints [27].

Finally, mobile and IoT applications often process sensitive data (e.g., face detection, speech recognition) whose transmission to the cloud may raise privacy concerns. Avoiding the transmission or performing a first preprocessing step locally would therefore also increase security.

5. Applications of deep learning at the edge

The increasing amount of smart devices caused a growth in the number of deep learning-based applications that would benefit from an execution at the edge. Tasks such as face recognition, object detection, and wakeword recognition are in fact already commonly executed in a decentralized fashion [26]. Some of the most relevant tasks that can benefit from edge computing are described in the following sections. While describing them, we try to stress the fact that, in principle, nothing prevents the implementation of any of these tasks using a standard cloud computing approach. However, that solution would yield suboptimal results in terms of responsiveness, security, and energy efficiency.

5.1 Computer vision

Computer vision is one of the most explored applications of deep learning due to the outstanding results obtained [1]. Tasks such as object detection and image recognition are common in a broad number of edge-oriented applications, such as autonomous driving, surveillance, vehicle, and person detection. These tasks often rely on collecting data from cameras and processing them immediately afterwards [26]. Therefore, the inference phase has real-time requirements (e.g., car detection in autonomous driving) which may be violated with a cloud approach due to connectivity problems. Moreover, vision-based applications often process large amounts of sensitive user data (e.g., pictures of faces), and therefore would benefit both from avoiding the energy-consuming transmission of these data to the cloud, and from the privacy-preserving properties of edge computing. Evidently, computer vision at the edge can only be realized if the model can be optimized to efficiently run a highly intensive processing of large videos on an embedded device, with an acceptable *frame rate*.

5.2 Language and speech processing

Deep learning has also obtained outstanding results in natural language processing (NLP) and speech recognition [1]. While language-based applications are typically not real-time, the perceived user experience (e.g., in an automatic translation system) would still benefit from the shorter response times that can be achieved thanks to edge computing. Similarly, privacy is again a relevant issue when dealing with user audio recordings or text conversations. Furthermore, some audio-based edge deep learning tasks, such as wakeword recognition for smart speakers, are "always-on" (i.e., the inference is performed constantly in the background). For these tasks, a cloud approach is even more suboptimal, as it requires a constant energy-hungry transmission of data to the cloud. Indeed, this is one of the most common tasks that are already performed at the edge [26]. As for computer vision, however, state-of-the-art language and speech models are typically of considerable size. In the last years, reduced versions of deep learning models have successfully been used for simple NLP tasks on embedded devices, avoiding any use of the network (e.g., the aforementioned wakeword recognition). However, the challenge is still open for more complex tasks such as machine translation or summarization, which require far more computing power.

5.3 Time series processing

Smartphones and wearables continuously collect time series of data from tens of sensors. These data can be processed to recognize the activity performed by the device owner and react to it in various ways, from simple logging [28] to power optimization [29]. Once again, analyzing this flow of data directly on the device would avoid energy-consuming transmission of private data (e.g., the biometric measurements of a smart watch) to the cloud. Moreover, it would allow a prompter response for the applications.

Similarly, ubiquitous sensors in smart cities and smart factories can also greatly benefit from deep learning-based inference on time series. Among the most popular applications in these domains there are electrical load prediction and balancing in smart grids [30], traffic and pollution monitoring in cities [31], soil monitoring for agriculture [5] and predictive maintenance of industrial equipment in factories [2]. Even more than for smartphones and wearables, energy reduction is critical for these kinds of sensors, which are typically expected to operate on battery for months or years [26]. Therefore, avoiding raw data transmission and performing at least a partial local processing is critical for these scenarios as well.

6. Hardware support for deep learning inference at the edge

The broad diffusion of deep learning has led to two main trends related to hardware [9]. On the one hand, a lot of custom accelerators for deep learning inference have been proposed, implemented either as application-specific integrated circuits (ASICs) or using field-programmable gate arrays (FPGAs). On the other hand, researchers and companies have also worked in the direction of improving the efficiency of deep learning processing on general purpose hardware, using highly optimized software libraries, some of which have been mentioned in Section 3.

In this section, we briefly overview the available platforms for deep learning at the edge, highlighting their advantages and disadvantages.

6.1 Custom accelerators

Resorting to hardware specialization to implement DNN inference leads to the highest energy efficiency and throughput. This comes at the cost of high design and manufacturing costs, especially for ASIC implementations. FPGAs, on the other hand, allow to drastically cut costs due to their field programmability, which enables easier flows for designers and dilutes manufacturing costs over large volumes. However, they are significantly more power hungry than modern ASICs.

Regardless of the target technology, DNN accelerators typically use spatial or systolic architectures to optimize the highly parallel MAC kernels that are at the core of most key DNN layers (such as FC, Conv, and LSTM layers described in Section 2) [32–38]. A large number of such accelerators, sometimes called Neural Processing Units (NPUs), has been recently proposed both by companies and by academic researchers. In the commercial world, two of the most popular products are Google's Edge TPU [32] and Intel Movidius [33]. In the academic world, Eyeriss [34] and Envision [35] are examples of flexible and powerful accelerators for feed-forward NNs. One feature of the latter is the use of low-precision quantization (see Section 7.1) to reduce the memory bandwidth and improve arithmetic operations efficiency using dynamic voltage scaling [39–41]. The quantization concept is brought to the extreme by binary NN accelerators such as the XNE [36]. Besides these examples, many other designs are constantly being proposed by hardware designers. Since the focus of this chapter are model-level, semihardware-independent optimizations, we do not go into details of the most recent accelerators and refer the interested reader to the excellent survey in [9].

6.2 Embedded GPUs

GPUs played a fundamental role in the diffusion of deep learning, thanks to their excellent throughput and efficiency for parallel tasks such as DNN inference [1]. Most GPUs are high-performance computing devices, with a power consumption in the order of 100s of Watts, clearly inappropriate for edge systems. Recently, however, GPU manufacturers have started to design architectures that focus on efficiency. The NVIDIA Jetson family of GPUs [42], for example, target explicitly embedded applications and consume in the order of 10 W. While these power values remain excessive for sensors and battery-operated devices, such kind of embedded GPUs might be installed in grid-plugged intermediate edge servers. Furthermore, the high-level of parallelism of GPUs would not be fully exploited by end-nodes due to the aforementioned difficulties for applying batching in latency-sensitive inference tasks [37]. In contrast, edge servers can typically resort to batching by gathering data from multiple sources (e.g., multiple sensors) [43].

Commercial embedded GPUs come with associated libraries and tool-chains [42] that enable the full exploitation of the capabilities of their hardware for DNN processing. In particular, CuDNN [44] is the library used by NVIDIA GPUs to map DNN layers from various high-level deep learning frameworks (TensorFlow, PyTorch, etc.) to optimized implementations based on CUDA [45]. Compared to the libraries for MCUs described in the following section, CuDNN supports a very comprehensive set of deep learning primitives, for both convolutional and sequential models.

6.3 Embedded CPUs and MCUs

Compared to custom accelerators and GPUs, embedded CPUs and micro-controllers have cost as their main selling point. However, these devices also have orders of magnitude lower performance, as well as extremely limited memory spaces. Nonetheless, modern MCUs often support low-precision integer operations (e.g., 8-bit, 16-bit), sometimes with a certain amount of parallelism using single instruction multiple multiple data (SIMD) instructions. Therefore, deep learning libraries for these platforms exploit as much as possible these capabilities, while minimizing the on-chip memory footprint. To achieve both goals at the same time, many MCU deep learning libraries focus extensively on quantized models.

CMSIS-NN [18] is one of such libraries, targeting ARM Cortex-M series MCUs. It consists of highly optimized implementations of common DNN kernels, based on a clever use of the ARM instruction set architecture (ISA) to increase throughput. For instance, the lack of 8-bit SIMD operations on Cortex-M is overcome mapping them to 16-bit instructions and using a custom reorganization of the data in memory to minimize the instructions needed for sign extension and merging. CMSIS-NN also comes with a companion set of tools, which allow to convert pretrained models (e.g., in TensorFlow, PyTorch or ONNX) and to perform posttraining quantization. X-CUBE-AI [24] is a similar library from STMicroelectronics, again able to "translate" networks trained with high-level frameworks into optimized code for STM32 MCUs. Currently, one limitation of these and other similar libraries is the limited support in terms of types of layers, especially for sequential models.

In the academic domain, several embedded CPU architectures with specific hardware features that favor the implementation of deep learning algorithms have been proposed, extending open ISAs. One notable example is the parallel ultra-low-power (PULP) platform [46], which is at the basis of the GAP8 processor [47]. GAP8 is a scalable, clustered many-core architecture targeting low-voltage and low-energy processing. The extensible RISC-V ISA on which PULP is based allows the implementation of DNN-oriented operations directly in hardware, such as bit-extraction for subbyte quantization and popcount for binary NNs (see Section 7.1). These features are exploited by the PULP-NN library [25] to yield improved efficiency compared to that of other MCUs. Another peculiar feature of GAP8 which is of great interest for deep learning applications is the fact that the main MCU cluster does not contain data caches, and uses software-controllable scratchpad memories instead. This increases the efficiency for memory-intensive tasks such as DNN layers, at the cost of a greater effort from the programmer. The latter, however, can be eliminated by resorting to automatic tools that implement custom scratchpad management starting from a high-level DNN specification [48].

7. Static optimizations for deep learning inference at the edge

Design time optimizations to improve the energy efficiency of deep learning models have been actively researched in recent years [9, 49], with a few main families of methodologies being proposed. Before going into the details of such approaches, however, it is important to mention what is arguably the single best optimization strategy, i.e., the careful tuning of the hyper-parameters of the target DNN architecture. Indeed, state-of-the-art DNNs are often greatly over-parametrized, in order to achieve the best possible accuracy on complex tasks, such as ImageNet's 1000-classes classification [50]. Many researchers have shown that such architectures can be dramatically shrunk with negligible accuracy losses, even for the same application. For example, SqueezeNet [51] is a CNN that obtains the same ImageNet accuracy as the classical AlexNet [6] with 50x less parameters. Another notable example are MobileNets [52], a family of CNNs that substitute standard Conv layers with depth-wise convolutions, thus obtaining high accuracy with a significantly reduced amount of MAC operations.[a] Additionally, popular DNN architectures originally proposed for standard tasks (such as ImageNet) are often reused for much simpler applications, especially in the IoT domain [53, 54]. In those scenarios, hyper-parameters

---

[a]Although it must be noted that this reduction in the number of MACs does not always translate into hardware efficiency improvements, because depending on the adopted tensor layout, depth-wise Conv layers may have inefficient data access patterns.

simplifications can be even more dramatic, ranging from input size reduction to the complete removal of several layers. Clearly, the drawback of changing the DNN architecture is that it prevents the fine-tuning [1] of pre-trained models. We do not describe hyper-parameters optimizations in detail, since they are highly task-specific. However, we remark that, whenever they are possible for the task at hand, these should be the first optimizations to be considered, as they can yield the largest complexity reduction and efficiency improvement.

In contrast, in the rest of this section we focus on four families of *general* optimization strategies that can be applied successfully to many different DNNs: *quantization, pruning, distillation,* and *collaborative inference.* We select these four families as they are currently the most effective and widely used by researchers and industry. Two other interesting trends, not covered in this chapter but worth mentioning, are filter decomposition [55] and the use of approximate computing techniques (e.g., voltage over-scaling, approximate functional units) [27, 56].

Importantly, the four families of optimizations described in this section, as well as the dynamic ones treated in Section 8 are not only (almost) orthogonal to the details of the DNN used for inference, but also to the type of inference hardware. This means that, although with different benefits in terms of energy efficiency, these strategies can be applied to *any* of the families of platforms described in Section 6. As such, most of the approaches described in the following do not try to reduce energy by lowering the *power consumption* of each individual operation, which is strongly hardware-dependent.[b] Instead, they exploit the fact that energy is a time-integral quantity, and try to decrease it by reducing the *number* of operations performed, thus being effective regardless of the underlying hardware, even if the power of each operation remains constant. In practice, therefore, most of the techniques presented in the chapter reduce either the memory occupation and bandwidth of DNNs, thus cutting the energy associated with loading and storing data through memory hierarchy levels, or the number and complexity (precision) of the arithmetic (MAC) operations required for inference. We select these techniques exactly for their generality. Clearly, hardware-specific power optimizations can be combined to them, yielding even higher energy savings, and we will mention some scenarios when this combination has been realized in the following.

---

[b]There are clearly exceptions to this rule. For example, integer quantization reduces both time and power as integer ALUs are normally more power-efficient than floating point ones, for practically any hardware platform.

7.1 Quantization

One of the most widely diffused optimization techniques for deep learning models is *quantization*. It consists of reducing the precision of the DNN weights and possibly of the activations, exploiting the resilience of DNNs to small errors and noise [27]. Reduced precision formats for DNNs can leverage either floating point (such as the 16-bit *minifloat*, currently supported by many deep learning oriented GPUs [57]) or integer (i.e., fixed point) arithmetic. The former is sometimes used also to speedup training, as minifloat representation often does not yield any accuracy loss compared to standard 32-bit floats. Integer quantization, in contrast, is mostly used to improve the efficiency and speed of the *inference* phase [16]. At training time, integer quantization is only (optionally) *simulated*, in order to account for its impact on the network outputs, as explained in detail in the following. Since our main focus is inference, in the following, we concentrate on integer quantization techniques.

7.1.1 Quantization algorithms

Integer quantization algorithms can be divided into uniform and nonuniform based on the way in which representable numbers are distributed on the real axis. Uniform quantization is the most common of the two, and can be performed in several ways, among which the *affine, symmetric,* and *stochastic* quantizers are the most popular [16].

The *affine quantizer* maps numbers in the range (x_{min}, x_{max}) to the range $(0, N_{levels} - 1)$, where $N_{levels} = 2^{precision}$. The input range (x_{min}, x_{max}) is the set of values that can be assumed by a given set of weights or activations (relative to a channel, a layer, or to the entire network, as explained below). Therefore, the first step needed to apply quantization is to determine this range. The way to do so depends on whether quantization is being applied during or after training, and is explained in Section 7.1.2.

Next, the uniform quantization step (Δ) and the zero-point (z) can be derived based on x_{min}, x_{max}, and N_{levels}. In particular, z is an integer corresponding exactly to zero. Ensuring that zero is still represented exactly after quantization is important, as it avoids that common operations like zero padding introduce a quantization error. One-sided weights and activations distribution are therefore relaxed in order to include 0, before quantizing with the following operations:

$$x_{int} = round\left(\frac{x}{\Delta}\right) + z \tag{14}$$

$$x_Q = clamp(0, N_{levels} - 1, x_{int}) \tag{15}$$

with *clamp* defined as:

$$clamp(a, b, x) = \begin{cases} a & x \leq a \\ x & a \leq x \leq b \\ b & x \geq b \end{cases} \tag{16}$$

The dequantization can be instead performed in the following way:

$$x_{float} = (x_Q - z)\Delta \tag{17}$$

The *symmetric quantizer* is a simplified version of the affine quantizer, restricting z to 0. Eqs. (14) and (15) then become:

$$x_{int} = round\left(\frac{x}{\Delta}\right) \tag{18}$$

$$x_Q = clamp(-N_{levels}/2, N_{levels}/2 - 1, x_{int}) \quad \text{if signed} \tag{19}$$

$$x_Q = clamp(0, N_{levels} - 1, x_{int}) \quad \text{if unsigned} \tag{20}$$

with the reverse operation becoming:

$$x_{float} = (x_Q)\Delta \tag{21}$$

The *stochastic quantizer* adds noise to float numbers before rounding them to integer [16]. The quantization is then performed in the following way:

$$x_{int} = round\left(\frac{x+\epsilon}{\Delta}\right) + z \quad \epsilon \sim Unif(-1/2, 1/2) \tag{22}$$

$$x_Q = clamp(0, N_{levels} - 1, x_{int}) \tag{23}$$

with its reverse operation being (17). Although this quantization technique can be shown to yield improved accuracy for the same precision, it requires to generate a new random number for each quantization operation. When this has to be done at runtime, on the output activations of a layer, this adds a considerable overhead to the inference phase. Therefore, stochastic quantization is generally used only for data that can be quantized offline, such as model weights [58, 59].

Most real DNN weights and activations are not distributed uniformly. Therefore, nonuniform quantization might yield better representations [9, 60]. In particular, having finer levels close to zero would be ideal, since tensor values distributions tend to be bell-shaped [60]. On the other hand, given that most inference HW platforms are based on uniform number representations for integers, nonuniform quantizers have to be carefully designed to be made hardware-friendly [61].

One popular nonuniform quantization approach is based on a *logarithmic number system*. This yields the desired finer granularity for smaller magnitude data, while grouping "outliers" in few levels [62]. In particular, using base-2 powers for levels yields a very efficient hardware implementation of logarithmic quantization, allowing multiplications to be implemented as bitwise shifts [63, 64].

The quantizers described above (both uniform and nonuniform) can be applied with different granularities to a DNN. The three most common granularities are:

- *Network-wise*: this is the simplest approach, with a single bit-width used for the whole network. Range and zero-value parameters of quantizers might still be different for different tensor (e.g., different layers weights). This is also the easiest solution to support from a HW point of view, since data widths remain constant throughout the inference, simplifying both memory accesses and MAC operations. However, for a given accuracy level, network-wise quantization might lead to significantly larger total weights and activation sizes compared to finer granularity approaches.
- *Layer-wise*: in this approach, each layer has a different quantizer, handling not only different ranges but also possibly different precisions (i.e., bit-widths). While this approach usually outperforms the previous one, it requires a more complex support from the underlying hardware.
- *Channel-wise*: this is an extension of the layer-wise approach tailored for convolutional neural networks. It consists of adapting the parameters of the quantizer to each convolutional kernel in a tensor. This strategy usually leads to an improved final accuracy, although it can only be applied to weights. In fact, channel-wise quantization of activations would greatly complicate data alignment for MAC operations in Conv layers [16].

Regardless of the quantization granularity, one important detail to underline is that weights and activations quantization are significantly different at runtime. In fact, while weights can be quantized offline, once and for all, the output of large MAC loops during inference must be requantized at runtime by the inference platform to produce new activations [65].

7.1.2 Quantization and training

The straightforward way to apply quantization to a DNN is to do it *post-training*, i.e., applying quantizers to an already trained model. This requires minimal setup, while leading to significant energy savings. The main operation required to implement posttraining quantization consists in determining the value range for each group of tensors to be quantized, i.e.,

x_{min} and x_{max}. For model weights, this range can be obtained immediately, since their value is fixed posttraining. Quantizing activations, instead, requires a *calibration* phase to determine data ranges to be used at runtime. This is normally done running forward passes on a subset of the dataset and storing the maximum and minimum values encountered for each activation tensor. This does not guarantee to find the absolute minimum and maximum values assumed by a given tensor, but is often sufficient. Values exceeding the expected range will be simply clipped by the *clamp*() function in (15), (19), and (23). The risk of clipping is partially mitigated by batch normalization, which keeps the values in a stable interval [16].

Posttraining quantization does not require a time-consuming training, nor the availability of training data, except the few samples used for calibration. However, it may cause noticeable accuracy degradations on complex tasks [66], although networks with more parameters tend to be less affected [16].

A very effective solution to cope with these accuracy degradations is *quantization-aware training*. In this approach, quantization is *simulated* while the model is being trained, thus giving the DNN the opportunity to learn how to compensate for the loss of precision. Clearly, the drawback of this approach is that it requires an expensive training run and is only feasible when training data are available.

During quantization-aware training, weights (and activations) are "fake-quantized". This means that their values are *rounded* in the forward pass, to mimic low precision, but internal computations are still performed in floating point. In the backward pass, quantized operations can be dealt with in different ways. One classical approach is to approximate them with a straight-through estimator [67]. This ensures that DNN outputs are produced as if data had been quantized, while still allowing the back-propagation of small gradients, which is fundamental for training convergence. At inference time, fake quantization operations are then removed, and the model uses actual integer weights and activations.

Quantization-aware training also requires a different procedure for estimating ranges of tensors. For example, since activation values change depending on the input, in [68] it is proposed to use an exponential moving average to estimate their range during training. Moreover, in order to avoid rapidly shifting activation values, their quantization is usually performed only after a considerable number of initial training steps, so that the network has reached a more stable state [68].

Besides the aforementioned BatchNorm layers, there are also other DNN architecture elements that can favor the application of quantization, especially during training. An important one is the use of *bounded* activation functions, such as the PArametrized Clipping acTivation (PACT) [69]. PACT is a bounded ReLU variant that follows this equation:

$$pact(x) = \begin{cases} 0 & x \leq 0 \\ x & 0 \leq x \leq k \\ k & x \geq k \end{cases} \qquad (24)$$

where k is a parameter learned during training. Clipping the maximum activation output to k has been shown to yield significant improvements in accuracy for a given quantization precision [69].

7.1.3 Binarization

Binarization is an extreme form of quantization, in which precision is reduced to 1 bit. Initially focusing only on weights [70], this technique has then been extended also to activations [59]. The most common form of binarization consists in using binary values 0 and 1 to represent integer values -1 and 1, respectively. The conversion of a floating point number to this format is simply obtained with the *sign*() function.

Binarization of both weights and activations yields extreme complexity reductions for DNN inference, since MAC operations can be completely eliminated and replaced by binary operations [59, 70, 71]. In particular, using the aforementioned semantic for binary values, multiplications can be replaced by bitwise XNORs, as shown in Table 1. The accumulation of the binarized elements of a tensor X with N elements, instead, can be computed as follows:

$$s = 2 \cdot popcount(X) - N \qquad (25)$$

where *popcount*() is a function that counts the number of bits at 1 in X.

Table 1 Equivalence between MUL and XNOR ($\odot$) when the values -1 and 1 are represented by the binary 0 and 1.

$X1_{val}$	$X2_{val}$	$X1_{binary}$	$X2_{binary}$	$X1_{val} * X2_{val}$	$X1_{binary} \odot X2_{binary}$
-1	-1	0	0	1	1
-1	1	0	1	-1	0
1	-1	1	0	-1	0
1	1	1	1	1	1

The dramatic impact of Binarization on model size and operations complexity is paid with significant accuracy degradations for complex tasks (e.g., ImageNet classification [50]). However, this approach is of extreme interest for simpler tasks, such as hand-written digits classification and human activity recognition [71, 72].

7.1.4 Benefits of quantization

Quantization brings significant advantages in terms of memory occupation, speed and energy consumption. First and foremost, low-precision data not only reduces the storage occupation of the DNN, but even more importantly, it limits the memory bandwidth required to bring weights and activations on-chip, which is often the dominant contributor to inference time and energy [13]. Moreover, integer operations are also faster and efficient than floating point ones on virtually all hardware platforms.

Both energy and speed should ideally improve at least *linearly* with respect to bit-width [9]. However, while this is true for what concerns memory bandwidth and energy, the real trend for inference speedup and total energy is often very different, depending on the target platform. In particular, on general purpose hardware such as CPUs and MCUs, the linear trend definitely stops for *subbyte quantization*. For these platforms, in fact, bytes are the atomic load and store elements, and lower precision quantized data have to be "packed" together in a single memory location. As a consequence, MAC operations on subbyte values require a set of additional operations to extract packed data into separate registers and then perform the reverse operation on results. This overhead greatly affects the speed of inference, so that, for example, 4-bit quantized DNNs are often slower than 8-bit ones [25]. Furthermore, some popular CPU architectures such as ARM Cortex-M, only support 16-bit Single Instruction Multiple Register (SIMD) MAC instructions. Therefore, even 8-bit values have to be moved to 16-bit registers and sign-extended before executing a MAC. Despite thorough ISA-dependent optimizations [18], this overhead inevitably limits the speedup and energy gain obtained at 8-bit or lower quantization.

The scenario is clearly different for custom hardware accelerators. Indeed, accelerator architectures specifically designed to handle subbyte quantizations have been proposed in literature. These designs avoid unpacking quantized values and thus benefit from the full gains achievable thanks to quantization. One notable example is the Envision accelerator [35], which combines quantization with voltage and frequency scaling to achieve much more than linear savings down to 4-bit.

Contrary to other types of quantization, binarization does not just reduce the precision of operations, but radically changes them. Therefore, as anticipated, the benefits that can be derived from it are even higher than the intuitive 32x reduction in model size and memory bandwidth compared to floating point. However, obtaining these gains on general purpose hardware is again not trivial, mainly because commercial CPUs do not offer an efficient way to implement the *popcount()* operation. In contrast, recent academic processor platforms [25] have added dedicated hardware and a corresponding instruction for popcount. Custom accelerators for binary NNs are also quite explored, due to their extreme compactness and efficiency [36, 73].

Regardless of the target hardware and precision, a considerable advantage of quantization lies in its orthogonality to the DNN architecture. Indeed, although there are architectural elements that favor its application, such as BatchNorm and bounded activations, quantization does not *require* any particular model characteristic in order to work. This is one of the main reasons why this technique has become popular and is now widely supported by the major deep learning frameworks, both in its training-aware and post-training forms [21, 22], making it easier to implement for developers. Clearly, what does depend on the platform are the energy and time benefits of quantization, as described above. Therefore, the same quantized model might lead to very different efficiency on two different hardware targets.

Quantization has been found very successful on convolutional neural networks, allowing to reduce significantly the precision with negligible accuracy loss [16, 71]. However, sequential models are a much harder challenge [74–76]. While the research on this type of networks has not been as extensive as the one on CNNs, current results are definitely less outstanding. In particular, while the easiest tasks and datasets benefit from quantization, the accuracy deterioration is noticeable on harder ones [75].

7.2 Pruning

It has been known for some time that, due to their overparametrization, deep learning models can tolerate high levels of sparsity in their weights [77]. This means that a large portion of the weights can assume value 0, while still producing accurate results. Furthermore, modern DNN activations are also inherently sparse, due to the use of functions such as ReLU (see Fig. 9), which turn all negative inputs to 0.

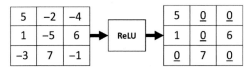

Fig. 9 Activations sparsity deriving from the application of a ReLU, one of the most frequently used activation functions for DNN hidden layers.

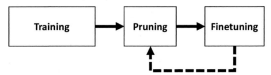

Fig. 10 The common weight pruning workflow.

Sparsity can be exploited to optimize the memory requirements of a DNN model by *compressing* it, which is particularly useful in memory-constrained edge devices. Moreover, if the target hardware offers the required support, sparsity can also be exploited by recognizing and skipping operations (i.e., MACs) on zero-valued weights/activations, thus improving inference speed and energy [9].

The sparsity of a model, and in particular of its weights, can be artificially increased by so-called *pruning* algorithms. These techniques identify the "least important" weights of a model and replace them with 0, in order to increase sparsity with the least possible impact on DNN accuracy.

7.2.1 Pruning algorithms

The great majority of pruning algorithms operate after an initial standard training, and iteratively eliminate connections and fine-tune the model in order to recover the drop in accuracy, as shown in Fig. 10. In particular, one of the first published techniques was based on eliminating weights with the smallest *saliency*, i.e., the smallest impact on the training loss [77]. The process was repeated until the desired weight reduction or accuracy were reached.

Unfortunately, the computation of weights saliency has become too expensive for modern DNNs, due to their increasing depth and total number of parameters, thus giving birth to a new family of so-called *magnitude-based* pruning approaches [78]. This family of techniques prunes the weights simply according to their *magnitude*, under the assumption that the smallest weights have the least impact on accuracy. Clearly, computing the magnitude of all weights is much simpler than evaluating their saliency, thus making these approaches much more computationally efficient at training time. With magnitude-based pruning, the majority of the weights that can be safely pruned with negligible impact on the final accuracy is found in fully connected layers.

Both magnitude- and saliency-based pruning simply try to maximize the number of 0-weights while minimizing the accuracy drop. Other works, however, have shown that this does not always correspond to the optimal solution when the target is energy minimization. For instance, the authors of [79] have shown that, in classical models such as AlexNet [6], most of the energy for inference is consumed by convolutional layers, and not by fully connected ones. Therefore, they have introduced *energy-driven* pruning, in which the energy impact of each weight is estimated to select the optimal pruning location, based on the consumption of different layers [79].

7.2.2 Benefits of pruning

The immediate benefit that can be derived from pruning techniques is a reduction in the total storage size of a DNN, obtained by storing sparse model parameters in *compressed* formats. By itself, however, compression does not reduce the inference energy consumption. Vice versa, it might actually *increase* it, due to the need of decompressing weights after loading them. Therefore, compression formats specifically tailored for DNN inference have been proposed, which try to simultaneously provide a low-cost decoding algorithm and a way to exploit sparsity for *skipping computations.*

One simple format is the compressed sparse row (CSR) [80]. CSR uses three vectors to store the nonzero values of a DNN weights matrix, and to recover their original location, as shown in Fig. 11. In particular, the lowermost vector contains the values of all the nonzero elements, while the middle one stores the indexes of these elements in the corresponding matrix row. Finally, the topmost vector contains pointers to the locations of the other two arrays where each matrix row starts. To clarify, in the example of the figure, the topmost vector is interpreted as follows:

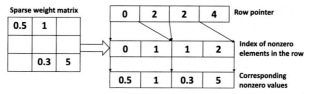

Fig. 11 Compressed sparse row (CSR) format.

- Nonzero elements relative to the 1st matrix row start at index 0 of the other two arrays and end before index 2.
- Elements relative to the 2nd matrix row start at index 2 and end before index 2 (meaning that this row does not contain any nonzero value).
- Elements relative to the 3rd matrix row start at index 2 and end before index 4.

CSR decoding is efficient if the matrix is read in row-major order. In fact, row pointers can be accessed in constant time based on the row index, and reconstructing the entire row is linear in the number of nonzero elements. The problem of this format, however, occurs when trying to skip computations related to zero-weights, and in particular when accessing the activations tensor with which the sparse matrix is multiplied [81]. In fact, since each CSR matrix row has nonzero elements in different positions, the corresponding activations must be accessed multiple times with a sparse pattern. Alternatively, the whole vector has to be loaded at once, but depending on its size, this might not be feasible for memory-constrained devices [80].

Compressed sparse column (CSC) solves this problem using the opposite approach with respect to CSR, i.e., storing row indices and column pointers. This allows to read the matrix by column when performing multiplications, and eliminates the problem of multiple accesses to the input activations. In fact, in a matrix-vector product, each matrix column is multiplied with the same input element, thus the input activations are guaranteed to be read at most once and in order. However, it creates an analogous problem for the *output* vector, which has to be either stored as a whole at the end of the product, or accessed multiple times in a sparse way [80]. Nonetheless, CSC becomes preferable to CSR when the size of the output is smaller than the size of the input [80], which is often true for deep learning models.

Both CSC and CSR matrices are built based on the output of *unstructured* pruning algorithms, in which weights can be zeroed-out at arbitrary locations in the weights matrix. This complicates the optimization from a hardware point of view. In fact, virtually all hardware platforms for deep learning use some form of parallel computation, from SIMD/SIMT operations in CPUs and GPUs, to systolic processing elements in accelerators [34]. With unstructured pruned formats such as CSC/CSR, however, each atomic computation step (e.g., the multiplication of a portion of a weights matrix row with a portion of the activation vector) may require a different number of operations, depending on the number of nonzero values involved. Therefore, if computations involving zero-values are skipped, it becomes hard to fully exploit the available parallel hardware [80].

Structured pruning strategies and compression formats have been introduced to improve hardware utilization. These approaches use the same pruning algorithms described above, but force zeroed-weights to respect certain patterns. Typically, they constrain given portions of the matrix (e.g., fixed size subsets of a row or a column) to contain *exactly* the same number of nonzero weights. For CNNs, the easiest way to perform structured pruning consists in eliminating entire convolutional filters from layers [82]. For fully connected and sequential models, instead, more elaborate structured pruning approaches are needed.

As a representative example, *bank-balanced sparsity* [83], is a structured pruning algorithm that splits the weights in subrows (banks), letting each bank have the same number of pruned weights (see Fig. 12). The resulting matrix can then be stored in a format called *compressed sparse banks* (CSB), as shown in Fig. 13. CSB uses only two vectors, storing the nonzero values and their indexes in the bank respectively. As shown in the figure, elements are re-arranged so that the first positions of the two vectors contain the first elements of each bank, and so on. Since each bank contains the same number of elements, row pointers are not needed, as they can be automatically inferred from the sparsity. Assuming that the entire activation vector can be loaded at

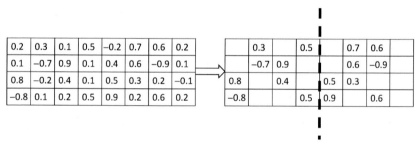

Fig. 12 An example of bank-balanced sparsity with a 50% pruning and 1 × 4 banks.

	0	1	2	3	4	5	6	7	8	9	10	11	12	13	14	15
0	K	L				M	N			O		P		Q		R
1			S	T	U			V		W	X				Y	Z

Data rearranged for interbank parallelization

	0	1	2	3	4	5	6	7	8	9	10	11	12	13	14	15
Values	K	M	O	Q	L	N	P	R	S	U	W	Y	T	V	X	Z
Bank internal indices	0	1	1	1	1	2	3	3	2	0	1	2	3	3	2	3

Fig. 13 An example of the compressed sparse banks (CSB) format.

one time and split into banks as well, this reorganization allows to perform *interbank parallelization*, where weights from different banks are simultaneously multiplied with the corresponding activations. The fact that each bank contains exactly the same number of elements ensures that parallel hardware is fully utilized. This improved parallelism comes with minimal costs in terms of accuracy compared to an unstructured CSR approach, for the same sparsity level [83].

7.3 Knowledge distillation

Knowledge distillation is a *model compression* method whose goal is deriving small but highly accurate networks from ones with far larger sizes. These models can be then deployed on the edge due to their reduced requirements and sizes. Specifically, this approach consists of training a small network (student) starting from one or more large pretrained DNNs (teachers) [84, 85]. In this scheme, the student learn directly from the teacher rather than just from data.

As shown in Fig. 14, two different types of predictions are derived from the networks: *hard* and *soft*. Both are typically obtained with a softmax on the output layer (see Eq. 5). However, hard predictions are obtained setting T to 1, as for standard classification tasks, while soft predictions are obtained with $T > 1$.

Distillation is then performed training the student network on a reduced dataset using two separate losses, as shown in Fig. 14. The *student loss*

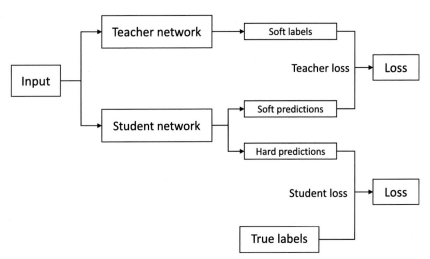

Fig. 14 A simplified overview of knowledge distillation.

measures the difference between the student network's hard predictions and the reduced dataset's true labels. The *teacher loss*, instead, measures the distance between the soft predictions of the student and those of the teachers. Therefore, it measures how different the predictions of the student are from the ones of the bigger network. When teachers are an ensemble of models, the teacher loss usually employs the geometric mean of their predictions. This second loss is computed using soft predictions with $T > 1$ because this causes the probabilities to have a more compact distribution. Less extreme values are in fact more informative and can be more easily learned by the student model [84].

Recent works have proposed variations to the basic architecture of Fig. 14, such as using multiple connections (or bridges) to enforce similar outputs between student and teacher at different layers, aside from the output [86]. Specifically, some hidden layers from the teacher network are chosen to "guide" the learning of others belonging to the student model. If the two layer sizes are different, an additional linear layer is added in-between the two to match the dimensions. More advanced applications of distillation have also been proposed in literature [87, 88].

The impact of distillation on energy efficiency is evident. At inference time, only the distilled student network will be used to process inputs, thus significantly reducing both the model size and the number of operations per input. However, the degree at which this network shrinking can be applied clearly depends on the complexity of the problem. Recent works have shown that when the sizes of student and teachers differ greatly, the performance drops significantly [89, 90]. In fact, the student network can only learn up to a certain extent from the teacher, becoming unable to mimic networks with too many parameters [90]. To tackle this problem, an additional intermediate size network called *teacher assistant* has been proposed in [89] in order to have a multiple-step distillation.

Network distillation has been shown to work equally effectively on both feed-forward and sequential models. For example, a distilled version of BERT [7], one of the state of the art NLP models, called DistilBERT [19] has been recently proposed. It manages to obtain 97% of the accuracy of the full model while performing inference 60% faster.

7.4 Collaborative inference

While the optimizations mentioned above and the availability of custom hardware help running deep learning models *entirely* at the edge, for large networks this may still be unfeasible due to memory and performance limitations [26]. Therefore, an interesting research branch seeks for a compromise between the benefits of edge and cloud computing by means of so-called *collaborative inference*, which consists in distributing the inference computation among multiple devices (e.g., edge nodes and cloud servers) [91].

One basic form of collaborative inference is proposed in [93], where the authors suggest to preprocess images used as inputs for a CNN at the edge, before running the actual inference in the cloud. Specifically, they propose to discard blurry images, since they will not be useful for the neural network, thus reducing the total time and energy spent transmitting raw data. A more advanced collaborative inference framework is presented in [91], where the authors propose to split the execution of a CNN between edge and cloud in a layer-wise fashion as shown in Fig. 15A. Specifically, the first layers are executed on the edge device, while the last ones are computed in the cloud, based on the observation that intermediate layers outputs (e.g., after a pooling) are smaller in size compared to raw inputs in many DNNs. Therefore, they show that computing a few layers at the edge and sending the resulting activations to the cloud is often the optimal approach in terms

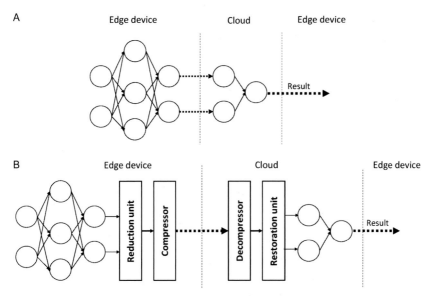

Fig. 15 An overview of Neurosurgeon [91] (A) and BottleNet [92] (B), two collaborative inference frameworks. Both compute the first layers locally, transmitting their output to the cloud where the final result is calculated and then sent back.

of balance between computation and transmission time and/or energy. The best split point for a given DNN is found at runtime, based on the edge-cloud connectivity conditions and on the load of the servers.

An extension of this approach is proposed in [92], where the DNN architecture is slightly modified to make layer-wise partitioning even more convenient, as shown in Fig. 15B. In particular, a pair of so-called reduction (compression) and restoration (decompression) layers are added at the selected DNN split point, in order to further reduce the size of the transmitted data. Compression and decompression may use standard algorithms such as JPEG. At training time, similar to quantization, they are approximated by a straight-through estimator. Finally, the work of [94] considers the case of multiple split points, for DNNs where feature sizes are not monotonically decreasing, such as autoencoders.

As an alternative to layer-wise partitioning, other authors propose to distribute the inference among a number of small IoT devices, each of which processes only a part of the input (e.g., some rows of an image), since it would not be able to handle a full layer [95]. However, this scheme introduces an additional data dependency. It is in fact necessary to have the results of adjacent partitions before being able to compute the following layer.

In [96] a three-level hierarchical framework for deep learning applications that process data from multiple sources (e.g., multiple sensors) is proposed. In this solution, sensors, edge servers and cloud perform a separate inference on the locally available data, aggregating the results of the previous level and forwarding theirs to the following one. This approach drastically reduces the volume of data transmitted, saving energy and reducing the latency, but may affect the accuracy. In fact, "higher-level" devices only have access to the final aggregated outputs of lower-level inference. Therefore, the authors of [43] propose an evolution of this approach, in which the architecture of a *single* DNN is modified to favor distributed processing for multiple-source tasks. In particular, the first layers of this DNN process the data from each source separately (i.e., there are no weights connecting features relative to different sources). These layers are processed locally by each device, which then transmits the result to the cloud. There, features from multiple sources are concatenated before executing the remaining layers. With respect to the previous approach, using a single DNN enables to train the entire system with an end-to-end approach based on standard back-propagation, which improves the accuracy.

7.5 Limitations of static optimizations

Static design time optimizations of deep learning models have been extensively studied due to their effectiveness in reducing the time, memory, and energy requirements for inference. However, recent works have pointed out that static optimizations may be suboptimal in many applications [27, 97–103]. In particular, the main limitation of these approaches comes from the fact that they are input-independent: that is, since optimizations are fixed at design time, they cannot be tuned based on the currently processed input. In contrast, for many realistic applications, inputs are not all equally "difficult" to process for a DNN. As an intuitive example, for an image classification task, a blurry image where the subject is small compared to the frame and has been captured from an unconventional angle might be much more difficult to classify than one where the subject is clear, large and well-positioned in the center of the frame. Similarly, a long and ambiguous sentence might represent a much more challenging task for a translation model than the sentence: "The cat is near the window."

In these scenarios, an aggressively optimized network (e.g., using a low bit-width quantization, a high-sparsity pruning) will likely miss-classify difficult inputs, whereas a less aggressive optimization would cause unnecessary time, memory and energy wastes for easy inputs. This has spurred the birth of optimization strategies that permit the tuning of the complexity versus accuracy trade-off at runtime, depending on the currently processed input. These strategies will be analyzed in the next section.

8. Dynamic (input-dependent) optimizations for deep learning inference at the edge

The limitations of static optimizations have brought to development of dynamic (or *adaptive*) deep learning techniques, where inference execution time and energy consumption (and the corresponding accuracy) can be tuned at runtime instead of design time. Such tuning can be implemented in a wide variety of ways, ranging from using two or more completely separate DNNs based on the difficulty of the input [97, 98], to a different quantization bit-width [27], a different number of layers [99, 100], or a different hardware platform [101]. In the following, we review the main research directions in this sense, focusing in particular on approaches that target edge inference.

8.1 Ensemble learning

Ensemble learning is typically viewed as an approach that exploits multiple machine learning models to improve accuracy. However, in recent years, a new sort of ensemble learning has been proposed, where two or more models are instead used to tune the trade-off between inference complexity and accuracy at runtime [97, 98]. One of the first embodiments of this idea are *big/little* DNNs, shown in Fig. 16 [98]. This approach is based on using two networks of different size at inference time for a classification task. The "little" model is always executed first, and its output is evaluated by the *success checker* block, which estimates its classification *confidence*. If the confidence exceeds a threshold, meaning that the little DNN was "sure" about its prediction, the inference is stopped and the prediction is simply forwarded to the output. In the opposite case, the "big" model is run on the input, and its output is used for the final classification.

The entire big/little scheme is based on the assumption that, for most applications, easy inputs are more frequent than hard ones [98]. Under this assumption, for the majority of inputs, only the "little" DNN will be executed. This allows to absorb the overheads accumulated in the opposite cases, i.e., when both models are run on the same input, clearly causing a larger execution time and energy consumption compared to using only the "big" model, for the same accuracy. In other words, big/little schemes yield energy savings only as long as "big" models are rarely activated.

Clearly, the function performed by the success checker, which controls the activation of the second network, plays a fundamental role. If not carefully tuned, this block could activate the "big" network even when the "little" had classified correctly, or vice versa, it could label wrong classifications from the

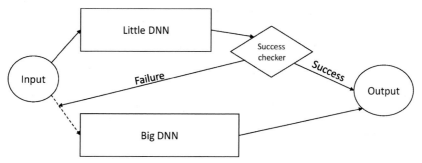

Fig. 16 General scheme of a big/little DNN.

first network as correct. In the first case, this would result in an energy waste, while in the second it would lower the accuracy of the system.

The *score margin* [98] has been proposed as an effective classification confidence estimate, based on the class probabilities produced by the network's output layer. Specifically, the score margin computes the difference between the largest two of these probabilities. If this difference is large, it means that the network produced a high probability *only for one class*, and therefore it is highly confident that the input belongs to it. On the other hand, a small difference means that there are *at least two classes* to which the input could belong with similar probability, according to the prediction of the DNN, hence the confidence is low. In summary, the score margin method activates the "big" network using the following equation:

$$p_{largest} - p_{2nd\_largest} < th \tag{26}$$

that is, whenever the difference between the top-2 probabilities is smaller than a threshold th. Finding the ideal threshold for a given accuracy level is not an easy task. The authors of [98] propose to use a fine-tuning step to find the best th for a given pair of networks and a dataset [98]. Alternatively [27], th can also be tuned at runtime based on external conditions, e.g., increasing it when the battery level is low to save more energy.

The score margin method is effective as long as the "little" model's outputs actually resemble the probability of an input belonging to a given class, which is not guaranteed for black-box models such as DNNs. In particular, it has been shown that modern DNNs estimate confidence probabilities less reliably [104]. In fact, their increased depth positively impacts their accuracy, but negatively affects their capability to predict the likelihood with which a given input belongs to a class. In practice, modern models tend to be *overconfident* even for inputs that are not actually classified correctly. For a big/little system, this makes it harder to find the optimal score margin threshold. To mitigate this problem, one solution is to use so-called *calibration* techniques, that make DNN scores more similar to actual probabilities, at the cost of some accuracy degradation [104].

The major disadvantage of big/little DNNs, however, is that they require a double effort at training time and they result in an increased model size. It is in fact necessary to separately train two different models, each one with its own hyper-parameters to be selected. After training, then, the weights of both models have to be stored on the inference device, which might not be possible on memory-constrained edge platforms.

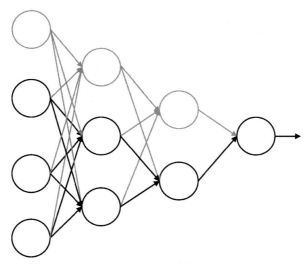

Fig. 17 Simplified view of the dynamic inference technique proposed in [97]. The *black portion* of the network is executed for all inputs, while the *gray part* is only activated for difficult data.

One solution to the storage size issue consists in building an ensemble in which the "little" network is a part of the "big" one, as shown in Fig. 17. This approach activates a portion of each layer (shown in black in the figure) for all inputs, whereas the remaining computations (gray part) are only performed for difficult data. Layers can be split at the level of individual neurons, as shown in the figure for fully connected architectures, while channel-wise partitioning is the approach proposed in [97] for CNNs. In this way, there is no need for two separate set of weights (and inference executions) for the "little" and "big" models, since the latter reuses the weights and computations of the former. Moreover, the scheme can be easily extended to more than two "submodels". Clearly, building a network like the one of Fig. 17 requires a custom training algorithm. Specifically, the DNN has to be trained *incrementally*, starting from the smallest submodel (the black part in Fig. 17). Initially, that portion has to be trained by itself, while disabling the rest of the model. Then, the next set of neurons/channels has to be added and trained, while keeping the weights of the previous portion fixed, and so on.

The work of [27] has proposed a similar way to trade-off complexity and accuracy using multiple "versions" of a DNN. In this case, however, the only difference between the versions is the bit-width used for quantization. The authors perform an offline characterization to identify the set of *relevant* quantization configurations for a given DNN, i.e., those that yield an interesting trade-off in terms of accuracy and energy. Then, a single posttraining quantization is performed, targeting the largest bit-width, and lower precision weights are simply obtained by truncation/rounding. This removes the need for multiple sets of weights, while also not requiring any training. Therefore, the approach is also applicable when training data are not available. Moreover, it is complementary to the previous one and can be used in conjunction with it to generate even more variants of the same DNN.

8.2 Conditional inference and fast exiting

The term *conditional inference* indicates a family of dynamic optimization techniques that change the portion of the DNN graph which is executed at runtime, depending on the input. This approach is based on the observation that neural networks are trained to learn complex nonlinear decision boundaries, in order to correctly classify as many inputs as possible. While this is indeed required for hard inputs, easy inputs often need a much simpler decision boundary, and therefore a much less deep network is sufficient to correctly classify them. In order to save energy and time, part of the DNN graph can then be "switched off" during inference, when an input is detected as easy. One popular form of conditional inference is the so-called *fast exiting*, in which the execution of a DNN is stopped early, avoiding the processing of the last layers.

BranchyNet [99] implements conditional inference adding *branches* to CNNs. Each branch is a possible exit point, which allows to complete the inference using only a portion of the layers (see Fig. 18 for an example with two branches). After each branch, the *entropy* of the softmax output is calculated to estimate the classification confidence. Being a measure of the "uncertainty" of a distribution, a larger entropy indicates a lower confidence. Based on a threshold on the entropy, the execution is either stopped at that branch or continued.

Similar to the "big/little" approach, BranchyNet reduces the overall energy consumption and number of computations as long as the full model or the deeper branches are rarely used. The authors of [99] suggest inserting branches deeper in the network for more difficult datasets. Easier tasks will in fact benefit greatly from branching earlier. Furthermore, the lateral part of each branch (i.e., the one not shared with the main model) may be composed of more than one layer, for example including one or more Conv layers before the final fully connected classifier. While this increases the

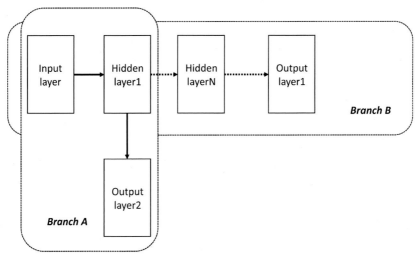

Fig. 18 An example of the BranchyNet architecture proposed in [99].

number of computations, it has been shown to improve the accuracy of the branch as well [99]. As for the score margin in big/little systems, the entropy threshold on which this approach is based to decide whether to stop the execution or not usually requires a fine-tuning step, exploring the effects of different values on the final accuracy and energy. The network and its branches require a custom training procedure, in which the loss functions is a weighted sum of the outputs of each branch. In the forward pass the output of each branch is calculated and then used to update the weights with the standard backward pass. However, this procedure can be optionally based on a pretrained network with fixed weights for the main model (Branch B in Fig. 18) which represents the "backbone" of the system. This makes the training of other branches significantly faster. The main drawback of BranchyNet is an increase in the memory occupation of the network, although smaller than for big/little architectures, due to the new weight tensors of the lateral layers [99].

A different approach to conditional inference is proposed in SkipNet [100]. In this case, instead of adding branches, some of the layers of the DNN are optionally *skipped* for easy inputs. For each set of "skippable" layers, so-called *gates* are added to the base network. Gates are neural networks themselves, although significantly smaller in size compared to the corresponding set of main network layers. They are run first, and based on their output, a policy decides whether to execute or skip the corresponding portion of the main network. Specifically, such policy is based on a combination of supervised and reinforcement learning, and tries to find an acceptable trade-off between the prediction accuracy and the number of layers skipped [100]. The SkipNet system requires a custom training procedure, which includes a first pretraining phase, in which the gate outputs are allowed to assume continuous values. This enables the gates' weights to start from suitable values before the start of the reinforcement learning phase, in which the network is updated in optimize the aforementioned policy. During inference, the model is then able to turn off layers depending on the complexity of the input, as shown in Fig. 19, saving computations with a negligible accuracy loss.

Conditional inference has been also applied to sequential models in [105], but with more modest results than those obtained with feed-forward networks.

8.3 Hierarchical inference

Hierarchical (or staged) inference is an application of a *divide et impera* paradigm to DNN execution. An inference task (e.g., a classification) is split into multiple subtasks, with the easiest being the most common. These multiple tasks are then sequentially executed in increasing complexity order, with the chance of stopping earlier to save energy. Optionally, the subinferences can be performed on different devices, such as edge nodes for the simplest tasks and cloud server for the most computationally intensive ones, in a collaborative fashion. The main difficulty in this approach is being able to split a single machine learning task into multiple subinferences, which is a strongly application-dependent problem. Therefore, this is a somewhat less generally applicable method compared to ensemble and conditional systems.

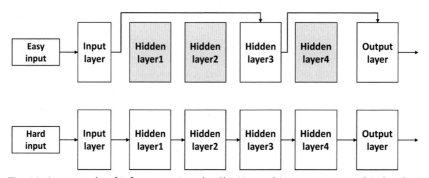

Fig. 19 An example of inference using the SkipNet architecture proposed in [100].

One domain where this methodology has been successfully applied is speech recognition for voice assistants such as Apple Siri, and in particular the implementation for the Apple Watch [26]. In general, speech recognition requires large and power hungry DNNs, unable to fit on a constrained smartwatch. On the other hand, the latency and energy requirements for interacting with a voice assistant are strict, in order to achieve a good user experience, and may not be obtained by a system that constantly transmits to the cloud. The task is then split in two parts: wakeword recognition and speech recognition. A reduced-size RNN is deployed on the edge device (the smart watch), constantly running in the background but with very low-power consumption. This network performs wakeword recognition, which being a much easier task than voice recognition can be handled by simple and low-power models. When a wakeword is detected, the edge device offloads the rest of the audio data to the cloud, where a computationally expensive model is used to interpret them. This division in subtasks permits both a consistent reduction of the data transmitted to the cloud, and a smaller response latency.

A similar hierarchical approach is proposed in [35], where the authors hierarchically split a face recognition task for personal devices such as smartphones. In particular, they propose to run separate inferences to understand: (1) whether the input picture contains a face or not; if it does, (2) whether the face belongs to the device owner or to someone else; if it belongs to someone else, (3) whether it is one of the owner's favorite contacts or not; etc. All steps are executed by increasingly complex CNNs on the same custom hardware accelerator.

Finally, an interesting combination of hierarchical and conditional inference is proposed in [4], for the classification of patient medical issues based on wearable sensors data. In that work, a CNN is split in two parts: a small one deployed on the edge device and a bigger one on the cloud. The local network only tries to predict whether the patient is sick or healthy with a minimal number of layers, while the remote one performs a more in depth analysis to understand the nature of the sickness. The latter model is clearly only invoked when the edge device predicts that the patient is sick. Moreover, the hierarchical splitting into multiple tasks is combined with conditional inference concepts. Indeed, the remote network is fed with the output of the local one, rather than with raw input data. This prevents the repetition of similar computations both locally and remotely to extract basic features. The overall architecture is summarized in Fig. 20.

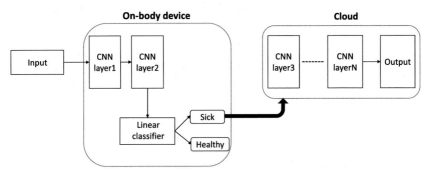

Fig. 20 Staged inference medical diagnosis based on wearable sensors data proposed in [4]. Easy inputs are classified directly on the edge device, while harder ones are sent to the cloud for further (and more computationally expensive) analysis.

8.4 Input-dependent collaborative inference

Collaborative inference, described in Section 7.4 is a very effective approach to optimize the execution of deep learning models by splitting them between edge and cloud. The standard approach to collaborative inference is, however, input-independent.[c] For example, the layer on which a DNN is partitioned in Neurosurgeon [91] depends only on the connectivity conditions and on the load of the cloud server, not on the processed input. This makes sense for some kinds of models, like CNNs, which typically process fixed-size inputs. However, it is not ideal for sequential models, where the *length* of the input time-sequence influences significantly the complexity of an inference.

Indeed, the works of [101] and [106] demonstrate through a characterization that, for RNNs, inference execution time and energy increase *linearly* with the length of the input, due to the dependencies between subsequent steps which prevent interstep parallelism. Therefore, the authors propose a dynamic framework for energy-efficient input-dependent RNN collaborative inference. The framework deploys a copy of the same RNN both on the edge device and in the cloud. It then uses a runtime *mapping engine* to determine the optimal platform where to execute inference for a given input sequence. The mapping engine bases its decision on an estimate of the edge and cloud execution time (or energy consumption) for the current input length, and on the current status of the connection between the two devices.

---

[c]Notice that the approach of Fig. 20 is also collaborative, as it involves edge and cloud. However, in that case, the two devices execute different (portions of) models, while standard collaborative inference is based on a single model.

Inference time and energy are estimated via linear regression models, based on the results of the aforementioned characterization. Such dynamic partitioning significantly outperforms both edge-only and cloud-only inference in terms of energy efficiency, for several NLP applications.

Importantly, the engine in [101, 106] maps the *entire* RNN execution on one of the two platforms, rather than partitioning it as typically done for CNNs [91, 92]. This is because, for most RNN applications, input sizes are much smaller than for CNNs, and most importantly they are smaller than hidden layer outputs. This eliminates the data compression advantage deriving from partial local processing described in Section 7.4. Indeed, the authors show that, for NLP tasks, the total communication time is dominated by the round-trip network latency, which is independent from data size.

8.5 Dynamic tuning of inference algorithm parameters

In this section, we show how deep learning-based tasks at the edge can be optimized by tuning some parameters of the inference algorithm, not directly related with the DNN. While this approach can in principle be applied to many different domains, it is still largely unexplored, therefore we present it using a recent example taken from NLP with sequential models (RNNs, transformers, etc.).

For basic classification or regression tasks, deep learning inference simply consists of the execution of a DNN forward pass. More complex tasks, such as machine translation or reinforcement learning, instead, use the DNN as part of a larger algorithm. Similar to the hyper-parameters of the network, the configurations of these algorithms influence both the inference accuracy and the processing complexity. Such configurations are usually chosen statically at design time, in order to obtain an acceptable accuracy on average. Exactly as detailed before for DNN architectures, this typically corresponds to an over-design for easy inputs.

Recently, examples of dynamic input-dependent tuning of inference algorithm parameters have been proposed in [102, 103, 107, 108] for sequential NLP models, in particular those based on the encoder–decoder architecture. In this architecture, the *decoder* DNN takes as input the fixed-length representation of an input sequence (e.g., a sentence in English) generated by the encoder, and produces an output sequence (e.g., a translation in German). At each step, the decoder outputs the likelihood of all possible outputs (e.g., all words in the German vocabulary), given the input representation and the previous outputs [1]. Since greedily selecting the most likely output at each step generates suboptimal translations, decoding is typically performed using *beam search*. With this algorithm the *BW* most likely partial sentences are expanded in each step, where *BW* is a parameter called *beam width*, and the final overall most likely translation is selected at the end of decoding. The beam width influences both accuracy and energy consumption, since *BW* forward passes of the decoder DNN must be executed in each step [1]. Classically, BW is statically chosen at design time, generating the scenario depicted in Fig. 21A. In contrast, multiple works have recently proposed to adapt this value dynamically at runtime, lowering it for easier inputs and increasing it for harder ones [102, 103, 107, 108]. Specifically, the authors of [103] developed an entropy-based policy to estimate the complexity of a

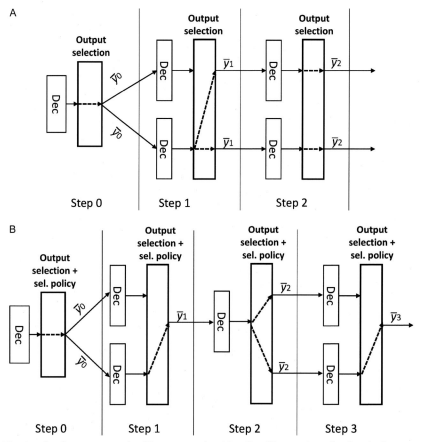

Fig. 21 On the top a standard beam search with a fixed beam size of 2. On the bottom its dynamic version, where the Sel. Policy chooses the best beam width to be used in the following step.

given sequence at each decoding step and change BW accordingly. An example of the result is shown in Fig. 21B. With experiments on translation and summarization tasks, they have demonstrated that this approach yields significant time and energy reductions, especially for single-core MCUs, where the different decoder execution must happen sequentially with one another.

9. Open challenges and future directions

Despite the constantly increasing interest for implementations of deep learning-based applications at the edge, there are still several open challenges that are just starting to be addressed.

Among the optimization methods presented in this chapter, quantization (especially to 8-bit integer) is by far the most widely supported by commercial products and frameworks. Additional efforts on other techniques such as (structured) pruning, collaborative inference, etc., are required to complete the transition from research prototypes and narrow-scoped implementations that work on a single combination of hardware platform and task, to industry-ready methodologies. The same is true for dynamic/adaptive models, which are potentially *the* main direction for edge deep learning in the future, but are still far from being extensively used by industry.

In terms of models, the great majority of the research works in the past have been focusing on CNNs, while other models (e.g., sequential ones) are comparatively much less studied. Nonetheless, sequential models are potentially even more relevant for edge devices, with applications in smart devices (voice recognition, translation, image description, etc.) as well as in IoT systems deployed in cities and factories to perform time series processing.

Finally, as pointed out in [109], deep learning on edge devices still lacks a standard and comprehensive set of benchmarks on which to perform meaningful and fair comparisons among different optimization techniques. These benchmarks should be flexible enough to support the highly heterogeneous platforms on which edge deep learning can be performed, ranging from MCUs to custom accelerators.

References

[1] Y. LeCun, Y. Bengio, G. Hinton, Deep learning, Nature 521 (7553) (2015) 436–444. https://doi.org/10.1038/nature14539, http://www.nature.com/articles/nature14539.
[2] J. Wang, Y. Ma, L. Zhang, R.X. Gao, D. Wu, Deep learning for smart manufacturing: methods and applications, J. Manufact. Syst. 48 (2018) 144–156.
[3] J. Ker, L. Wang, J. Rao, T. Lim, Deep learning applications in medical image analysis, IEEE Access 6 (2017) 9375–9389.

[4] M. Parsa, P. Panda, S. Sen, K. Roy, Staged inference using conditional deep learning for energy efficient real-time smart diagnosis, in: 2017 39th Annual International Conference of the IEEE Engineering in Medicine and Biology Society (EMBC), IEEE, 2017, pp. 78–81.

[5] A. Kamilaris, F.X. Prenafeta-Boldú, Deep learning in agriculture: a survey, Comput. Electron. Agric. 147 (2018) 70–90.

[6] A. Krizhevsky, I. Sutskever, G.E. Hinton, Imagenet classification with deep convolutional neural networks, in: Advances in Neural Information Processing Systems, 2012, pp. 1097–1105.

[7] J. Devlin, M.-W. Chang, K. Lee, K. Toutanova, BERT: pre-training of deep bidirectional transformers for language understanding, Proceedings of the 2019 Conference of the North American Chapter of the Association for Computational Linguistics: Human Language Technologies 1 (2019) 4171–4186.

[8] L. Deng, G. Hinton, B. Kingsbury, New types of deep neural network learning for speech recognition and related applications: an overview, in: 2013 IEEE International Conference on Acoustics, Speech and Signal Processing, IEEE, 2013, pp. 8599–8603.

[9] V. Sze, Y.H. Chen, T.J. Yang, J.S. Emer, Efficient processing of deep neural networks: a tutorial and survey. Proc. IEEE 105 (12) (2017) 2295–2329. ISSN: 15582256. https://doi.org/10.1109/JPROC.2017.2761740.

[10] I. Goodfellow, Y. Bengio, A. Courville, Deep Learning, MIT Press, 2016.

[11] V. Nair, G.E. Hinton, Rectified linear units improve restricted Boltzmann machines, in: Proceedings of the 27th International Conference on Machine Learning (ICML-10), 2010, pp. 807–814.

[12] A.H. Namin, K. Leboeuf, R. Muscedere, H. Wu, M. Ahmadi, Efficient hardware implementation of the hyperbolic tangent sigmoid function. in: Proceedings—IEEE International Symposium on Circuits and Systems, ISSN 02714310, IEEE, 2009, ISBN: 9781424438280, pp. 2117–2120. https://doi.org/10.1109/ISCAS.2009.5118213.

[13] L. Benini, Plenty of room at the bottom? Micropower deep learning for cognitive cyber physical systems. in: 2017 7th IEEE International Workshop on Advances in Sensors and Interfaces (IWASI), IEEE, 2017, p. 165. https://doi.org/10.1109/iwasi. 2017.7974239. 165.

[14] Y. LeCun, B. Boser, J.S. Denker, D. Henderson, R.E. Howard, W. Hubbard, L.D. Jackel, Backpropagation applied to handwritten zip code recognition. Neural Comput. 1 (4) (1989) 541–551. https://doi.org/10.1162/neco.1989.1.4.541.

[15] S. Ioffe, C. Szegedy, Batch normalization: accelerating deep network training by reducing internal covariate shift, in: Proceedings of the 32nd International Conference on Machine Learning (ICML), 2015, pp. 448–456.

[16] R. Krishnamoorthi, Quantizing deep convolutional networks for efficient inference: a whitepaper, arXiv preprint arXiv:1806.08342 (2018), http://arxiv.org/abs/1806.08342.

[17] S. Santurkar, D. Tsipras, A. Ilyas, A. Madry, How does batch normalization help optimization? in: Advances in Neural Information Processing Systems, ISSN 10495258, vol. 2018, 2018, pp. 2483–2493.

[18] L. Lai, N. Suda, V. Chandra, CMSIS-NN: efficient neural network kernels for arm cortex-M CPUs, arXiv preprint arXiv:1801.06601 (2018), http://arxiv.org/abs/1801.06601.

[19] V. Sanh, L. Debut, J. Chaumond, T. Wolf, DistilBERT, a distilled version of BERT: smaller, faster, cheaper and lighter, in: Proceedings of the 5th Workshop on Energy Efficient Machine Learning and Cognitive Computing (EMC2), 2019, pp. 1–5.

[20] A. Vaswani, N. Shazeer, N. Parmar, J. Uszkoreit, L. Jones, A.N. Gomez, Ł. Kaiser, I. Polosukhin, Attention is all you need, in: Advances in Neural Information Processing Systems, ISSN 10495258, vol. 2017-Decem, 2017, pp. 5999–6009.

[21] A. Paszke, S. Gross, F. Massa, A. Lerer, J. Bradbury, G. Chanan, T. Killeen, Z. Lin, N. Gimelshein, L. Antiga, A. Desmaison, A. Köpf, E. Yang, Z. DeVito, M. Raison, A. Tejani, S. Chilamkurthy, B. Steiner, L. Fang, J. Bai, S. Chintala, PyTorch: an imperative style, high-performance deep learning library, in: H. Wallach, H. Larochelle, A. Beygelzimer, F. d'Alché-Buc, E. Fox, R. Garnett (Eds.), Advances in Neural Information Processing Systems 32, Curran Associates, Inc., 2019, pp. 8024–8035. http://arxiv.org/abs/1912.01703

[22] M. Abadi, A. Agarwal, P. Barham, E. Brevdo, Z. Chen, C. Citro, G.S. Corrado, A. Davis, J. Dean, M. Devin, S. Ghemawat, I. Goodfellow, A. Harp, G. Irving, M. Isard, Y. Jia, R. Jozefowicz, L. Kaiser, M. Kudlur, J. Levenberg, D. Mane, R. Monga, S. Moore, D. Murray, C. Olah, M. Schuster, J. Shlens, B. Steiner, I. Sutskever, K. Talwar, P. Tucker, V. Vanhoucke, V. Vasudevan, F. Viegas, O. Vinyals, P. Warden, M. Wattenberg, M. Wicke, Y. Yu, X. Zheng, TensorFlow: large-scale machine learning on heterogeneous distributed systems, arXiv preprint arXiv:1603.04467 (2016), http://arxiv.org/abs/1603.04467.

[23] ONNX, https://onnx.ai/.

[24] ST, STM32Cube-AI, https://www.st.com/en/ecosystems/stm32cube.html.

[25] A. Garofalo, M. Rusci, F. Conti, D. Rossi, L. Benini, Pulp-NN: accelerating quantized neural networks on parallel ultra-low-power RISC-V processors. in: Philosophical Transactions of the Royal Society A: Mathematical, Physical and Engineering Sciences, ISSN 1364503X, vol. 378, 2020, https://doi.org/10.1098/rsta.2019.0155.

[26] J. Chen, X. Ran, Deep learning with edge computing: a review. Proc. IEEE 107 (8) (2019) 1655–1674. https://doi.org/10.1109/JPROC.2019.2921977.

[27] D. Jahier Pagliari, E. Macii, M. Poncino, Dynamic bit-width reconfiguration for energy-efficient deep learning hardware. in: Proceedings of the International Symposium on Low Power Electronics and Design, ISSN 15334678, 2018, ISBN: 9781450357043, pp. 1–6. https://doi.org/10.1145/3218603.3218611.

[28] ST, iNemo, https://www.st.com/en/mems-and-sensors/lsm6dsox.html.

[29] D. Jahier Pagliari, M. Ansaldi, E. Macii, M. Poncino, CNN-based camera-less user attention detection for smartphone power management, in: 2019 IEEE/ACM International Symposium on Low Power Electronics and Design (ISLPED), IEEE, 2019, pp. 1–6.

[30] L. Li, K. Ota, M. Dong, When weather matters: IoT-based electrical load forecasting for smart grid. IEEE Commun. Mag. 55 (10) (2017) 46–51. https://doi.org/10.1109/MCOM.2017.1700168.

[31] Y. Duan, Y. Lv, Y.L. Liu, F.Y. Wang, An efficient realization of deep learning for traffic data imputation. Transp. Res. C Emerg. Technol. 72 (2016) 168–181. ISSN: 0968090X. https://doi.org/10.1016/j.trc.2016.09.015.

[32] Google, Edge TPU, https://cloud.google.com/edge-tpu/.

[33] Intel, Movidius, https://software.intel.com/content/www/us/en/develop/articles/intel-movidius-neural-compute-stick.html.

[34] Y.-H. Chen, T. Krishna, J.S. Emer, V. Sze, Eyeriss: an energy-efficient reconfigurable accelerator for deep convolutional neural networks, IEEE J. Solid-State Circuits 52 (1) (2016) 127–138.

[35] B. Moons, R. Uytterhoeven, W. Dehaene, M. Verhelst, Envision: a 0.26-to-10TOPS/W subword-parallel dynamic-voltage-accuracy-frequency-scalable Convolutional Neural Network processor in 28nm FDSOI. in: Digest of Technical Papers—IEEE International Solid-State Circuits Conference, ISSN 01936530, vol. 60, IEEE, 2017, ISBN: 9781509037575, pp. 246–247. https://doi.org/10.1109/ISSCC.2017.7870353.

[36] F. Conti, P.D. Schiavone, L. Benini, XNOR Neural engine: a hardware accelerator IP for 21.6-fJ/op binary neural network inference, IEEE Transactions on Computer-Aided Design of Integrated Circuits and Systems 37 (11) (2018) 2940–2951. https://doi.org/10.1109/TCAD.2018.2857019. http://arxiv.org/abs/1807.03010.

[37] A. Shawahna, S.M. Sait, A. El-Maleh, FPGA-based accelerators of deep learning networks for learning and classification: a review. IEEE Access 7 (2019) 7823–7859. ISSN: 21693536. https://doi.org/10.1109/ACCESS.2018.2890150.

[38] J.E. Stone, D. Gohara, G. Shi, OpenCL: a parallel programming standard for heterogeneous computing systems, Comput. Sci. Eng. 12 (3) (2010) 66–73.

[39] B. Moons, R. Uytterhoeven, W. Dehaene, M. Verhelst, DVAFS: trading computational accuracy for energy through dynamic-voltage-accuracy-frequency-scaling. in: Proceedings of the 2017 Design, Automation and Test in Europe, DATE 2017, IEEE, 2017, ISBN: 9783981537093, pp. 488–493. https://doi.org/10.23919/DATE.2017.7927038.

[40] D. Jahier Pagliari, E. Macii, M. Poncino, Automated synthesis of energy-efficient reconfigurable-precision circuits, IEEE Access 7 (2019) 172030–172044.

[41] D. Jahier Pagliari, M. Poncino, Application-driven synthesis of energy-efficient reconfigurable-precision operators, in: 2018 IEEE International Symposium on Circuits and Systems (ISCAS), IEEE, 2018, pp. 1–5.

[42] J.T.X. Nvidia, Developer Kit, 2015, https://www.nvidia.com/it-it/autonomous-machines/embedded-systems/.

[43] A. Thomas, Y. Guo, Y. Kim, B. Aksanli, A. Kumar, T.S. Rosing, Hierarchical and distributed machine learning inference beyond the edge. in: Proceedings of the 2019 IEEE 16th International Conference on Networking, Sensing and Control, ICNSC 2019, IEEE, 2019, ISBN: 9781728100838, pp. 18–23. https://doi.org/10.1109/ICNSC.2019.8743164.

[44] S. Chetlur, C. Woolley, P. Vandermersch, J. Cohen, J. Tran, B. Catanzaro, E. Shelhamer, cuDNN: efficient primitives for deep learning, arXiv preprint arXiv:1410.0759 (2014), http://arxiv.org/abs/1410.0759.

[45] D. Kirk, NVIDIA CUDA software and GPU parallel computing architecture. in: International Symposium on Memory Management, ISMM, Vol. 7, 2007, ISBN: 9781595938930, p. 103. https://doi.org/10.1145/1296907.1296909.

[46] D. Rossi, I. Loi, F. Conti, G. Tagliavini, A. Pullini, A. Marongiu, Energy efficient parallel computing on the PULP platform with support for OpenMP. in: 2014 IEEE 28th Convention of Electrical and Electronics Engineers in Israel, IEEEI 2014, 2014, ISBN: 9781479959877, pp. 1–5. https://doi.org/10.1109/EEEI.2014.7005803.

[47] GAP8—The IoT Application Processor, https://greenwaves-technologies.com/ai_processor_gap8/, (Accessed May, 2020).

[48] A. Burrello, F. Conti, A. Garofalo, D. Rossi, L. Benini, Work-in-progress: dory: lightweight memory hierarchy management for deep NN inference on iot endnodes. in: Proceedings of the International Conference on Hardware/Software Codesign and System Synthesis Companion, CODES/ISSS 2019, IEEE, 2019, ISBN: 9781450369237, pp. 1–2. https://doi.org/10.1145/3349567.3351726.

[49] H. Ismail Fawaz, G. Forestier, J. Weber, L. Idoumghar, P.A. Muller, Deep learning for time series classification: a review. Data Mining and Knowledge Discovery , 33 (4) (2019) 917–963. ISSN: 1573756X. https://doi.org/10.1007/s10618-019-00619-1.

[50] J. Deng, W. Dong, R. Socher, L.-J. Li, K. Li, L. Fei-Fei, ImageNet: a large-scale hierarchical image database. in: 2009 IEEE Conference on Computer Vision and Pattern Recognition, IEEE, 2009, pp. 248–255. https://doi.org/10.1109/cvprw.2009.5206848.

[51] F.N. Iandola, S. Han, M.W. Moskewicz, K. Ashraf, W.J. Dally, K. Keutzer, SqueezeNet: AlexNet-level accuracy with 50x fewer parameters and <0.5 MB model size, arXiv preprint arXiv:1602.07360 (2016). http://arxiv.org/abs/1602.07360.

[52] A.G. Howard, M. Zhu, B. Chen, D. Kalenichenko, W. Wang, T. Weyand, M. Andreetto, H. Adam, MobileNets: efficient convolutional neural networks for mobile vision applications, arXiv preprint arXiv:1704.04861 (2017), http://arxiv.org/abs/1704.04861.

[53] M. Min, L. Xiao, Y. Chen, P. Cheng, D. Wu, W. Zhuang, Learning-based compu-
    tation offloading for IoT devices with energy harvesting. IEEE Trans. Veh. Technol.,
    68 (2) (2019) 1930–1941. ISSN: 00189545. https://doi.org/10.1109/TVT.2018.
    2890685.

[54] C.M.J.M. Dourado, S.P.P. da Silva, R.V.M. da Nóbrega, A.C. Antonio, P.P. Filho,
    V.H.C. de Albuquerque, Deep learning IoT system for online stroke detection in skull
    computed tomography images. Comput. Netwk. 152 (2019) 25–39. https://doi.org/
    10.1016/j.comnet.2019.01.019.

[55] Y.D. Kim, E. Park, S. Yoo, T. Choi, L. Yang, D. Shin, Compression of deep con-
    volutional neural networks for fast and low power mobile applications, in: 4th
    International Conference on Learning Representations, ICLR 2016—Conference
    Track Proceedings, 2016.

[56] D. Jahir Pagliari, M. Poncino, E. Macii, Energy-efficient digital processing via
    approximate computing. in: Smart Systems Integration and Simulation, Springer,
    2016, ISBN: 9783319273921, pp. 55–89. https://doi.org/10.1007/978-3-319-
    27392-1_4.

[57] P. Micikevicius, S. Narang, J. Alben, G. Diamos, E. Elsen, D. Garcia, B. Ginsburg,
    M. Houston, O. Kuchaiev, G. Venkatesh, H. Wu, Others, Mixed precision training.,
    in: Proceedings of the 6th International Conference on Learning Representations
    (ICLR), 2018, pp. 1–12.

[58] M. Courbariaux, I. Hubara, D. Soudry, R. El-Yaniv, Y. Bengio, Binarized neural
    networks: training deep neural networks with weights and activations constrained
    to +1 or −1, arXiv preprint arXiv:1602.02830 (2016), http://arxiv.org/abs/1602.
    02830.

[59] M. Courbariaux, Y. Bengio, J.P. David, Binaryconnect: training deep neural networks
    with binary weights during propagations, in: Advances in Neural Information
    Processing Systems, ISSN 10495258, vol. 2015, 2015, pp. 3123–3131.

[60] S. Han, H. Mao, W.J. Dally, Deep compression: compressing deep neural networks
    with pruning, trained quantization and huffman coding, in: Proceedings of the 4th
    International Conference on Learning Representations (ICLR), 2016, pp. 1–14.

[61] S. Zhou, Y. Wu, Z. Ni, X. Zhou, H. Wen, Y. Zou, Dorefa-net: training low bitwidth
    convolutional neural networks with low bitwidth gradients, arXiv preprint arXiv:
    1606.06160 (2016).

[62] E.H. Lee, D. Miyashita, E. Chai, B. Murmann, S.S. Wong, LogNet: energy-efficient
    neural networks using logarithmic computation, in: 2017 IEEE International
    Conference on Acoustics, Speech and Signal Processing (ICASSP), IEEE, 2017,
    pp. 5900–5904.

[63] P. Gysel, J. Pimentel, M. Motamedi, S. Ghiasi, Ristretto: A framework for empirical
    study of resource-efficient inference in convolutional neural networks, IEEE Trans.
    Neural Netw. Learn. Syst. 29 (11) (2018) 5784–5789.

[64] D. Miyashita, E.H. Lee, B. Murmann, Convolutional neural networks using logarith-
    mic data representation, arXiv preprint arXiv:1603.01025 (2016).

[65] S. Gupta, A. Agrawal, K. Gopalakrishnan, P. Narayanan, Deep learning with limited
    numerical precision, in: International Conference on Machine Learning, 2015,
    pp. 1737–1746.

[66] P. Gysel, J. Pimentel, M. Motamedi, S. Ghiasi, Ristretto: a framework for empirical
    study of resource-efficient inference in convolutional neural networks. IEEE Trans.
    Neural Netw. Learning Syst. 29 (11) (2018) 5784–5789. ISSN: 21622388. https://
    doi.org/10.1109/TNNLS.2018.2808319.

[67] Y. Bengio, N. Léonard, A. Courville, Estimating or propagating gradients through
    stochastic neurons for conditional computation, arXiv preprint arXiv:1308.3432
    (2013).

[68] B. Jacob, S. Kligys, B. Chen, M. Zhu, M. Tang, A. Howard, H. Adam, D. Kalenichenko, Quantization and training of neural networks for efficient integer-arithmetic-only inference. in: Proceedings of the IEEE Computer Society Conference on Computer Vision and Pattern Recognition, ISSN 10636919, 2018, ISBN: 9781538664209, pp. 2704–2713. https://doi.org/10.1109/CVPR.2018.00286.

[69] J. Choi, Z. Wang, S. Venkataramani, P.I.-J. Chuang, V. Srinivasan, K. Gopalakrishnan, PACT: parameterized clipping activation for quantized neural networks, arXiv preprint arXiv:1805.06085 (2018), http://arxiv.org/abs/1805.06085.

[70] M. Rastegari, V. Ordonez, J. Redmon, A. Farhadi, Xnor-net: imagenet classification using binary convolutional neural networks, in: European Conference on Computer Vision, Springer, 2016, pp. 525–542.

[71] I. Hubara, M. Courbariaux, D. Soudry, R. El-Yaniv, Y. Bengio, Binarized neural networks, in: Advances in Neural Information Processing Systems, ISSN 10495258, 2016, pp. 4114–4122.

[72] M. Edel, E. Köppe, Binarized-blstm-rnn based human activity recognition, in: 2016 International Conference on Indoor Positioning and Indoor Navigation (IPIN), IEEE, 2016, pp. 1–7.

[73] R. Andri, L. Cavigelli, D. Rossi, L. Benini, YodaNN: an ultra-low power convolutional neural network accelerator based on binary weights, in: 2016 IEEE Computer Society Annual Symposium on VLSI (ISVLSI), IEEE, 2016, pp. 236–241.

[74] J. Ott, Z. Lin, Y. Zhang, S.-C. Liu, Y. Bengio, Recurrent neural networks with limited numerical precision, arXiv preprint arXiv:1608.06902 (2016), http://arxiv.org/abs/1611.07065.

[75] Q. He, H. Wen, S. Zhou, Y. Wu, C. Yao, X. Zhou, Y. Zou, Effective quantization methods for recurrent neural networks, arXiv preprint arXiv:1611.10176 (2016), http://arxiv.org/abs/1611.10176.

[76] S. Shin, K. Hwang, W. Sung, Fixed-point performance analysis of recurrent neural networks. in: ICASSP, IEEE International Conference on Acoustics, Speech and Signal Processing—Proceedings, ISSN 15206149, vol. 2016, IEEE, 2016, ISBN: 9781479999880, pp. 976–980. https://doi.org/10.1109/ICASSP.2016.7471821.

[77] Y. LeCun, J.S. Denker, S.A. Solla, Optimal brain damage, in: Advances in Neural Information Processing Systems, 1990, pp. 598–605.

[78] S. Han, J. Pool, J. Tran, W.J. Dally, Learning both weights and connections for efficient neural networks, in: Advances in Neural Information Processing Systems, ISSN 10495258, vol. 2015, 2015, pp. 1135–1143.

[79] T.-J. Yang, Y.-H. Chen, V. Sze, Designing energy-efficient convolutional neural networks using energy-aware pruning, in: Proceedings of the IEEE Conference on Computer Vision and Pattern Recognition, 2017, pp. 5687–5695.

[80] R. Dorrance, F. Ren, D. Marković, A scalable sparse matrix-vector multiplication kernel for energy-efficient sparse-blas on FPGAs. in: ACM/SIGDA International Symposium on Field Programmable Gate Arrays—FPGA, 2014, ISBN: 9781450 326711, pp. 161–169. https://doi.org/10.1145/2554688.2554785.

[81] G. Goumas, K. Kourtis, N. Anastopoulos, V. Karakasis, N. Koziris, Understanding the performance of sparse matrix-vector multiplication. in: Proceedings of the 16th Euromicro Conference on Parallel, Distributed and Network-Based Processing, PDP, 2008, IEEE, 2008, ISBN: 0769530893, pp. 283–292. https://doi.org/10.1109/PDP.2008.41.

[82] S. Anwar, K. Hwang, W. Sung, Structured pruning of deep convolutional neural networks. ACM J. Emerg. Technol. Comput. Syst. 13 (3) (2017) 1–18. ISSN: 15504840. https://doi.org/10.1145/3005348.

[83] S. Cao, C. Zhang, Z. Yao, W. Xiao, L. Nie, D. Zhan, Y. Liu, M. Wu, L. Zhang, Efficient and effective sparse LSTM on FPGA with bank-balanced sparsity.

in: FPGA 2019—Proceedings of the 2019 ACM/SIGDA International Symposium on Field-Programmable Gate Arrays, 2019, ISBN: 9781450361378, pp. 63–72. https://doi.org/10.1145/3289602.3293898.

[84] G. Hinton, O. Vinyals, J. Dean, Distilling the knowledge in a neural network, in: NIPS Deep Learning and Representation Learning Workshop, 2015. http://arxiv.org/abs/1503.02531.

[85] C. Bucilă, R. Caruana, A. Niculescu-Mizil, Model compression. in: Proceedings of the ACM SIGKDD International Conference on Knowledge Discovery and Data Mining, vol. 2006, 2006, ISBN: 1595933395, pp. 535–541. https://doi.org/10.1145/1150402.1150464.

[86] A. Romero, N. Ballas, S.E. Kahou, A. Chassang, C. Gatta, Y. Bengio, Fitnets: hints for thin deep nets, in: Proceedings of the 3rd International Conference on Learning Representations (ICLR), 2015, pp. 1–13.

[87] Y. Tian, D. Krishnan, P. Isola, Contrastive representation distillation, in: Proceedings of the 8th International Conference on Learning Representations (ICLR), 2020, pp. 1–19.

[88] B.B. Sau, V.N. Balasubramanian, Deep model compression: distilling knowledge from noisy teachers, arXiv preprint arXiv:1610.09650 (2016), http://arxiv.org/abs/1610.09650.

[89] S.-I. Mirzadeh, M. Farajtabar, A. Li, N. Levine, A. Matsukawa, H. Ghasemzadeh, Improved knowledge distillation via teacher assistant, in: Proceedings of the 34th Conference on Artificial Intelligence (AAAI), 2020, pp. 5191–5198.

[90] J.H. Cho, B. Hariharan, On the efficacy of knowledge distillation. in: Proceedings of the IEEE International Conference on Computer Vision, ISSN 15505499, vol. 2019, 2019, ISBN: 9781728148038, pp. 4793–4801. https://doi.org/10.1109/ICCV.2019.00489.

[91] Y. Kang, J. Hauswald, C. Gao, A. Rovinski, T. Mudge, J. Mars, L. Tang, Neurosurgeon: collaborative intelligence between the cloud and mobile edge. International Conference on Architectural Support for Programming Languages and Operating Systems—ASPLOS Part F1271 (1) (2017) 615–629. https://doi.org/10.1145/3037697.3037698.

[92] A.E. Eshratifar, A. Esmaili, M. Pedram, BottleNet: a deep learning architecture for intelligent mobile cloud computing services. in: Proceedings of the International Symposium on Low Power Electronics and Design, ISSN 15334678, vol. 2019, IEEE, 2019, ISBN: 9781728129549, pp. 1–6. https://doi.org/10.1109/ISLPED.2019.8824955.

[93] C. Liu, Y. Cao, Y. Luo, G. Chen, V. Vokkarane, M. Yunsheng, S. Chen, P. Hou, A new deep learning-based food recognition system for dietary assessment on an edge computing service infrastructure. IEEE Trans. Services Comput. 11 (2) (2017) 249–261. https://doi.org/10.1109/TSC.2017.2662008.

[94] A.E. Eshratifar, M.S. Abrishami, M. Pedram, JointDNN: an efficient training and inference engine for intelligent mobile cloud computing services. IEEE Trans. Mob. Comput. Early Access (2019) 1. https://doi.org/10.1109/tmc.2019.2947893.

[95] Z. Zhao, K.M. Barijough, A. Gerstlauer, DeepThings: distributed adaptive deep learning inference on resource-constrained IoT edge clusters. IEEE Trans. Comput. Aided Des. Integrated Circuits Syst. 37 (11) (2018) 2348–2359. ISSN: 02780070. https://doi.org/10.1109/TCAD.2018.2858384.

[96] H. Yin, Z. Wang, N.K. Jha, A hierarchical inference model for internet-of-things, IEEE Trans. Multi-Scale Comput. Syst. 4 (3) (2018) 260–271.

[97] H. Tann, S. Hashemi, R.I. Bahar, S. Reda, Runtime configurable deep neural networks for energy-accuracy trade-off. in: 2016 International Conference on Hardware/Software Codesign and System Synthesis, CODES+ISSS 2016, IEEE, 2016, ISBN: 9781450330503, pp. 1–10. https://doi.org/10.1145/2968456.2968458.

[98] E. Park, D. Kim, S. Kim, Y.D. Kim, G. Kim, S. Yoon, S. Yoo, Big/little deep neural network for ultra low power inference. in: 2015 International Conference on Hardware/Software Codesign and System Synthesis, CODES+ISSS 2015, IEEE, 2015, ISBN: 9781467383219, pp. 124–132. https://doi.org/10.1109/CODESISSS. 2015.7331375.

[99] S. Teerapittayanon, B. McDanel, H.T. Kung, BranchyNet: fast inference via early exiting from deep neural networks. in: Proceedings—International Conference on Pattern Recognition, ISSN 10514651, IEEE, 2016, ISBN: 9781509048472, pp. 2464–2469. https://doi.org/10.1109/ICPR.2016.7900006.

[100] X. Wang, F. Yu, Z.Y. Dou, T. Darrell, J.E. Gonzalez, SkipNet: learning dynamic routing in convolutional networks. in: Lecture Notes in Computer Science (Including Subseries Lecture Notes in Artificial Intelligence and Lecture Notes in Bioinformatics), ISSN 16113349, vol. 11217 LNCS, 2018, ISBN: 9783030012601, pp. 420–436. https://doi.org/10.1007/978-3-030-01261-8_25.

[101] D. Jahier Pagliari, R. Chiaro, Y. Chen, E. Macii, M. Poncino, Optimal input-dependent edge-cloud partitioning for RNN inference, in: 2019 26th IEEE International Conference on Electronics, Circuits and Systems (ICECS), IEEE, 2019, pp. 442–445.

[102] D. Jahier Pagliari, F. Panini, E. Macii, M. Poncino, Dynamic beam width tuning for energy-efficient recurrent neural networks, in: Proceedings of the 2019 on Great Lakes Symposium on VLSI, 2019, pp. 69–74.

[103] D. Jahier Pagliari, F. Daghero, M. Poncino, Sequence-to-sequence neural networks inference on embedded processors using dynamic beam search. Electronics (Switzerland) , 9 (2) (2020) 337. ISSN: 20799292. https://doi.org/10.3390/electronics 9020337.

[104] C. Guo, G. Pleiss, Y. Sun, K.Q. Weinberger, On calibration of modern neural networks, in: 34th International Conference on Machine Learning, ICML 2017, vol. 3, JMLR.org, 2017, ISBN: 9781510855144, pp. 2130–2143.

[105] A. Graves, Adaptive computation time for recurrent neural networks, arXiv preprint arXiv:1603.08983 (2016).

[106] D. Jahier Pagliari, R. Chiaro, Y. Chen, S. Vinco, E. Macii, M. Poncino, Input-dependent edge-cloud mapping of recurrent neural networks inference, in: 2020 57th ACM/EDAC/IEEE Design Automation Conference (DAC), 2020, pp. 1–6.

[107] M. Mejia-Lavalle, C.G.P. Ramos, Beam search with dynamic pruning for artificial intelligence hard problems, in: 2013 International Conference on Mechatronics, Electronics and Automotive Engineering, IEEE, 2013, pp. 59–64.

[108] M. Freitag, Y. Al-Onaizan, Beam search strategies for neural machine translation, in: Proceedings of the First Workshop on Neural Machine Translation, 2017, pp. 56–60.

[109] C.R. Banbury, V.J. Reddi, M. Lam, W. Fu, A. Fazel, J. Holleman, X. Huang, R. Hurtado, D. Kanter, A. Lokhmotov, D. Patterson, D. Pau, J.-s. Seo, J. Sieracki, U. Thakker, M. Verhelst, P. Yadav, Benchmarking TinyML systems: challenges and direction, arxiv, preprint: arXiv:2003.04821 (2020).

About the authors

Francesco Daghero is a PhD student at Politecnico di Torino. He received a M.Sc. degree in computer engineering from Politecnico di Torino, Italy, in 2019. His research interests concern embedded machine learning and Industry 4.0.

Daniele Jahier Pagliari received the M.Sc. and Ph.D. degrees in computer engineering from Politecnico di Torino, Italy, in 2014 and 2018, respectively. He is currently an Assistant Professor in the same institution. His research interests include computer-aided design of digital systemsand low-power optimization for embedded systems, with particular focus on embedded machine learning.

Massimo Poncino is a Full Professor of Computer Engineering with the Politecnico di Torino, Italy. His current research interests include several aspects of design automation of digital systems, with emphasis on the modeling and optimization of energy-efficient systems. He received a PhD in computer engineering and a Dr.Eng. in electrical engineering from Politecnico di Torino.

CHAPTER NINE

"Last mile" optimization of edge computing ecosystem with deep learning models and specialized tensor processing architectures

Yuri Gordienko, Yuriy Kochura, Vlad Taran, Nikita Gordienko, Oleksandr Rokovyi, Oleg Alienin, and Sergii Stirenko
National Technical University of Ukraine "Igor Sikorsky Kyiv Polytechnic Institute", Kyiv, Ukraine

Contents

Abstract

In the context of edge computing (EC) paradigm the new type of specific System on a Chip (SoC) devices with tensor processing architectures (TPAs) appeared for running deep learning (DL) models efficiently on edge computing accelerators (ECAs). Despite availability of numerous benchmarks of ECAs on the real world applications the contribution of the overheads under different available configurations and actual EC performance is not covered thoroughly at the moment. The matter is the high performance hardware (ECAs of the latest releases) can be applied for the highly optimized DL models, but all these accelerations can be diminished by the overheads at the "last mile" where the end-users can use inefficient support provided by the default OS, applicability of slow connectivity modes, and ignorance of some specific manufacturer recommendations like overclocking regimes and so on. In this work several different TPA hardware implementations were considered, including Coral Edge TPU by Google, and Movidius Neural Compute Stick by Intel. The comparative analysis of their performance (inference time, iteration time, speedup due to optimization measures) was performed including comparison with the traditional computing architectures on the basis of central processing unit (CPU) and general purpose graphic processing unit (GPU). These results give the opportunity to estimate the performance of ECAs for the specific use cases (the human pose recognition practical use case is considered here) and find the optimal environment configurations for the better speedup, low overhead and high efficiency of ECAs.

1. Introduction

During the last years the progress of machine learning (ML), especially deep learning (DL) and deep neural networks (DNNs) [1–3], is closely related to recent advances in high performance computing (HPC), distributed computing and specific computing architectures.

The requirements for higher accuracy, faster learning and smaller inference time are also closely related to requirements for more computing power, more memory, more network bandwidth, etc. Currently, various IT infrastructures are used for this purpose: from desktop computers (shared memory model) to parallel hardware architectures such as HPC (like clusters with shared and distributed memory models), cloud computing (systems with distributed memory models), fog computing [4], and dew computing [5,6] which are especially popular and widely used today [5–7] for peripheral computing devices in the framework of Edge Computing (EC) paradigm [8,9].

The evolution of these infrastructures follows the current trend of hardware accelerated DL applications. Over the past decade, this has been closely related to the development of graphics processing units (GPUs) or general purpose graphics processing units (GPGPUs). Recently, however, a wide variety of old and new alternative platforms have emerged, from digital signal processors (DSPs) and field programmable gate-arrays

(FPGAs) [10,11] to a completely new architecture for applied integrated circuits (ASICs) [12], including neuromorphic hardware [13,14], tensor processing architectures (TPA) such as like TCs (Tensor Cores) from NVIDIA [15], TPUs (tensor processing units) from Google [16], designed for specialized tensor processing tasks that are widely used in machine learning applications.

In the context of EC paradigm the new type of specific System on a Chip (SoC) devices appeared for running deep learning models efficiently on edge computing devices. These devices (edge computing accelerators—ECAs) have evident advantages like low latency, high energy efficiency, security, locality, etc. that allow for numerous opportunities in the real world applications [17].

During the last years numerous edge computing accelerators with the correspondent SDK appeared on the market including NVidia Jetson Nano [18], Google Coral [19], Intel NPU [20], and their derivatives like Horned Sungem (based on Intel Movidius NPU) [21], HiSilicon Kirin 970 SoC with HiAI Architecture and a dedicated NPU [22].

Usually they allow to use CNN-based deep learning inference at the edge layer, supports heterogeneous execution across various computer vision accelerators (including GPU, NPU, NPU, VPU, etc.), apply some internal optimization to get maximal inference performance, and quickly deploy applications and solutions for vision inference by leveraging corresponding libraries.

In this study, we will briefly discuss some of the specialized ECAs, highlight their implementation in the EC context, present the results of research on the performance of some popular hardware implementations in EC context, and suggest an assessment method for evaluating their effective use in this context.

The main goal of this work is to study the inference performance and efficiency of existing specialized hardware on the popular use case of human pose recognition, allowing to predict performance in the view of the hidden overheads for the proprietary solutions even.

The rest of this work is organized as follows. In Section 2, we summarize the state of the art in specialized hardware implementations for EC. Section 3 describes the experimental part concerning the selected equipment, data, models and indicators used. Section 4 summarizes the experimental results obtained, Section 5 discusses these results, and Section 6 summarizes the lessons learned.

2. State of the art
2.1 Background of edge computing

In the context of the distributed computing, fog computing (FC) is usually considered as an architecture that uses edge level computing devices to perform a significant portion of computation, storage, and communication locally and route the summarized data over the internet connection. Such a paradigm can efficiently support huge volume of emerging IoT applications that demand real-time working mode, predictable performance and latency [4].

Dew computing (DC) paradigm goes beyond the concept of a Cloud-Fog computing to a micro-service concept in vertically distributed computing hierarchy. In comparison to FC, DC paradigm further shifts the limits to computing applications, data, and low level services away from centralized virtual nodes (from the FC level even) to the end users. A cloud-fog-dew type of architecture allows to distribute and share the information among the end-user devices (e.g., laptops, smartphones, smart watches, and other types of smart gadgets, especially wearable electronics in health care applications) thus enabling data access even when the global network connection is not available. DC paradigm also leverages the local resources that are continuously connected to a network, such as various smart gadgets and sensors. In fact, DC covers a wide range of technologies that allow for everyday, general purpose usage in working and living environments, down to the level of the simple and special purpose equipment and micro-services [5–7].

Recently, the edge computing (EC) paradigm becomes very popular. FC-DC and EC provide nearly the same functionalities by using the computing capabilities within a local area network (LAN) to perform computations and store data that would ordinarily have been carried out in the cloud. The main difference between FC-DC and EC relates to the place where data processing occurs: EC usually takes place directly on the edge devices to which sensors/actuators are connected or some local gateway devices nearby to the sensors/actuators. But FC-DC moves the data processing to processing units that are connected to the LAN or into the LAN hardware itself so usually they are more distant from the sensors/actuators.

EC paradigm allow for storing and processing data on edge devices without uploading to CC-FC-DC platform with the following advantages: faster (real-time) data processing and analysis; higher security; low cost of ownership and support; low energy consumption; low networking cost; better scalability; higher independence and others [8,9].

Due to these advantages, and because the resources of EC intelligence equipment are usually limited, and AI algorithms based on DL often need huge computing and storage resources, expensive power supply and support, so the lightweight AI solutions for EC applications are very actively investigated now [23]. Implementation of DL applications for EC devices differs from DL applications on cloud computing where computationally intensive DL calculations can be provided as a service. Several studies were dedicated to investigation of combined EC/CC approaches, so-called hybrid edge-cloud architectures [23].

It should be noted that DL computational demands for resource-constrained EC devices are also related with the various additional operations and correspondent overhead expenses (input/output, OS support, connectivity, shared communication, etc.) which are rarely considered, measured and discussed on the real world application (except for some simple illustrative examples on the standard datasets or under simplified conditions). In this context, the performance of these EC systems are need to be estimated, or benchmarked, for the proper real world application deployment, resource optimization, and efficient decision-making. Despite availability of numerous benchmarks of EC devices (usually provided by manufacturers of EC devices), infrastructures, approaches [24] and the tests on the real world applications [17] the contribution of the overheads under different available configurations and actual EC performance is not covered thoroughly at the moment.

2.2 Edge computing hardware implementations

Now healthcare organizations can provide the better services for onsite computer-aided diagnostics (CADe/CADx) by the combination of emerging information technologies (ITs) such as artificial intelligence (including DL [25–28]), decentralized infrastructures (including blockchain [29,30]) and distributed computing (including CC-FC-DC). By means of edge computing the healthcare end users (e.g., patients, doctors, hospitals, etc.) can leverage the potential of the IT combination at their premises.

This work is motivated by the works dedicated to medical applications where DL approaches were used for measuring and estimating health state by CADe/CADx of various diseases (like cancer [31–34], tuberculosis [35–37], several other lung abnormalities [27,38,39], etc.) and could be easily ported to EC for the healthcare end users. It has become evident during the current COVID-2019 pandemia when the huge volume of the related DL content had appeared and has been reviewed and criticized thoroughly [40].

Moreover, EC with DL becomes more and more popular due to usage of wearable electronics (WE) for onsite and real time health monitoring not only in medicine, but also in other fields like sport, transport, manufacturing, leisure, etc. [41–46], where numerous wearable sensors are used and some summary should be obtained. Here, the human pose monitoring task is considered a simulated use case for time performance estimation by some available EC devices under different configurations of end user environments. The thorough investigations of other optimization parameters and their impact on the accuracy and loss are under work right now and will be published elsewhere.

Now many personal gadgets have potential to perform the inference by many DL models (after the proper adaptation) on EC devices such as IoT-controllers, smartphones, WE, etc. In addition to them low-power and obsolete laptops and desktops without powerful CPU-chips or built-in hardware accelerators like GPU-cards can be equipped by the new category of chips (ECA [17]) aimed at inferences at the EC level by means of their ability to perform calculations for DL models with the high time performance (latency, inference, iteration, and other times) and without compromising prediction performance (accuracy, loss, etc.). Despite several attempts to measure the performance characteristics of such ECA [17,24,47], their characterization is not finished on the real world applications and influence of the overheads under different available configurations and actual ECA performance is not covered thoroughly at the moment.

Among various previously mentioned ECAs and the correspondent SDKs, here the two widely used ECAs (actually, Google Coral [19] and Horned Sungem based on Intel Movidius NPU [21]) were investigated because of their applicability not only for IoT and WE devices, but also for the low-end end user PCs like laptops and desktops.

To provide the proper basis for comparison this investigation was performed for several different software/hardware configurations which are most widely available for end users with the low-end end user PCs like laptops and desktops without powerful CPU-chips or built-in hardware accelerators like GPU-cards. We report on their time performance concerning inference time and iteration time which is the inference time and all related overhead time (data input and results output). These results and further insights can be used for the predictable and low-cost optimization of the end-user software/hardware configuration to get the better inference performance by means of cheap and widely available ECAs on the basis of Google Coral, Intel Movidius, and other solutions.

It should be noted that various aspects DL models were thoroughly investigated and benchmarked with varying architectures and tasks (i.e., motion, audio, and vision) across seven platform configurations with three different ECAs including Intel Neural Compute Stick, Google Coral, NVidia Jetson Nano [17,48–54], and some brand new implementations [55]. Authors reported on their execution performance concerning various aspects like latency, memory, power consumption.

2.3 Practical edge computing use cases

Many real world applications can leverage the potential of DL on ECA, especially for "real time" mode of operation with some privacy concerns that limit inference to the local ECAs [23]. Below they are shortly listed inside the quite vague and intersecting categories and very briefly characterized for outlining the general application context.

2.3.1 Computer vision

Classification, detection, and segmentation tasks are used in many kinds of specific applications like security surveillance, traffic management, health care CADx/CADe, and others. Now, video data from edge layer cameras can be pre-processed by built-in computing units and the current trend is to incorporate DL computing units also [56–58].

2.3.2 Network management

Recently, DL approaches are widely used to support some network security functions like scheduling, intrusion detection, IoT security, and others, has been proposed [59–61]. In the context of network applications the natural requirement is to provide the lowest possible latency and real-time operation for automatic decision making.

2.3.3 Human–computer interaction

DL on ECAs has also become popular for the better human–computer interaction tasks with usage of natural language processing techniques implemented in many the ECAs as voice-manipulated assistants like Microsoft Cortana, Google Assistant, Amazon Alexa, Apple Siri, and others [62–64]. In combination with virtual reality (VR) and augmented reality (AR) technologies DL has been used in real-time regime with low latency to predict the field of view of the user and detect objects of interest in the user's field of view with adding virtual overlays [65–67].

2.3.4 Internet of things

In numerous applications in healthcare, manufacture, city, and power grid management the huge number of IoT sensors/actuators should be orchestrated in automatic regime with real-time decision making for IoT actuators on the basis of IoT sensor data. Many intersections of this application domain with the other listed above concern data stream operation and time correlation that could be leveraged by the DL on ECAs [68,69].

2.3.5 Human activity recognition

This domain include computer vision and sensor-based WE for health care, sport, pedestrian analysis, security, assisted living, intelligent driver systems, augmented reality, mechanical robotics, and many others [70–72]. In the context of this work the human pose estimation by DL is of the great interest. It is targeted on prediction or estimation of location of several key-points of human body (like the knees, elbows, shoulders, back, etc.) [73–76]. It is considered as a challenging DL task because some key-points may not be visible because of loose clothing, low lighting, shadowing by other objects, improper camera position, etc.

2.4 Challenges in edge computing

Despite the recent progress in AI, DL, and EC the significant challenges remain in deploying DL models and applications on EC infrastructure (accelerators, controllers, servers and their combination). They usually related to latency and inference time, energy, migration, optimization. Despite many excellent studies described above have considered thoroughly on reducing latency and inference time, the problem remains especially for processing high-quality input images and videos. The lowest energy consumption is desirable for DL applications on ECA, because they are battery-powered often [77]. It can be reached by reducing the number of DL-related computations, but it can decrease the prediction performance. Optimization of the specific hardware chips (e.g., GPUs and TPUs) by decrease of energy consumption without detriment to accuracy is one of the hottest topics in this field [23,24,78]. Due to variety of available and future ECAs the problem of smooth migration appears for the large DL models targeted for EC infrastructure. Many studies were devoted to this problems and some effective solutions were proposed on the basis of virtual machines, containers, and specific network techniques [79–82]. Addressing these challenges requires system measurements and experiments to gain an empirical understanding of the migration challenges.

But impact of environment overheads on the performance of ECAs for the specific use cases is not investigated yet and search for the optimal configurations for low overhead and high utilization of ECAs is of great practical interest.

In logistics, transportation, telecommunications the "last mile" problem means the difficulty in delivering the goods from the high-speed connections to the final destinations over the low-speed last connections when the whole efficiency suffers from this slow-down. The similar situation can appear when the high performance hardware (for example, ECAs of the latest releases) can be applied for the highly optimized DL models, but all these accelerations can be diminished by the overheads at the "last mile," when the end-users use inefficient support provided by the default OS, applicability of slow connectivity modes, and ignorance of some specific manufacturer recommendations like overclocking regimes and so on. The problem is aggravated by the usage of some obsolete components into EC ecosystem, for example, non-efficient OSs, slow USB connectivity, ignorance of overclocking regimes proposed by manufacturers. The effect of the proper optimization on the performance is intuitively evident, but the concrete speedup analysis was not performed before for the specific ECA types, DL models and applications.

3. Methodology

3.1 Hardware used

Among many hardware implementations proposed by different manufacturers to accelerate the execution of deep learning models at EC layer two popular types of ECAs were selected (Google Coral and Horned Sungem) and investigated in different environment configurations with comparison of their usage on CPU and GPU platforms. In this work the motivation for their selection was dictated by their wide availability, low price, reach support, and current use in the applications from our actual experience: in health care for onsite Chest X-Ray image inference and lung disease prediction [32,35,39], in sport and everyday life for activity recognition, and elderly care for human pose and activity recognition [7,43,45–47].

3.1.1 Google TPUs

Briefly, during the last years Google released several versions of their Tensor Processing Unit (TPU) platform which can be used in cloud-based environment, known as Google Cloud TPU v1, v2, and v3, (https://cloud. google.com/tpu/) [16] and in EC environment, known as EdgeTPU under the brand name Google Coral (https://www.coral.ai).

The TPUs become popular for increasing the efficiency and speed of DNNs, because according to the available benchmarks, TPU can process DNNs up to 30× faster and can be up to 80× more energetically efficient than CPUs or GPUs [11], [83]. It is possible, because the TPU is specifically adapted to process DNNs with more instructions per cycle than CPU and GPU. Moreover TPU is the unique TPA because of its current availability as a cloud service resource only in the following configurations: Cloud TPU v2 (180 teraflops, 64 GB High Bandwidth Memory (HBM)), Cloud TPU v3 (420 teraflops, 128 GB HBM), Cloud TPU v2 Pod Alpha (11.5 petaflops, 4 TB HBM, 2-D toroidal mesh network) [11]. The EdgeTPU is an application specific integrated circuit (ASIC) that supports only signed integer operations (8/16 bits) and TensorFlow Lite models with the specific requirements [4]. Among two variations of the EdgeTPU (Coral Dev Board and as Coral Accelerator) the USB-based standalone version (known as Coral Accelerator) was used.

3.1.2 Intel VPUs

Intel has released several modifications of Vision Processing Units (VPUs) which are effectively the low-power System-on-Chips (SoCs) designed to accelerate deep-learning deployments and computer vision applications. For example, Intel Movidius Myriad X contains several processors and computing units optimized for high parallelism and DNN inference. The VPU is available in different configurations. In the context of the current work USB-based dongle versions (known as Intel Neural Compute Sticks—NCS) is of interest. For example, NCS hosts the Movidius Myriad X VPU with 4GBit of RAM. This USB stick can be plugged to any USB-equipped computing unit (like standalone laptop or PC) as a co-processor to speed-up the inference of DNNs. Several manufacturers propose the similar solution equipped by the same VPU, for example, in this work the popular Horned Sungem solution [21] is considered which is based on Intel Movidius MA245x chip (Table 1).

Recently, the cloud-based version of Google TPUs [83] and Coral Dev Board version of EdgeTPU [17] were tested and analyzed thoroughly, and this work is dedicated to USB-based Coral Accelerator in different environment configurations including various OSs, connectivities, and possible overclocking. Despite availability of some performance tests and extensive

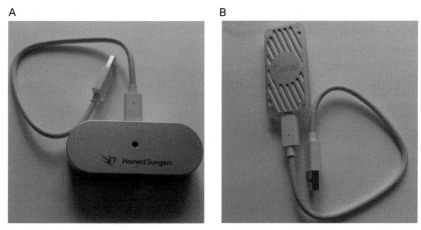

Fig. 1 Examples of hardware used in the work: Horned Sungem (A) and Google Coral (B).

Table 1 Parameters of the hardware used (Fig. 1).

Hardware	GPU	Horned Sangem/ usb2	Horned Sangem/ usb3	Coral stick/ usb2/ standard	Coral stick/ usb3/ standard
Chip	NVIDIA 1080 Ti	Intel Movidius MA245x	Intel Movidius MA245x	Google Edge TPU	Google Edge TPU
OPs	0.6 TFLOPs	0.1 TOPs	0.1 TOPs	4 TOPs	4 TOPs
Memory	11 GB	512 MB	512 MB	8 MB	8 MB
I/O interface (rate)	PCIe 2.0 ×16 (64 Gbps)	USB 2.0 (480 Mbps)	USB 3.0 (5 Gbps)	USB 2.0 (480 Mbps)	USB 3.0 (5 Gbps)
OS	Linux	Linux	Linux	Linux/ Windows	Linux/ Windows
Overclocking	No	No	No	No/Max	No/Max

reviews of the available TPAs, like Coral Dev Board version of EdgeTPU [17], Google TPU [16], TPUv2 [83–85], the systematic studies on their performance with regard to different end user environments (OSs, connectivity type, overclocking, etc.) are absent.

That is why in this work the specific attention was dedicated to different environment configurations (OSs, connectivity type, overclocking, etc.)

that can be easily changed by end users for obtaining the better performance as it is demonstrated here for the proposed use case of human pose recognition task for sport, health care, manufacture, and other applications.

Two the most popular OSs were considered (Windows 10 and Ubuntu 18.0) along with the two popular input/output connectivity interfaces, namely, USB 2.0 (with a maximum rate of 480 megabits per second) and USB 3.0 (with a maximum rate of 5 gigabits per second), and two overclocking regimes (standard—actually no overclocking, and maximal) proposed by Horned Sungem SDK are considered.

3.2 Use case and data used

As a continuation of the previous works targeted on the human activity recognition, classification, load and fatigue prediction in applications for health care, sport, pedestrian analysis, assisted living and others [7,44–46], the use case of human pose recognition was selected. Human pose estimation relates to computer vision approaches to detect human posture in images and video sequences (frames) and, moreover, to determine locations of the targeted body limbs and joints between them. It is one of the most fundamental and challenging problems in computer vision as a whole with the numerous mentioned above important applications. PoseNet [86] is one of the most popular computer vision models [87–89] that can be effectively used to estimate the human pose from image or video frames by estimating locations of key body joints. The PoseNet model's performance varies depending on the device and output, but it is invariant to the size of the image, thus it can predict pose positions in the wide scale range [86].

After numerous inference runs on different videos the following video with 389 frames was proposed in the work as an example of presence of the quite different content in the same video. The video included three various views: the close view without humans as targets (Fig. 2, left column), the middle view with several clear seen humans (Fig. 2, central column), and the far one with several unclear seen humans (Fig. 2, right column). These three fragments illustrate the qualitatively different situations when targets are absent (Fig. 2, left column), or targets are present, but humans are not recognized and human poses are not determined (Fig. 2, central column), or targets are present, but humans and their poses are not determined (Fig. 2, right column). It should be noted that the middle view fragment can contain the very different numbers of persons with varying human poses (Fig. 3).

Fig. 2 Examples of the three various views (close—left column, middle—central column, far—right column) and correspondent frames (frame 10—left column, frame 90—central column, and frame 300—right column) with various frame sizes (1280*720—top row, 640*480—central row, and 480*360—bottom row).

3.3 Optimization methods

Usually to improve DNN inference performance of the EC applications several main approaches are used: (1) on-device computation of DNNs; (2) edge server-based computation, where data from the ECA are sent to edge server for computation; and (3) combined computation among ECAs, edge servers, and the cloud [23].

As far as ECAs have different constraints and requirements, different optimization techniques are used to obtain the actual hardware acceleration. They are usually divided into some categories like optimized implementation, quantization, and structured simplification that convert a DNN into a compact one [90–92] by approaches like tensor factorization [92], sparse connection [93], channel pruning [94,95], and others. The extensive reviews on optimization of DNNs are usually related with various general perspectives, models, optimization algorithms, and datasets [94].

Several aspects of the actual calculations with quantization on ECAs with TPA should be considered here. Sometimes various manufacturers of ECAs propose their TPA supporting frameworks where different quantization scenarios can be used in various TPAs. As far as Google Coral EdgeTPU does not support floating-point operations, quantization is used when the model weights are represented as signed-integer numbers. For example, some of the TPA like EdgeTPU supports quantization-aware training

A

B

Fig. 3 Comparison analysis of pose detection for various devices: Coral (A) and GPU (B).

which performs parameter quantization during training time, then freezing the model, and converting to TensorFlow Lite format. The post-training quantization is performed for all the parameters and activations without re-training the model. The EdgeTPU compiler verifies if the quantized TensorFlow Lite model meets the requirements, statically allocates the model weights on the TPU on-chip memory and defines the execution on the acceleration hardware. If some operations could not be supported by the EdgeTPU runtime, then the compiler forces their execution on the CPU instead of TPU. If model's weights do not fit in the TPU on-chip memory, the weights are dynamically streamed from off-chip memory, e.g., RAM, to the on-chip memory, introducing additional latency [17].

In the context of this work, on-device computation was investigated by the considered ECAs with various optimization techniques applied to resolve the "last mile" problem. The idea is to estimate effect of the potential overheads on performance that were not taken into account by ECA manufacturers (hardware optimization) of DL model developers (software optimization) at the "last mile" where the end-users can use inefficient support provided by the default OS, applicability of slow connectivity modes, and ignorance of some specific manufacturer recommendations like overclocking regimes and so on.

For performance analysis, the benchmark script was developed that run the selected model and measure execution time. Among several steps in the model lifetime, two stages were selected for timing: inference stage and iteration stage that included the whole frame processing including all related overheads including data input (frame loading), inference, result output. As far as this work is dedicated to impact of some infrastructure optimization measures on possible speedup improvements that could not influence the accuracy, the investigation of similar impact on the accuracy performance by other optimization measures (for example, by modification of the model itself by pruning or quantization) is out of scope of this work. But these aspects were considered and analyzed by us and other authors before for other models and published elsewhere [91–95].

3.4 Models and metrics used

Among a wide range of DL models designed for computer vision tasks MobileNet network architecture was selected as specially designed for portable devices such as smartphones, robots and other low power ECAs. The main component of the network is depth-wise convolution, which applies single filter to each input channel. Then point-wise 1×1 convolution combines output from previous step. Such two-step convolution decreases computational overheads, which is important for embedded devices. Despite availability of many layers, this network is small enough and already optimized for low latency, developers introduce width multiplier and resolution multiplier hyperparameters which allow to decrease the number of computations. MobileNet can be used for object detection and localization, face embeddings, fine grained recognition and other computer vision tasks [96]. Here, we used DL models that have been implemented by means of TensorFlow or TensorFlow + Keras frameworks [97].

With regard to the problem of the highest efficiency of ECAs like the before mentioned USB-dongle solutions on the low-power laptops and desktops without powerful CPU–chips or built-in hardware accelerators like GPU–cards, raw ($t^{\mathrm{raw}}_{\mathrm{inf}}$) inference time (i.e., the execution time of the model inference for each frame) and raw ($t^{\mathrm{raw}}_{\mathrm{over}}$) iteration time (the whole duration of the frame processing loop including data input, inference time, results output) were the key metrics here. Also several derivative characteristics were calculated for the subset of 50 frames in the moving window: mean inference time (t_{inf}) and iteration time (t_{over}) obtained, standard deviation (t^{std}), skewness (t^{skew}), and kurtosis (t^{kurt}).

For example, for Horned Sungem device, the mean inference time (t_{inf}) and iteration time (t_{over}) dependencies vs the frame size under Linux OS were calculated for various USB interfaces like USB 2.0 and 3.0 (Fig. 8A), where the correspondent inference time increase with the frame size (Fig. 8B) was defined as:

$$ti_{\mathrm{inf}}(\mathrm{size}) = t_{\mathrm{inf}}\,(\mathrm{size})/t_{\mathrm{inf}}\,(480 \times 360),$$

the iteration time increase with the frame size (Fig. 8B) was defined as:

$$ti_{\mathrm{over}}(\mathrm{size}) = t_{\mathrm{over}}\,(\mathrm{size})/t_{\mathrm{over}}\,(480 \times 360),$$

the actual inference USB2-to-USB3-related speedup (Fig. 8B) was defined as:

$$sm_{\mathrm{inf}}(\mathrm{size}) = t^{\mathrm{usb2}}_{\mathrm{inf}}(\mathrm{size})/t^{\mathrm{usb3}}_{\mathrm{inf}}\,(\mathrm{size}),$$

and the iteration USB2-to-USB3-related speedup (Fig. 8B) was defined as:

$$sm_{\mathrm{over}}(\mathrm{size}) = t^{\mathrm{usb2}}_{\mathrm{over}}(\mathrm{size})/t^{\mathrm{usb3}}_{\mathrm{over}}\,(\mathrm{size}).$$

Then, for Google Coral device, the following additional metrics were determined with regard of the availability of various configuration parameters like OSs, types of connectivity, overclocking regimes. The performance speedup for each configuration (su) due to usage of USB 3.0 in comparison to usage of USB 2.0 (the USB2-to-USB3 related speedup) was measured as ratio of the mean (inference or iteration) time measured for the configuration with USB 2.0 (t^{usb2}) and the correspondent time measured for the configuration with USB 3.0 (t^{usb3}):

$$su = t^{\mathrm{usb2}}/t^{\mathrm{usb3}}.$$

The performance speedup for each configuration (so) due to usage of Ubuntu OS in comparison to usage of Windows OS for all configurations (frame size, USB2/USB3, maximal overclocking/no overclocking) and running stages (inference/iteration) was measured as ratio of the time measured for the configuration under Windows (t_w) and the correspondent time measured for the configuration under Ubuntu (t_u):

$$so = t_w/t_u.$$

The performance speedup for each configuration (sc) due to usage of maximal overclocking regime in comparison to usage of standard regime without overclocking was measured as ratio of the time measured for the configuration in standard regime without overclocking (t_{std}) and the correspondent time measured for the configuration in maximal overclocking regime (t_{max}):

$$sc = t_{std}/t_{max}.$$

The efficiency (e) of devices was measured as ratio of the mean inference time (t_{inf}) and mean iteration time (inference time with all overheads) (t_{over}):

$$e = t_{inf}/t_{over}.$$

The set of benchmark tests was carried out to characterize the performance of the ECAs and the results are presented in the next section. Because of the proprietary nature of the ECAs and the limited availability of APIs in the correspondent SDKs, some important accelerator-related metrics (for example, TPU/VPU actual utilization and others) cannot be obtained at the moment.

4. Results

The full set of the tested configurations with the correspondent references on the results obtained is given in Table 2. Along with tests on the ECAs (Horned Sungem and Google Coral) the similar runs were carried out for the full versions of the same model on the available CPU and GPU as theoretically slowest and fastest available infrastructures, respectively, for the final comparative analysis of the optimization effect.

4.1 Frame sequences

4.1.1 CPU

For the laptop equipped by CPU (Intel Core i5-3337U) the presented sequences of raw (t^{raw}) inference and iteration times demonstrate the very volatile behavior that can be explained by the very different content of the frames (Fig. 4). Moreover, their moving window subsets have the very

Table 2 Configurations of CPU/GPU/TPU/VPU devices used for inference/iteration runs with the correspondent references on the results obtained.

Device	OS	I/O connectivity	Overclocking	Results (figure)
CPU Intel Core i5-3337U	Windows 10	FCBGA1023	–	Figs. 4 and 13
GPU NVIDIA 1080 Ti	Ubuntu 18	PCIe 2.0 x16	–	Figs. 5 and 13
Horned Sungem	Ubuntu 18	USB 2.0	–	Figs. 6, 8, and 12A
Horned Sungem	Ubuntu 18	USB 3.0	–	Figs. 8, 12A and 13
Google Coral	Ubuntu 18	USB 2.0	Std	Figs. 7, 9–11, 12B and 14
Google Coral	Ubuntu 18	USB 2.0	Max	Figs. 9–11 and 14
Google Coral	Ubuntu 18	USB 3.0	Std	Figs. 9–11, 12B and 14
Google Coral	Ubuntu 18	USB 3.0	Max	Figs. 9–11, 13 and 14
Google Coral	Windows 10	USB 2.0	Std	Figs. 9–11 and 14
Google Coral	Windows 10	USB 2.0	Max	Figs. 9–11 and 14
Google Coral	Windows 10	USB 3.0	Std	Figs. 9–11 and 14
Google Coral	Windows 10	USB 3.0	Max	Figs. 9–11 and 14

volatile mean (t^{mean}), standard deviation (t^{std}), skewness (t^{skew}), kurtosis (t^{kurt}) values also. This volatility do not allow correctly compare performance metrics for different datasets (for example, different videos). That is why the following comparison analysis (see below) was performed for the same video with the same frame-to-frame performance characteristics. The presented sequences for inference times (t_{inf}) (Fig. 4A and C) and iteration times (t_{over}) (Fig. 4B and D) are visually very similar because of the very small time expenses on overhead operations for CPU-based inference. It should be noted that the moving window subsets have the distributions that are very close to the normal distribution (by close to zero values of skewness and kurtosis in Fig. 4C and D) in the middle view zone only where the number of the detected humans was not so volatile like in the other views.

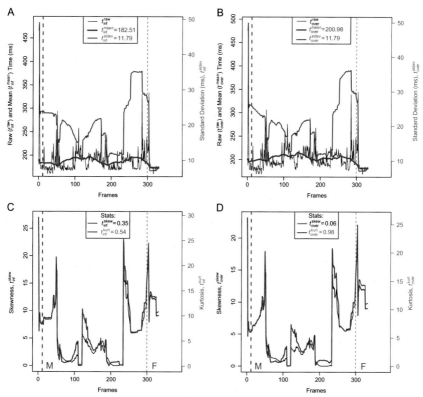

Fig. 4 The raw values (t^{raw}) for all frames, their moving average (t^{mean}), standard deviation (t^{std}), skewness (t^{skew}), and kurtosis (t^{kurt}) for inference (t_{inf}) and iteration (t_{over}) times for the laptop equipped by CPU (Intel Core i5-3337U). The vertical dashed lines denotes starts of the middle (red M) and the far (blue F) view zones. In the legend the numerical values of the correspondent characteristics for frame 200 are given.

4.1.2 GPU

All measured values were compared with the same values after running the same scripts on the GPU device used (NVIDIA 1080) to obtain the general impression as to the progress of the arts with the standalone non–edge computing infrastructures and estimate their performance. From the presented time dependencies (Fig. 5) of raw (t^{raw}) inference and iteration times, one can observe the very volatile behavior that can be explained by the very different content of the frames also. Again, their moving window subsets have the very volatile mean (t^{mean}), standard deviation (t^{std}), skewness (t^{skew}), kurtosis (t^{kurt}) values. In contrast to CPU these sequences for inference times (t_{inf}) (Fig. 5A and C) and iteration times (t_{over}) (Fig. 5B and D) are visually

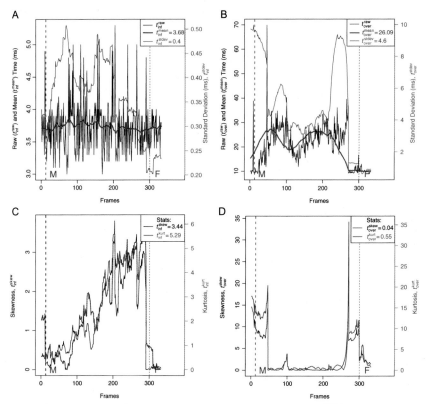

Fig. 5 The raw values (t^{raw}) for all frames, their moving average (t^{mean}), standard deviation (t^{std}), skewness (t^{skew}), and kurtosis (t^{kurt}) for inference (t_{inf}) and iteration (t_{over}) times for the laptop equipped by GPU (NVIDIA 1080). The vertical dashed lines denotes starts of the middle (red M) and the far (blue F) view zones. In the legend the numerical values of the correspondent characteristics for frame 200 are given.

very different due to the extremely high time expenses on overhead operations for GPU-based inference. In fact, the inference times (t_{inf}) are nearly by the degree of order lower than the iteration times (t_{over}). And the moving window subsets have the distributions that are very close to the normal distribution (by close to zero values of skewness and kurtosis) in the middle view zone and for the iteration times (t_{over}) only (Fig. 5D) where the number of the detected humans was not so volatile like in the other views.

4.1.3 Horned Sungem (Intel Movidius)

From the observed time dependencies (Fig. 6) of raw (t^{raw}) inference and iteration times, one can note again the very volatile behavior that can be

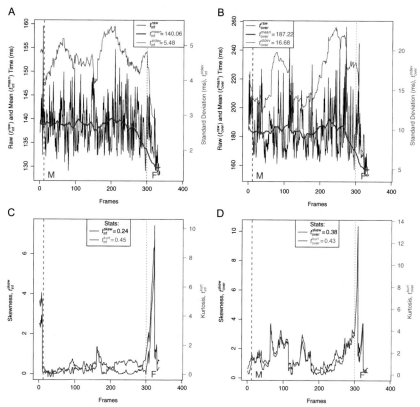

Fig. 6 The raw values (t^{raw}) for all frames, their moving average (t^{mean}), standard deviation (t^{std}), skewness (t^{skew}), and kurtosis (t^{kurt}) for inference (t_{inf}) and iteration (t_{over}) times for the laptop equipped by Horned Sungem (Intel Movidius). The vertical dashed lines denotes starts of the middle (red M) and the far (blue F) view zones. In the legend the numerical values of the correspondent characteristics for frame 200 are given.

explained by the very different content of the frames also. Their moving window subsets have the very volatile mean (t^{mean}), standard deviation (t^{std}), skewness (t^{skew}), kurtosis (t^{kurt}) values. In contrast to GPU these sequences for inference times (t_{inf}) (Fig. 6A and C) and iteration times (t_{over}) (Fig. 6B and D) are visually very similar due to the relatively small time expenses on overhead operations for TPU-based inference. In fact, the inference times (t_{inf}) are only by 20%–25% lower than the iteration times (t_{over}). The moving window subsets have the distributions that are very close to the normal distribution (by close to zero values of skewness and kurtosis) in the middle view zone for the inference and iteration (t_{over}) times (Fig. 6C and D), but not so close in the near and far views. It should be noted that for Horned Sungem the distribution of measured times is very sensitive to the near and far views, i.e., to the unreliable content without clearly recognizable targeted objects where skewness (Fig. 6C) and kurtosis (Fig. 6D) change sharply exactly at the division lines (M and F on Fig. 6C and D) between these zones.

4.1.4 Google Coral

For Google Coral device, the observed time dependencies (Fig. 7) of raw (t^{raw}) inference and iteration times, and all statistical parameters of their moving window subsets have the more irregular behavior even that can be

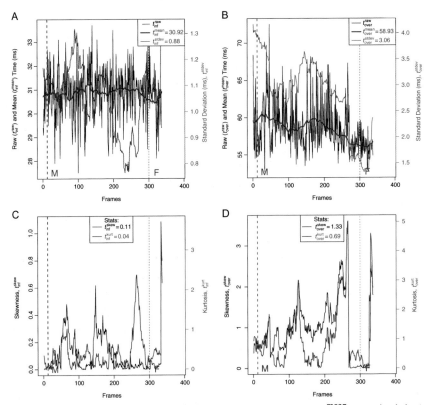

Fig. 7 The raw values (t^{raw}) for all frames, their moving average (t^{mean}), standard deviation (t^{std}), skewness (t^{skew}), and kurtosis (t^{kurt}) for inference (t_{inf}) and iteration (t_{over}) times for the laptop equipped by Google Coral. The vertical dashed lines denotes starts of the middle (red M) and the far (blue F) view zones. In the legend the numerical values of the correspondent characteristics for frame 200 are given.

explained by the very different content of the frames. The moving window subsets have the distributions that are very far from the normal distribution (estimated by high values of skewness and kurtosis, at least) in all view zones (Fig. 7C and D). It means that for Google Coral the distribution of measured times is much more sensitive to any changes of the content with or without targeted objects where skewness (Fig. 7C) and kurtosis (Fig. 7D) change sharply in all view zones.

4.2 Optimization effect

4.2.1 Horned Sungem

From these results the question arises about influence of USB2/USB3 usage on the overall performance of the systems considered here. For this purpose the inference and iteration time dependencies vs the frame size were investigated for Horned Sungem under Linux OS (Fig. 8) and Google Coral under Linux/Windows OSs (Fig. 9). For this purpose the mean values of the inference time (t_{inf}) and iteration time (t_{over}) were calculated for the moving window subset with 50 measured raw values at frame 200 and speedup comparison was calculated for USB2/USB3 on the frame-to-frame basis.

Despite the natural general trend of increase of mean inference time (t_{inf}) and iteration time (t_{over}) with the frame size for both USB interfaces

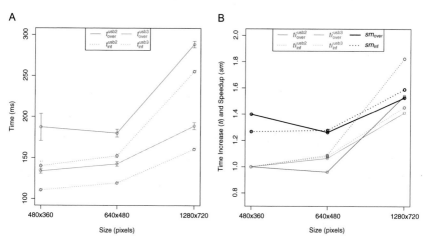

Fig. 8 The mean inference time (t_{inf}) and iteration time (t_{over}) dependencies versus the frame size under Linux OS (A); the correspondent inference time increase (ti_{inf}) and iteration time increase (ti_{over}), and the instant (after moving average) inference (sm_{inf}) and iteration (sm_{over}) time speedup with the frame size (B).

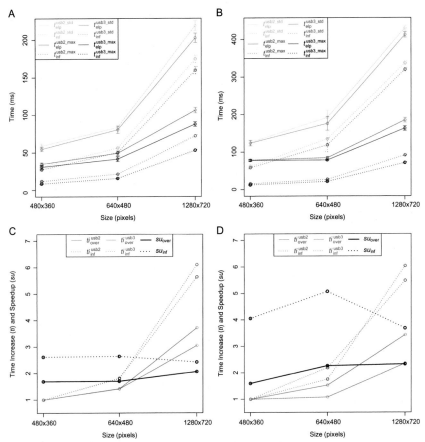

Fig. 9 The mean inference time (t_{inf}) and iteration time (t_{over}) dependencies vs the frame size under Linux (A) and Windows (B) OS; the correspondent inference time increase (ti_{inf}) and iteration time increase (ti_{over}); and the instant (after moving average) inference (su_{inf}) and iteration (su_{over}) time speedup with the frame size under Linux (C) and Windows (D) OS.

(Fig. 8A), the actual inference (sm_{inf}) and iteration (sm_{over}) time speedups are nearly the same (1.4 ± 0.1) for all frame sizes in the limits of standard deviations.

4.2.2 Google Coral

Due to the wider availability of tools for adapting the TF-model for various hardware/software configurations for Google Coral device it was possible to perform the additional inference runs.

The mean inference time (t_{inf}) and iteration time (t_{over}) dependencies vs the frame size were investigated for Google Coral for various USB interfaces under Linux/Windows OSs (Fig. 9). They demonstrate the same tendency as for Horned Sungem under Linux. It should be noted the significant (nearly two times) higher performance of Google Coral under Linux in comparison to Windows. The instant (after moving average) inference (su_{inf}) and iteration (su_{over}) time speedups with the frame size under Ubuntu (Fig. 9C) and Windows (Fig. 9D) OS, are little bit higher than for Horned Sungem. The USB2-to-USB3 related speedup is a little bit higher for Windows than for Ubuntu also, but hardly reaches the value of 2 ± 0.1.

The performance speedup values (su) due to usage of USB 3.0 in comparison to usage of USB 2.0 for all configurations (frame size, maximal overclocking/no overclocking), running stages (inference/iteration) and OS (Windows/Ubuntu) demonstrate increase in the range from 1.25 to 2.5 (Fig. 9C and D).

4.2.3 Time vs frame size dependence for various USB interfaces and OSs

The comparative performance analysis of the Coral device was performed for various types of connectivity (like USB 2.0 and USB 3.0) and types of overclocking (like standard and maximal) (Fig. 10). It was shown that efficiencies for all hardware combinations increases with an increase of the frame size despite the increase of the inference and iteration times. For all hardware configurations the efficiencies are higher for Ubuntu OS than for Windows OS.

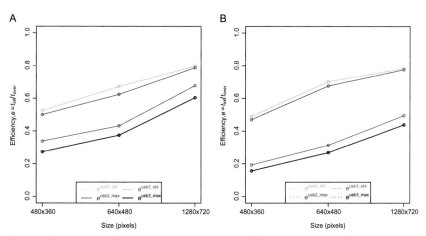

Fig. 10 Efficiency (*e*) of Google Coral device under different operating systems: Ubuntu (A) and Windows (B).

It means that under other equal conditions usage of Ubuntu OS is more efficient and preferable in comparison with Windows OS. The configurations with the more powerful types of connectivity (USB 3.0) and type of overclocking (maximal) demonstrate the lower efficiency than their less powerful counterparts (USB 2.0 and standard, respectively). It can be explained by nearly the same values of overhead times ($t_{oi} = t_{over} - t_{inf}$) for all configurations, and the less different inference times for the more powerful configurations. It means that the further improvement of efficiency can be obtained not only by progress of inference algorithms, but also by decrease of overheads.

The performance speedup values (so) due to usage of Ubuntu OS in comparison to usage of Windows OS for all configurations (frame size, USB2/USB3, maximal overclocking/no overclocking) and running stages (inference/iteration) demonstrate increase in the range from 1.25 to 2.5 (Fig. 11A).

The performance speedup values (sc) due to usage of maximal overclocking regime in comparison to usage of standard regime without overclocking for all configurations (frame size, Windows/Ubuntu, USB2/ USB3) and running stages (inference/iteration) demonstrate increase in the range from 1.01 to 1.4 (Fig. 11B).

Previous results were obtained for the middle view zones where the targeted objects (humans) were clearly presented and their poses can be reliably identified. But similar performance tendencies was observed for other

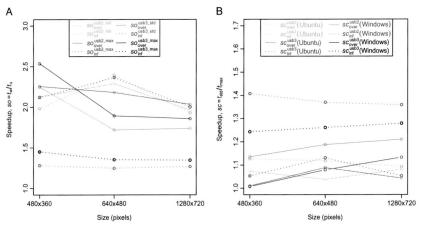

Fig. 11 The performance speedup due to usage of Ubuntu OS in comparison to usage of Windows OS (A) and due to usage of maximal overclocking regime in comparison to usage of standard regime without overclocking (B).

(near and far) views without the targeted objects or with objects of low quality for reliable human pose identification. For this purpose frame–to–frame performance analysis was carried out also when inference and iteration times for each frame from two different were compared. For example, the frame–to–frame inference (s^f_{inf}) and iteration (s^f_{over}) USB2-to-USB3-related speedup with time were calculated for Horned Sungem (Fig. 12A) and Google Coral (Fig. 12B) with the results presented in Table 3. The results on Horned Sungem (Fig. 12A) are in good correlation with the results presented earlier for the same USB2-to-USB3-related speedup (Fig. 8B) which were calculated by means of moving window in the middle view region of the video.

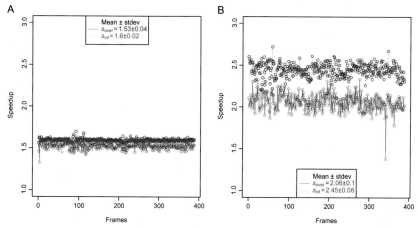

Fig. 12 The frame-to-frame inference (s^f_{inf}) and iteration (s^f_{over}) USB2-to-USB3-related speedups with time for Horned Sungem (A) and Google Coral (B) for frame size 1280 × 720.

Table 3 Actual frame-to-frame USB2-to-USB3-related speedup for Horned Sungem and Google Coral devices.

Frame size	360×480	640×480	1280×720
Horned Sungem (Iteration/Inference)	1.35 ± 0.15/ 1.24 ± 0.05	1.26 ± 0.04/ 1.28 ± 0.02	1.53 ± 0.04/ 1.60 ± 0.02
Google Coral (Iteration/Inference)	1.67 ± 0.13/ 2.61 ± 0.13	1.78 ± 0.14/ 2.63 ± 0.11	2.06 ± 0.10/ 2.45 ± 0.08

5. Discussion

In general, the evident improvement of the performance characteristics (inference and iteration times) due to application of the better connectivity (USB 3.0 instead USB 2.0). overclocking, and more OS was observed and measured for the concrete practical use case. The results of the comparative analysis for inference and iteration stages are shown below for inference and iteration times in linear (Fig. 13A) and logarithmic scale (Fig. 13B). The linear representation gives the visual comprehension of the available server-side fastest (GPU) and slowest (CPU) devices in comparison to edge computing devices, which are very close to the fastest server-side devices (at least, in this use case and for Google Coral device).

The logarithmic representation (Fig. 13B) also gives the visual impression about the high efficiency of some server-based devices (CPU) and some edge computing devices (Horned Sungem), but also demonstrates the low efficiency of the some fastest server-side (GPU) and edge computing (Google Coral) devices.

As to the possible optimization measures in the context of the considered configurations (Table 4), the following speedup values for various configurations and running stages for Google Coral device were summarized in Table 4 and Fig. 14B. For example, in the current use case and under condition of the availability of edge computing device like Google Coral

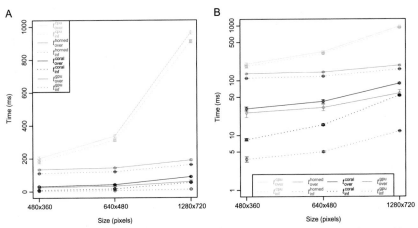

Fig. 13 The comparative analysis of inference and iteration times in linear (A) and logarithmic (B) scale.

Table 4 Speedup values for various configurations, running stages, and frame sizes for Google Coral device (in comparison to the baseline configuration with Windows, USB 2.0, and absence of overclocking).

Configurations				Frame size		
OS	I/O	Overclocking	Running stage	Small (480×360)	Middle (640×480)	Large (1280×720)
Windows	USB 2.0	Std	Over	1	1	1
Windows	USB 2.0	Std	Inf	1	1	1
Ubuntu	USB 2.0	Std	Over	2.13	2.29	1.96
Ubuntu	USB 2.0	Std	Inf	1.98	2.39	1.93
Ubuntu	USB 3.0	Std	Over	3.59	3.89	4.04
Ubuntu	USB 3.0	Std	Inf	5.19	6.32	4.67
Ubuntu	USB 3.0	Max	Over	4.08	4.62	4.90
Ubuntu	USB 3.0	Max	Inf	7.31	8.66	6.35

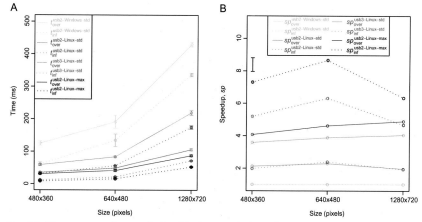

Fig. 14 The comparative analysis of inference (t_{inf}) and iteration (t_{over}) times (A) and speedup (sp) values (B) after various optimization procedures for Google Coral device.

connected by USB 2.0 interface to personal computer under Windows OS the simplest optimization procedure will consists in the transition from Windows to Ubuntu OS (iteration speedup in the range of 1.96–2.13 times), then transition from USB 2.0 to USB 3.0 (cumulative iteration speedup in the range of 3.59–4.04), and then usage of overclocking regime (cumulative iteration speedup in the range of 4.08–4.90). The additional finding is related with the non-monotonous dependence of the inference time speedup with the frame size, for example, the largest values were observed for the middle size frame. It should be noted the maximal cumulative inference speedup can reach the higher values even (in the range of 6.35–8.66) that opens opportunities to get the possible additional boosts related with the better optimization of overhead operations in addition to optimization of software (DL models) and number crunching hardware (TPU, VPU, TC) themselves.

The similar results were reported for the inference time speedup for Google Coral ECA, where USB 3.0 accelerates the execution time (actually the averaged inference time without taking into account the first warm-up iteration) by three to seven times compared to USB 2.0 [17]. And the similar tendencies were reported for other ECAs like Jetson Nano [17].

Recently, in the context of implementing AI and EC combination, the "last mile" problem was formulated on the basis of the other criteria like performance, cost, and privacy [98]. For example, applications of AI techniques to medicine, particularly those medical disciplines that extensively rely on digital imaging, some other difficulties related with "the hiatus of human trust and the hiatus of machine experience" under the title of "last mile of implementation" are reported and discussed [99]. The other recent systematic review of DL applications to medicine also highlighted the need to focus on the last mile of implementation from the end user point of view with difficulties arising by other reasons when the direct clinical impact, deployment and automation of deep learning models are not sufficiently considered [100]. This work emphasizes the other highly nontrivial aspect of the "last mile" problem when the very efficient software (DL models) and hardware (ECAs) components of EC ecosystem can be very ineffective due to their non-optimal integration and configuration within EC ecosystem. At the same time, the proposed quantitative analysis of several possible configurations allows end users (without reach expertise in optimization of DL models and ECAs even) to estimate precisely the cumulative effect on performance after application of all these optimization measures.

6. Conclusions

The speedup of ECA performance for the various environment configurations were investigated and analyzed by several metrics based on inference and iteration time. In addition to quantitative estimations, this work offers useful insights for the improvement of the available configurations around the ECAs considered for the current DL application for edge computing. The essence of these improvements that they can be implemented by the end users even without the specific expertise in DL model and ECA optimizations.

The biggest speedup of deep learning models on edge computing accelerators (at least for two considered ECA families proposed by Google and Intel) can be obtained by transition from Windows to Linux (for example, Ubuntu 18) OS. Sometimes this effect is considered as evident by many IT experts, but the concrete gain is not known often for the specific ECA types, DL models and applications. This work gives such quantitative estimations with the correspondent mean and standard deviation values. For example, for Google Coral device transition from Windows 10 to Ubuntu 18 OS allows to obtain the iteration time speedup up to 2.13 times, transition from USB 2.0 to USB 3.0—up to 2 times, transition to overclocking regime—1.2 times with the cumulative iteration speedup up to 4.9 times. In the view of the availability of numerous versions of various DL models that proposed in the open source format and used freely by the end users the important question is the balance between the promised and the actual values of accuracy and speedup. That is why the future work should include the investigation of possible "last mile" optimization for various DL model versions. But the results obtained here can hopefully help to understand, benchmark, and implement the optimal configuration of EC ecosystem (combination of DL models, ECAs, and their integrating environment) with the expected speedup without additional investments.

Acknowledgments

The work was partially supported by the National Research Foundation of Ukraine in the framework of the grant 2020.01/0490 "Artificial Intelligence Platform for Distant Computer-Aided Detection (CADe) and Computer-Aided Diagnosis (CADx) of Human Diseases.

References

[1] Y. Bengio, Deep learning of representations: looking forward, in: International Conference on Statistical Language and Speech Processing, Springer, Berlin, Heidelberg, 2013, pp. 1–37.
[2] Y. LeCun, Y. Bengio, G. Hinton, Deep learning, Nature 521 (7553) (2015) 436–444.
[3] J. Schmidhuber, Deep learning in neural networks: an overview, Neural Netw. 61 (2015) 85–117.
[4] R. Buyya, S.N. Srirama (Eds.), Fog and Edge Computing: Principles and Paradigms, John Wiley & Sons, 2019.

[5] Y. Wang, Cloud-dew architecture, Int. J. Cloud Comput. 4 (3) (2015) 199–210.

[6] K. Skala, D. Davidovic, E. Afgan, I. Sovic, Z. Sojat, Scalable distributed computing hierarchy: cloud, fog and dew computing, Open J. Cloud Comput. 2 (1) (2015) 16–24.

[7] Y. Gordienko, S. Stirenko, O. Alienin, K. Skala, Z. Sojat, A. Rojbi, A.L. Coto, Augmented coaching ecosystem for non-obtrusive adaptive personalized elderly care on the basis of Cloud-Fog-Dew computing paradigm, in: 2017 40th International Convention on Information and Communication Technology, Electronics and Microelectronics (MIPRO), IEEE, 2017, pp. 359–364.

[8] F. Al-Turjman, Al-Turjman, Edge Computing, Springer International Publishing, 2019.

[9] K. Cao, Y. Liu, G. Meng, Q. Sun, An overview on edge computing research, IEEE Access 8 (2020) 85714–85728.

[10] G. Lacey, G.W. Taylor, S. Areibi, Deep Learning on FPGAs: Past, Present, and Future, arXiv preprint arXiv: 1602.04283, 2016.

[11] E. Nurvitadhi, et al., Can FPGAs beat GPUs in accelerating next-generation deep neural networks, in: Proceedings of the ACM/SIGDA International Symposium on Field-Programmable Gate Arrays (FPGA'17), 2017, pp. 5–14.

[12] T. Chen, Z. Du, N. Sun, J. Wang, C. Wu, Y. Chen, O. Temam, DianNao: a small-footprint high-throughput accelerator for ubiquitous machine learning, in: Proceedings of the 19th International Conference on ASPLOS, 2014, pp. 269–284.

[13] F. Akopyan, et al., TrueNorth: design and tool flow of a 65 mW 1 million neuron programmable neurosynaptic chip, IEEE Trans. Comput. Aided Des. Integr. Circuits Syst. 34 (10) (2015) 1537–1557.

[14] P. Ienne, Architectures for Neuro-Computers: Review and Performance Evaluation, Technical Report, EPFL, Lausanne, Switzerland, 1993.

[15] NVIDIA Corporation, Programming Tensor Cores in CUDA 9, 2017, (Accessed 2019) Available at https://devblogs.nvidia.com/programming-tensor-cores-cuda-9.

[16] N.P. Jouppi, et al., In-datacenter performance analysis of a tensor processing unit, Int. Symp. Comput. Archit. 45 (2) (2017) 1–12.

[17] M. Antonini, T.H. Vu, C. Min, A. Montanari, A. Mathur, F. Kawsar, Resource characterisation of personal-scale sensing models on edge accelerators, in: Proceedings of the First International Workshop on Challenges in Artificial Intelligence and Machine Learning for Internet of Things, 2019, pp. 49–55.

[18] NVIDIA AI IoT Repository. https://github.com/NVIDIA-AI-IOT.

[19] Open Source Repository for Coral.ai. https://github.com/google-coral.

[20] OpenVINO™ (Open Visual Inference and Neural Network optimization) Toolkit. https://github.com/intel/ros_openvino_toolkit.

[21] Horned Sungem for AI. https://github.com/HornedSungem.

[22] 96Boards.ai Initiative. https://github.com/96boards/documentation.

[23] J. Chen, X. Ran, Deep learning with edge computing: a review, Proc. IEEE 107 (8) (2019) 1655–1674.

[24] B. Varghese, N. Wang, D. Bermbach, C.H. Hong, E. de Lara, W. Shi, C. Stewart, A Survey on Edge Benchmarking, 2020. arXiv preprint arXiv:2004.11725.

[25] Y.W. Chen, L.C. Jain, Deep Learning in Healthcare, Springer, 2020.

[26] A. Esteva, A. Robicquet, B. Ramsundar, V. Kuleshov, M. DePristo, K. Chou, J. Dean, A guide to deep learning in healthcare, Nat. Med. 25 (1) (2019) 24–29.

[27] J. Irvin, P. Rajpurkar, M. Ko, Y. Yu, S. Ciurea-Ilcus, C. Chute, J. Seekins, Chexpert: A large chest radiograph dataset with uncertainty labels and expert comparison, in: Proceedings of the AAAI Conference on Artificial Intelligence, vol. 33, 2019, pp. 590–597.

[28] W. Cao, J. Su, Z. Peng, W. Xu, Q. Liu, Electronic medical record of university hospital based on deep learning, in: IOP Conference Series: Materials Science and Engineering, vol. 569, IOP Publishing, 2019, July, p. 052110. No. 5.

[29] H. Rathore, A. Mohamed, M. Guizani, Blockchain applications for healthcare, in: Energy Efficiency of Medical Devices and Healthcare Applications, Academic Press, 2020, pp. 153–166.

[30] Y. Gordienko, O. Alienin, O. Rokovyi, V. Valko, S. Stirenko, G. Fedak, O. Lodygensky, Blockchain and Smart Contracts for Provenance of Deep Learning Content in Healthcare, Presented at 6th International Conference High Performance Computing (HPC-UA 2020) November 06 - 07, 2020 Kyiv / Ukraine (http://hpc.ugrid.org/#/program), HPC-2020, 2020. In press.

[31] M. Negassi, R. Suarez-Ibarrola, S. Hein, A. Miernik, A. Reiterer, Application of artificial neural networks for automated analysis of cystoscopic images: a review of the current status and future prospects, World J. Urol. 38 (2020) 2349–2358, https://doi.org/10.1007/s00345-019-03059-0.

[32] Y. Gordienko, P. Gang, J. Hui, W. Zeng, Y. Kochura, O. Alienin, S. Stirenko, Deep learning with lung segmentation and bone shadow exclusion techniques for chest x-ray analysis of lung cancer, in: International Conference on Computer Science, Engineering and Education Applications, Springer, Cham, 2018, pp. 638–647.

[33] Q. Hu, L.F.D.F. Souza, G.B. Holanda, S.S. Alves, F.H.D.S. Silva, T. Han, P.P. Reboucas Filho, An effective approach for CT lung segmentation using mask region-based convolutional neural networks, Artif. Intell. Med. 103 (2020) 101792.

[34] R.M. Sundhari, Enhanced histogram equalization based nodule enhancement and neural network based detection for chest x-ray radiographs, J. Ambient. Intell. Humaniz. Comput. (2020) 1–9, https://doi.org/10.1007/s12652-020-01701-z.

[35] S. Stirenko, Y. Kochura, O. Alienin, O. Rokovyi, Y. Gordienko, P. Gang, W. Zeng, Chest X-ray analysis of tuberculosis by deep learning with segmentation and augmentation, in: 2018 IEEE 38th International Conference on Electronics and Nanotechnology (ELNANO), IEEE, 2018, pp. 422–428.

[36] E. Tasci, Pre-processing effects of the tuberculosis chest X-ray images on pre-trained CNNs: an investigation, in: The International Conference on Artificial Intelligence and Applied Mathematics in Engineering, Springer, Cham, 2019, April, pp. 589–596.

[37] D. Verma, C. Bose, N. Tufchi, K. Pant, V. Tripathi, A. Thapliyal, An efficient framework for identification of tuberculosis and pneumonia in chest X-ray images using neural network, Prog. Comput. Sci. 171 (2020) 217–224.

[38] P. Rajpurkar, J. Irvin, K. Zhu, B. Yang, H. Mehta, T. Duan, M.P. Lungren, Chexnet: Radiologist-Level Pneumonia Detection on Chest X-Rays With Deep Learning, 2017. arXiv preprint arXiv:1711.05225.

[39] P. Gang, W. Zeng, Y. Gordienko, Y. Kochura, O. Alienin, O. Rokovyi, S. Stirenko, Effect of data augmentation and lung mask segmentation for automated chest radiograph interpretation of some lung diseases, in: International Conference on Neural Information Processing, Springer, Cham, 2019, pp. 333–340.

[40] P. Gang, W. Zhen, W. Zeng, Y. Gordienko, Y. Kochura, O. Alienin, S. Stirenko, Dimensionality reduction in deep learning for chest X-ray analysis of lung cancer, in: 2018 Tenth International Conference on Advanced Computational Intelligence (ICACI), IEEE, 2018, pp. 878–883.

[41] S.L. Halson, Monitoring training load to understand fatigue in athletes, Sports Med. 44 (2) (2014) 139–147.

[42] N. Singh, K.J. Moneghetti, J.W. Christle, D. Hadley, V. Froelicher, D. Plews, Heart rate variability: an old metric with new meaning in the era of using mhealth technologies for health and exercise training guidance. Part two: prognosis and training, Arrhythm. Electrophysiol. Rev. 7 (4) (2018) 247.

[43] S. Stirenko, P. Gang, W. Zeng, Y. Gordienko, O. Alienin, O. Rokovyi, A. Rojbi, Parallel statistical and machine learning methods for estimation of physical load, in: International Conference on Algorithms and Architectures for Parallel Processing, Springer, Cham, 2018, pp. 483–497.

[44] Y. Gordienko, S. Stirenko, Y. Kochura, O. Alienin, M. Novotarskiy, N. Gordienko, Deep Learning for Fatigue Estimation on the Basis of Multimodal Human-Machine Interactions, 2017. arXiv preprint arXiv:1801.06048.

[45] V. Vesterinen, K. Häkkinen, E. Hynynen, J. Mikkola, L. Hokka, A. Nummela, Heart rate variability in prediction of individual adaptation to endurance training in recreational endurance runners, Scand. J. Med. Sci. Sports 23 (2) (2013) 171–180.

[46] P. Gang, J. Hui, S. Stirenko, Y. Gordienko, T. Shemsedinov, O. Alienin, E.A. González, User-driven intelligent interface on the basis of multimodal augmented reality and brain-computer interaction for people with functional disabilities, in: Future of Information and Communication Conference, Springer, Cham, 2018, pp. 612–631.

[47] N. Gordienko, O. Lodygensky, G. Fedak, Y. Gordienko, Synergy of volunteer measurements and volunteer computing for effective data collecting, processing, simulating and analyzing on a worldwide scale, in: 2015 38th International Convention on Information and Communication Technology, Electronics and Microelectronics (MIPRO), IEEE, 2015, pp. 193–198.

[48] X. Zhou, R. Canady, S. Bao, A. Gokhale, Cost-effective hardware accelerator recommendation for edge computing, in: 3rd {USENIX} Workshop on Hot Topics in Edge Computing (HotEdge 20), 2020.

[49] G. Dinelli, G. Meoni, E. Rapuano, G. Benelli, L. Fanucci, An FPGA-based hardware accelerator for CNNS using on-chip memories only: design and benchmarking with intel movidius neural compute stick, Int. J. Reconfigurable Comput. 2019 (2019), https://doi.org/10.1155/2019/7218758, 7218758.

[50] A.M. Nguyen Hoang, Computer Vision Deployment on Edge Devices, 2019. https://www.theseus.fi/handle/10024/261705.

[51] M. Rubin, Evaluation of Machine Learning Inference Workloads on Heterogeneous Edge Devices, 2020, Thesis https://mediatum.ub.tum.de/doc/1540918/file.pdf.

[52] C. Wisultschew, A. Otero, J. Portilla, E. de la Torre, Artificial vision on edge IoT devices: a practical case for 3D data classification, in: 2019 XXXIV Conference on Design of Circuits and Integrated Systems (DCIS), IEEE, 2019, pp. 1–7.

[53] M. Norrgård, Using Computer Vision in Retail Analytics, 2020, Thesis https://www.doria.fi/bitstream/handle/10024/177198/norrgard_marcus.pdf.

[54] M. Almeida, S. Laskaridis, I. Leontiadis, S.I. Venieris, N.D. Lane, EmBench: quantifying performance variations of deep neural networks across modern commodity devices, in: The 3rd International Workshop on Deep Learning for Mobile Systems and Applications, ACM, 2019, pp. 1–6.

[55] L.F. Isikdogan, B.V. Nayak, C.T. Wu, J.P. Moreira, S. Rao, G. Michael, SemifreddoNets: Partially Frozen Neural Networks for Efficient Computer Vision Systems, 2020. arXiv preprint arXiv:2006.06888.

[56] M. Charkhabi, N. Rahurkar, Efficient training and inference in highly temporal activity recognition, in: 2019 International Conference on Image and Video Processing, and Artificial Intelligence, vol. 11321, International Society for Optics and Photonics, 2019, p. 113211N.

[57] V. Mittal, A. Tyagi, B. Bhushan, Smart surveillance systems with edge intelligence: convergence of deep learning and edge computing, Proceedings of the International Conference on Innovative Computing & Communications (ICICC) 2020 (2020), https://doi.org/10.2139/ssrn.3599865, 3599865.

[58] V. Mittal, B. Bhushan, Accelerated computer vision inference with AI on the edge, in: 2020 IEEE 9th International Conference on Communication Systems and Network Technologies (CSNT), IEEE, 2020, April, pp. 55–60.

[59] Y. Zhang, K. Xu, Network Management in Cloud and Edge Computing, Springer, Singapore, 2020.

[60] A.S. Almogren, Intrusion detection in edge-of-things computing, J. Parallel Distrib. Comput. 137 (2020) 259–265.

[61] J. Wang, L. Zhao, J. Liu, N. Kato, Smart resource allocation for mobile edge computing: a deep reinforcement learning approach, IEEE Trans. Emerg. Top. Comput. (2019) 1, https://doi.org/10.1109/TETC.2019.2902661.

[62] X. Wang, Y. Han, V.C. Leung, D. Niyato, X. Yan, X. Chen, Convergence of edge computing and deep learning: a comprehensive survey, IEEE Commun. Surv. Tutor. 22 (2) (2020) 869–904.

[63] T. Young, D. Hazarika, S. Poria, E. Cambria, Recent trends in deep learning based natural language processing, IEEE Comput. Intell. Mag. 13 (3) (2018) 55–75.

[64] M.B. Hoy, Alexa, Siri, Cortana, and more: an introduction to voice assistants, Med. Ref. Serv. Q. 37 (1) (2018) 81–88.

[65] J. Chakareski, S. Gupta, Multi-connectivity and edge computing for ultra-low-latency lifelike virtual reality, in: 2020 IEEE International Conference on Multimedia and Expo (ICME), IEEE, 2020, pp. 1–6.

[66] P. Ren, X. Qiao, Y. Huang, L. Liu, S. Dustdar, J. Chen, Edge-assisted distributed DNN collaborative computing approach for mobile web augmented reality in 5G networks, IEEE Netw. 34 (2) (2020) 254–261.

[67] M. Elawady, A. Sarhan, Mixed reality applications powered by IoE and edge computing: a survey, in: Ghalwash Atef Zaki (Ed.), Internet of Things—Applications and Future, Springer, Singapore, 2020, pp. 125–138.

[68] Z. Zhao, K.M. Barijough, A. Gerstlauer, DeepThings: distributed adaptive deep learning inference on resource-constrained IoT edge clusters, IEEE Trans. Comput. Aided Des. Integr. Circuits Syst. 37 (11) (2018) 2348–2359.

[69] M. Mohammadi, A. Al-Fuqaha, S. Sorour, M. Guizani, Deep learning for IoT big data and streaming analytics: a survey, IEEE Commun. Surv. Tutor. 20 (4) (2018) 2923–2960.

[70] P. Battistoni, D. Gregorio, M. Sebillo, M. Vitiello, AI at the edge for sign language learning support, in: 2019 IEEE International Conference on Humanized Computing and Communication, 2019, pp. 16–23.

[71] H. Liu, L. Wang, Collision-free human-robot collaboration based on context awareness, Robot. Comput. Integr. Manuf. 67 (2021) 101997.

[72] D. Dantas, C. Braun, K. Forte, F. Brito, A. Silva, S. Lins, A. Klautau, Testbed for Connected Artificial Intelligence Using Unmanned Aerial Vehicles and Convolutional Pose Machines, 2020. arXiv preprint arXiv:2001.04944.

[73] P. Agarwal, M. Alam, A lightweight deep learning model for human activity recognition on edge devices, Prog. Comput. Sci. 167 (2020) 2364–2373.

[74] A. De Vita, D. Pau, C. Parrella, L. Di Benedetto, A. Rubino, G.D. Licciardo, Low-power HW accelerator for AI edge-computing in human activity recognition systems, in: 2020 2nd IEEE International Conference on Artificial Intelligence Circuits and Systems (AICAS), IEEE, 2020, pp. 291–295.

[75] S. Salkic, B.C. Ustundag, T. Uzunovic, E. Golubovic, Edge computing framework for wearable sensor-based human activity recognition, in: International Symposium on Innovative and Interdisciplinary Applications of Advanced Technologies, Springer, Cham, 2019, June, pp. 376–387.

[76] E. Golubovic, Edge computing framework for wearable sensor-based human activity recognition, in: Advanced Technologies, Systems, and Applications IV-Proceedings of the International Symposium on Innovative and Interdisciplinary Applications of Advanced Technologies (IAT 2019), vol. 83, Springer, 2019, July, p. 376.

[77] C. Jiang, T. Fan, H. Gao, W. Shi, L. Liu, C. Cérin, J. Wan, Energy aware edge computing: a survey, Comput. Commun. 151 (2020) 556–580.

[78] E. Li, L. Zeng, Z. Zhou, X. Chen, Edge AI: on-demand accelerating deep neural network inference via edge computing, IEEE Trans. Wirel. Commun. 19 (1) (2019) 447–457.

[79] A. Yousefpour, C. Fung, T. Nguyen, K. Kadiyala, F. Jalali, A. Niakanlahiji, J.P. Jue, All one needs to know about fog computing and related edge computing paradigms: a complete survey, J. Syst. Archit. 98 (2019) 289–330.

[80] C. Puliafito, E. Mingozzi, F. Longo, A. Puliafito, O. Rana, Fog computing for the internet of things: a survey, ACM Trans. Internet Technol. 19 (2) (2019) 1–41.

[81] W. Zhang, S. Li, L. Liu, Z. Jia, Y. Zhang, D. Raychaudhuri, Hetero-edge: orchestration of real-time vision applications on heterogeneous edge clouds, in: IEEE INFOCOM 2019-IEEE Conference on Computer Communications, IEEE, 2019, pp. 1270–1278.

[82] C. Puliafito, E. Mingozzi, C. Vallati, F. Longo, G. Merlino, Virtualization and migration at the network edge: an overview, in: 2018 IEEE International Conference on Smart Computing (SMARTCOMP), IEEE, 2018, pp. 368–374.

[83] Y. Gordienko, Y. Kochura, V. Taran, N. Gordienko, A. Rokovyi, O. Alienin, S. Stirenko, Scaling analysis of specialized tensor processing architectures for deep learning models, in: Deep Learning: Concepts and Architectures, Springer, Cham, 2020, pp. 65–99, https://doi.org/10.1007/978-3-030-31756-0_3.

[84] R. Blog, E. Haußmann, Comparing Google's TPUv2 Against Nvidia's V100 on ResNet-50, 2019, Accessed 2019 https://www.hpcwire.com/2018/04/30/riseml-benchmarks-google-tpuv2-against-nvidia-v100-gpu.

[85] Y. Kochura, Y. Gordienko, V. Taran, N. Gordienko, A. Rokovyi, O. Alienin, S. Stirenko, Batch size influence on performance of graphic and tensor processing units during training and inference phases, in: Z. Hu, et al. (Eds.), Proceedings ICCSEEA 2019, AISC 938, 2019, pp. 1–11.

[86] PoseNet TensorFlow Lite model. https://www.tensorflow.org/lite/models/pose_estimation/overview.

[87] Q. Dang, J. Yin, B. Wang, W. Zheng, Deep learning based 2d human pose estimation: a survey, Tsinghua Sci. Technol. 24 (6) (2019) 663–676.

[88] S.N. Boualia, N.E.B. Amara, Pose-based human activity recognition: a review, in: 2019 15th International Wireless Communications & Mobile Computing Conference (IWCMC), IEEE, 2019, pp. 1468–1475.

[89] Y. Chen, Y. Tian, M. He, Monocular human pose estimation: a survey of deep learning-based methods, Comput. Vis. Image Underst. 192 (2020) 102897.

[90] B. Jacob, S. Kligys, B. Chen, M. Zhu, M. Tang, A.G. Howard, H. Adam, D. Kalenichenko, Quantization and Training of Neural Networks for Efficient Integer-Arithmetic-Only Inference, 2017, CoRR abs/1712.05877 (2017). arXiv:1712.05877 http://arxiv.org/abs/1712.05877.

[91] S. Han, H. Mao, W.J. Dally, Deep Compression: Compressing Deep Neural Networks With Pruning, Trained Quantization and Huffman Coding, arXiv preprint arXiv:1510.00149, 2015.

[92] S. Han, J. Pool, J. Tran, W. Dally, Learning both weights and connections for efficient neural network, in: C. Cortes (Ed.), Advances in Neural Information Processing Systems, 28, MIT Press, 2015, pp. 1135–1143.

[93] A. Mallya, S. Lazebnik, Packnet: adding multiple tasks to a single network by iterative pruning, in: Proceedings of the IEEE Conference on Computer Vision and Pattern Recognition, 2018, pp. 7765–7773.

[94] E. Wang, et al., Deep neural network approximation for custom hardware: where we've been, where we're going, ACM Comput. Surv. 52 (2019) 1–39. arXiv preprint arXiv:1901.06955.

[95] Y. Gordienko, Y. Kochura, V. Taran, N. Gordienko, A. Bugaiov, S. Stirenko, Adaptive iterative pruning for accelerating deep neural networks, in: 2019 XIth International Scientific and Practical Conference on Electronics and Information Technologies (ELIT), IEEE, 2019, pp. 173–178, https://doi.org/10.1109/ELIT.2019.8892346.

[96] A.G. Howard, M. Zhu, B. Chen, D. Kalenichenko, W. Wang, T. Weyand, M. Andreetto, H. Adam, MobileNets: Efficient Convolutional Neural Networks for Mobile Vision Applications, ArXiv:1704.04861, 2017.

[97] M. Abadi, et al., Tensorflow: a system for large-scale machine learning, in: 12th USENIX Symposium on Operating Systems Design and Implementation, 2016, pp. 265–283.

[98] Z. Zhou, X. Chen, E. Li, L. Zeng, K. Luo, J. Zhang, Edge intelligence: paving the last mile of artificial intelligence with edge computing, Proc. IEEE 107 (8) (2019) 1738–1762.

[99] F. Cabitza, A. Campagner, C. Balsano, Bridging the "last mile" gap between AI implementation and operation: "data awareness" that matters, Ann. Transl. Med. 8 (7) (2020) 501.

[100] M.G. Seneviratne, N.H. Shah, L. Chu, Bridging the implementation gap of machine learning in healthcare, BMJ Innov. 6 (2019) 45–47, https://doi.org/10.1136/bmjinnov-2019-000359.

Further reading

[101] L. Wynants, B. Van Calster, M.M. Bonten, G.S. Collins, T.P. Debray, M. De Vos, E. Schuit, Prediction models for diagnosis and prognosis of covid-19 infection: systematic review and critical appraisal, BMJ 369 (2020).

About the authors

Yuri Gordienko is NVIDIA Deep Learning Institute Ambassador, Head of KPI–Samsung R&D Lab, Coordinator and Principal Investigator in NVIDIA GPU Education and NVIDIA GPU Research Center, and Professor at National Technical University of Ukraine "Kyiv Polytechnic Institute." His research is mainly focused on artificial intelligence, computer vision, high-performance computing, cloud computing, distributed computing, parallel computing, eHealth, simulations, and statistical methods. He has published more than 60 papers in peer-reviewed international journals.

Yuriy Kochura has a MS in applied physics; now he is studying for a PhD's degree at Computer Engineering Department, National Technical University of Ukraine "Kyiv Polytechnic Institute." His research is mainly focused on deep learning, eHealth, and accelerated computing. He has published more than 10 papers in peer-reviewed international journals.

Vlad Taran has a MS in computer science; now he is studying for a PhD's degree at Computer Engineering Department, National Technical University of Ukraine "Kyiv Polytechnic Institute." His research is mainly focused on artificial intelligence, computer vision, embedded systems, high-performance computing, and distributed computing. He has published more than 10 papers in peer-reviewed international journals.

Nikita Gordienko has a BS in computer science; now he is studying for a master's degree at Computer Engineering Department, National Technical University of Ukraine "Kyiv Polytechnic Institute." His research is mainly focused on artificial intelligence, high-performance computing, distributed computing, mobile computing, eHealth, IoT, and brain-computer interface. He has published more than 10 papers in peer-reviewed international journals.

Oleksandr Rokovyi has a PhD in computer science, and he is an associate professor at National Technical University of Ukraine "Kyiv Polytechnic Institute." His research is mainly focused on artificial intelligence, computer vision, embedded systems, cyber security, parallel and distributed computing, and computer networks. He has published more than 10 papers in peer-reviewed international journals.

Oleg Alienin a M.S. in computer science, now he is Assistant Professor at National Technical University of Ukraine "Kyiv Polytechnic Institute". His research is mainly focused on artificial intelligence, computer vision, embedded systems, cyber security, parallel and distributed computing, computer networks. He has published more than 10 papers in peer reviewed international journals.

Sergii Stirenko is Head of Computer Engineering Department, Research Supervisor of KPI-Samsung R&D Lab, Head of NVIDIA GPU Education and NVIDIA GPU Research Center, and Professor at National Technical University of Ukraine "Kyiv Polytechnic Institute." His research is mainly focused on artificial intelligence, high-performance computing, cloud computing, distributed computing, parallel computing, eHealth, simulations, and statistical methods. He has published more than 60 papers in peer-reviewed international journals.

CHAPTER TEN

Hardware accelerator for training with integer backpropagation and probabilistic weight update

Hyunbin Park[a] and Shiho Kim[b]
[a]IT & Mobile Communications, Samsung Electronics, Seoul, South Korea
[b]School of Integrated Technology, Yonsei University, Seoul, South Korea

Contents

Abstract

Advances in the architecture of inference accelerators and quantization techniques of neural networks allow effective on-device inference in embedded devices. Privacy issues for user data, as well as increasing needs of user-specific services, have led to a need for on-device training. The dot product operation required in backpropagation can be computed efficiently by multiplier–accumulators (MACs) in the inference accelerator if forward and backward propagation of the neural network have the same precision. This chapter introduces a quantization technique to enable computation by the digital neuron inference accelerator with the same precision as that using the forward path. Updating the 5-bit weights with gradients of higher precision is challenging. To address this issue, this chapter also introduces a probabilistic weight update. It also describes the hardware implementation of the probabilistic weight-update scheme. The proposed training technique achieves 98.15% recognition accuracy on the MNIST dataset.

Advances in Computers, Volume 122
ISSN 0065-2458
https://doi.org/10.1016/bs.adcom.2020.11.006

343

1. Introduction

The requirements for training using private user data, such as face, voice, biometric information, and usage pattern, with deep learning in embedded devices is increasing for user-specific services. Along with advances in deep-learning techniques, implementation of deep-learning-based applications on embedded or edge devices has also been a focus area [1–5]. The user data stored in edge devices might be transmitted to servers to utilize the computational resources of the server. However, online transmission of personal data for training on servers is vulnerable to leakage and risks user privacy. In fact, most European countries restrict such utilization of user data through privacy regulations, such as the General Data Protection Regulation (GDPR) in Europe [6,7].

Therefore, several studies have adopted on-device training with a training accelerator in an embedded system [4,5,8–12], as illustrated in Fig. 1. Training requires computations of the forward path as well as the backward path of the training dataset with dozens of epochs. Therefore, on-device training of neural networks is challenging because there are constraints such as computing power and power consumption of edge devices.

Fig. 2 shows computational dataflow of two adjacent $(k-1)$th and kth layers during the inference and training (learning) phase [13]. The data are propagated through the convolutional layers in the forward direction, where h_k represents the kth layer's output activation, b_{k-1} is the bias unit, and $\varphi(\bullet)$ is an activation function. During the training step, the error for loss gradients, δ_{k-1}, is calculated by applying the chain rule, the error backpropagates

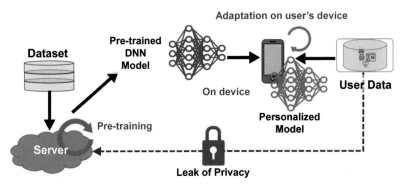

Fig. 1 Artificial intelligence at on embedded device or edge training with users' personal data.

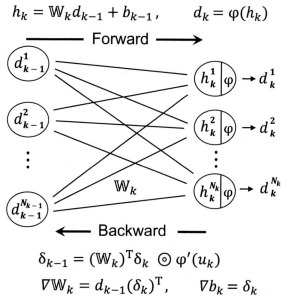

$$h_k = \mathbb{W}_k d_{k-1} + b_{k-1}, \qquad d_k = \varphi(h_k)$$

—— Forward ——→

←—— Backward ——

$$\delta_{k-1} = (\mathbb{W}_k)^{\mathrm{T}} \delta_k \odot \varphi'(u_k)$$
$$\nabla \mathbb{W}_k = d_{k-1}(\delta_k)^{\mathrm{T}}, \qquad \nabla b_k = \delta_k$$

Fig. 2 Forward and backward computational data flow of two adjacent layers for deep neural network. Here, $\odot$ indicates the Hadamard product, also known as element-wise multiplication between two matrices of the same dimension.

through the layer, and we apply the gradient-descent method and update the gradient of weights ($\nabla \mathbb{W}_k$) and bias parameter (∇b_k) of the neural network as shown in the equations in Fig. 1. The calculated gradient of weight ($\nabla \mathbb{W}_k$) are used to update their corresponding weights, where the gradient of weight is multiplied by the learning rate (η) and updating the previous weight as

$$\mathbb{W}_k \Leftarrow \mathbb{W}_k + \eta \nabla \mathbb{W}_k. \tag{1}$$

The training process requires more computation overhead and additional storage for the intermediate data used in error calculation and inference processing. All the data produced from the forward propagation must be re-accessed in the back-propagation phase to calculate the weight gradient. To enhance the efficiency of a hardware accelerator in an edge device, several studies have adopted strategies either to compute the only top-k gradient [14,15] or to quantize the gradient of neural networks with a negligible accuracy drop [16,17].

Prior studies [14,15] only compute the top-k of the derivative of the gradient loss of objective function L with respect to output $(\frac{\partial L}{\partial y})$ and weight $(\frac{\partial L}{\partial W})$

to reduce computation, where the precision of the gradient is FP32 (it is the standard representation of 32-bit single-precision floating point). The study proposed in [16] uses only the top 5% of gradients to update weights, and it achieves 99.39% recognition accuracy of the MNIST dataset. This result shows that usage of only a portion of gradients effectively reduces the computation of training without an accuracy drop.

To perform efficient training of neural networks on edge devices, back-propagated-gradients should be quantized and the gradient of weight should also be accumulated with a lower bit-width than a single precision, which reduces the hardware resources of the adder, multiplier, and storage element of a hardware training accelerator. Quantizing the forward path of neural networks effectively enhances the computational efficiency for inference performed in a hardware accelerator with negligible accuracy error [18–23].

Fig. 3 shows the forward and backward computational data flow of two adjacent layers for a method that consists of training a DNN (deep neural network) with binary weights during the forward and backward propagations, called the BinaryConnect scheme [24]. However, in the BinaryConnect scheme, techniques of quantizing only forward path propagate FP32 gradients backwards and accumulate gradients of weights in FP32 storage, as shown in Fig. 3. DoReFa-Net [16] first applies an affine transform upon the errors to map them into $[-1, 1]$, and the quantized error is still presented as FP32 numbers with discrete states. Wu et al. proposed a quantization technique [17] for training and inference with 8-bit integers. This technique propagates 8-bit integer gradients backwards and accumulates them on 8-bit integer weights.

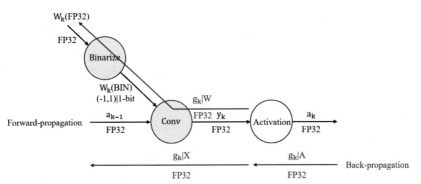

Fig. 3 Forward and backward computational data flow of two adjacent layers for the BinaryConnect scheme [24]. Here, a_k is the activation of the *kth* layer, y_k is the output of the convolution layer, and $(g_k|Z)$ is the gradient of backward propagation with respect to Z.

Four quantization operators—weight, activation, gradient, and error value (WAGE)—are added to reduce precision, and the bit width of signed integers are included in the forward and backward computation dataflow. The quantization function (WAGE) converts a floating-point number to its limited-bit-width signed integer representation. The DNN trained by the WAGE method achieves almost the same inference accuracy as a DNN represented by FP32.

This chapter introduces a back-propagation technique that can train neural networks on the Digital Neuron presented in Chapter 1 of this book. The digital neuron computes the dot product of 8-bit integer activations and 5-bit integer weights that are partitioned into two partial sub-integers. The proposed back-propagation technique, therefore, quantizes the gradients into an 8-bit integer and accumulates the gradients on 5-bit integer weights for weight update that are partitioned into two partial sub-integers. The weight change is limited to the range of $+1$ and -1 caused by weight update, and after each training iteration, the weight is increased by 1/32 in cases where the weight is a 5-bit integer. However, this change can lead to instability in training. To solve this problem, the proposed technique selects a weight to update probabilistically for each training iteration.

The rest of this chapter is divided into the following sections. In Section 2, we present the proposed integer backpropagation with probabilistic weight update. In Section 3, we describe the hardware implementation of the probabilistic weight update. In Section 4, we verify the proposed technique via Python simulation. In Section 5, we discuss the advantages of the proposed scheme. Finally, in Section 6, we summarize this chapter.

2. Integer back propagation with probabilistic weight update

The proposed integer backpropagation with a probabilistic weight update can be summarized into the following three steps.

(1) Converses the gradient from FP32 to integer (Section 2.1).

(2) Randomly selects indices of the weights to be updated (Section 2.2).

(3) In the weight-update process, probabilistically increases or decreases the weights of the randomly selected indices based on the gradient, where the weight is increased or decreased by 1 or 2 such that the updated weight is not ± 11 and ± 13 (Section 2.3).

These three steps will be explained in the following three subsections, respectively.

2.1 Conversion from FP32 to integer for integer back propagation

Fig. 4A shows the forward propagation of the proposed training scheme of a fully connected layer and softmax with cross entropy. Forward propagation refers to the forward data process for inference presented in Chapter 1 [25]. The digital neuron computes the inner product of the weight vector and activation of the previous layer, where the activations are unsigned

A

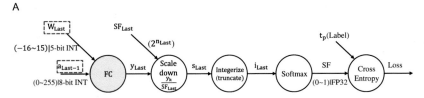

Forward propagation (inferencing) of the proposed scheme with 5-bit weight vector

B

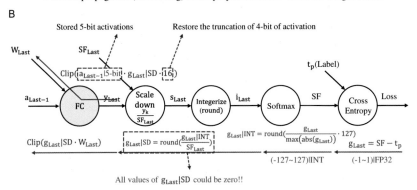

Dataflow for simple back propagation

C

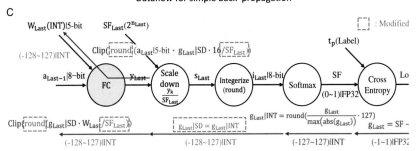

Modified back propagation with the 5-bit weight vector scheme

Fig. 4 Proposed inference and training scheme of a fully connected layer and softmax with cross entropy. (A) Forward propagation, (B) backpropagation and limitation, and (C) modified backpropagation.

8-bit integers ranged from 0 to 255, and the weights are 5-bit integers ranged from -16 to 15, where we exclude ± 11 and ± 13. All vectors are integerized after division [25,26]. The output of the FC layer, y_{Last}, is scaleddown by $2^{n_{Last}}$ and then integerized. The computations of the FC, scale-down, and integer layers are performed between the input of the NTs to the input of the output feature map presented in Chapter 1. The softmax computation of the integerized output, i_{Last}, is performed by the Softmax node. The output of the Softmax node (SF) is numbered from 0 to 1 with a precision of FP32. In the proposed scheme, FP32-based-computation is adopted only in the SF and cross-entropy nodes.

The 8-bit activations in the forward path should be stored for training. The proposed scheme allows the gradient of the FC node to be computed using the digital neuron that receives an 8-bit vector and a 5-bit vector. Therefore, the proposed scheme transforms the 8-bit activations to a 5-bit vector for delivering the activations to the input port for the 5-bit vector of the digital neuron. Therefore, the 8-bit activations are stored after truncating the LSB 4-bit vector to transform the unsigned 8-bit vector into a signed 5-bit vector.

Fig. 4B shows the backpropagation and limitation of the proposed training scheme in Fig. 4A. The gradient of the Softmax node and the cross-entropy node, g_{Last}, is derived from the subtraction of the output of the Softmax node SF and t_p label. Because g_{Last} comprises floating-point numbers from 0 to 1, to integerize g_{Last}, the proposed scheme multiplies g_{Last} by 127 after normalizing g_{Last} to $(-1, 1)$.

Fig. 5 shows why normalization is required. In the case where the most probable inference label matches the answer label, all values of SF-t_p are approximately zero. Therefore, all values of round $((SF-t_p) \times 127)$ could be zero. In this case, the back propagation of all zeros prohibits training. The higher the inference accuracy, the more frequent the backpropagation

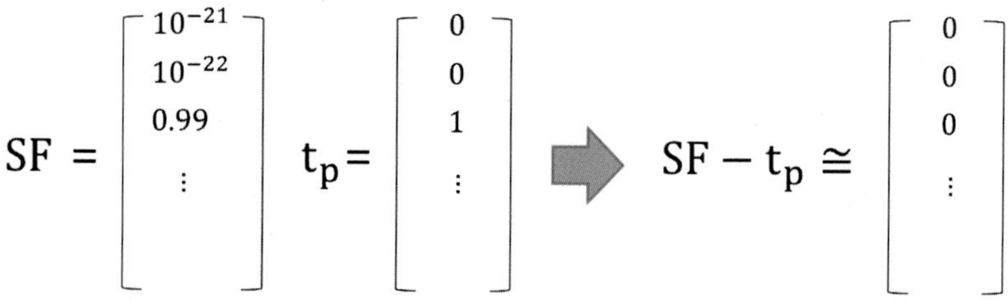

$$SF = \begin{bmatrix} 10^{-21} \\ 10^{-22} \\ 0.99 \\ \vdots \end{bmatrix} \quad t_p = \begin{bmatrix} 0 \\ 0 \\ 1 \\ \vdots \end{bmatrix} \quad \Rightarrow \quad SF - t_p \cong \begin{bmatrix} 0 \\ 0 \\ 0 \\ \vdots \end{bmatrix}$$

Fig. 5 If the most probable inference label matches the answer label, all values of SF-t_p are approximately zero.

of all-zero gradients. The normalization is executed by dividing g_{Last} by the maximum value of the magnitude of g_{Last}, which is expressed by

$$\frac{g_{Last}}{\max\left(abs(g_{Last})\right)} \tag{2}$$

The scale-down node divides $g_{Last}|INT$ by the scalability SF_{Last}, where $g_{Last}|INT$ is the normalized and integerized g_{Last}. The scaled-down $g_{Last}|INT$ is represented by $g_{Last}|SD$.

To obtain the gradient of weights W_{Last} of the FC layer, the computation of the inner product of the stored 5-bit activation of the forward propagation, $a_{Last-1}|5\text{-bit}$, and 8-bit $g_{Last}|SD$ is required. The digital neuron computes the inner product of $a_{Last-1}|5\text{-bit}$ and 8-bit $g_{Last}|SD$. The $a_{Last-1}|5\text{-bit}$ is approximately $\frac{1}{16}a_{Last-1}|8\text{-bit}$ because the 4-bit vector is truncated. Therefore, $a_{Last-1}|5\text{-bit} \cdot g_{Last}|SD$ is multiplied by 16 after clipping, where the clipping is to set the upper and lower limits of -128 and 127, respectively.

In Fig. 4B, however, all components of the round $(g_{Last}|INT/SF_{Last})$ could be all-zero vectors. This can prohibit training. Therefore, this section shows the modified backpropagation to resolve this problem in Fig. 4B. Fig. 4C shows the modified backpropagation of a fully connected layer and softmax with cross entropy. Because the division of SF_{Last} can make all components of the round $(g_{Last}|INT/SF_{Last})$ zero, the division of SF_{Last} and the round function are moved to the backpropagation of the FC layer. This allows integers from -128 to 127 to be propagated backward.

To obtain the gradient of weights W_{Last} of the FC layer, the computation of the inner product of the stored 5-bit activation of the forward propagation, $a_{Last-1}|5\text{-bit}$, and 8-bit $g_{Last}|SD$ is required. The digital neuron computes the inner product of $a_{Last-1}|5\text{-bit}$ and 8-bit $g_{Last}|SD$. The $a_{Last-1}|5\text{-bit}$ is approximately $\frac{1}{16}a_{Last-1}|8\text{-bit}$ because the 4-bit is truncated. Therefore, $a_{Last-1}|5\text{-bit} \cdot g_{Last}|SD$ is multiplied by 16 after clipping, where the clipping is to set the upper and lower limits of -128 and 127, respectively.

In Fig. 4B, however, all components of the round $(g_{Last}|INT/SF_{Last})$ could be all-zero vectors. This can prohibit training. Therefore, this section will show the modified backpropagation to resolve this problem in Fig. 4B. Fig. 4C shows the modified backpropagation of a fully connected layer and softmax with cross entropy. Because the division of SF_{Last} can make all components of the round $(g_{Last}|INT/SF_{Last})$ zero, the division of SF_{Last} and the round function are moved to the backpropagation of the FC layer. This allows integers from -128 to 127 to be propagated backward.

Fig. 6A shows the forward propagation of the proposed training scheme of a convolutional layer. The digital neuron computes the inner product of the weight vector and activation of the previous layer, where the activations are 8-bit integers from 0 to 255, and the weights are 5-bit integers from -16 to 15 except ± 11 and ± 13. The output of the FC layer, y_k, is scaled down by 2^{n_k}, and then integerized to generate i_k. After that, the activation node receives i_k the ReLU function with an upper limit of 255 and produces a_k 8-bit integers from 0 to 255. The a_k is stored in registers after truncating the LSB 4-bit to transform unsigned 8-bit to signed 5-bit for training. The pooling node receives a_k and produces p_k.

Fig. 6B shows the modified backpropagation of the proposed training scheme of a convolutional layer. Backpropagation of the pooling node unpools the gradient by storing the gradients in the expanded feature map in the maximum index in forward propagation, as in the method used in Ref. [27]. The estimator of the gradient method [28] is employed in the backpropagation of the rectified linear unit (ReLU). The estimator of the gradient approximates the gradient, as shown in Fig. 7. The activation function of the ReLU is actually stepwise in the domain between 0 and 255.

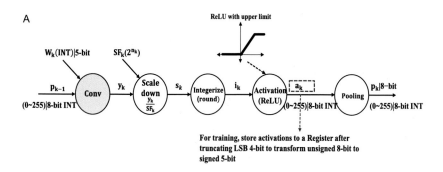

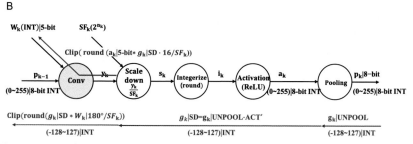

Fig. 6 Proposed training scheme of a convolutional layer. (A) Forward propagation and (B) modified backpropagation.

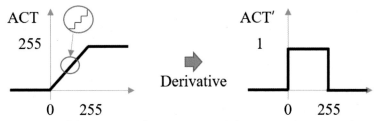

Fig. 7 Gradient of the activation function, ReLU of the proposed training scheme.

The activation gradients of the stepwise domain were approximated to be one. In backward propagation, $g_k | SD$ is produced by multiplying the unpooled gradient, $g_k | UNPOOL$, and the estimator of the gradient, ACT'.

The division of SF_k is performed in the convolution node in the backward propagation. The gradients of W_k and p_{k-1} can be expressed as.

$$\text{Gradient of } W_k(\text{INT}) = \text{Clip}\left(\text{round}\left(\frac{a_k | 5 - \text{bit} * g_k | SD \cdot 16}{SF_k}\right)\right), \quad (3)$$

$$\text{Gradient of } p_{k-1} = \text{Clip}\left(\text{round}\left(\frac{g_k | SD * W_k | 180°}{SF_k}\right)\right), \quad (4)$$

where the round and clipping are used for integerization and bitwise limitation.

2.2 Random selection of indices of weights to be updated

The representative training techniques, such as gradient descent [29], momentum [30], AdaGrad [31], ADADELTA [32], and Adam [33], all accumulate the gradients of weights in terms of weight loss. However, the accumulation of the gradient of noise is averaged out to zero [24]. Therefore, although the accumulation of the gradients of all weights occurs during every iteration of training, the accumulation of the gradients of noise does not actually occur from the perspective of overall training.

Therefore, the proposed training scheme does not update all weights in every iteration, but selects indices of weights to be updated in a probabilistic manner. To the author's knowledge, this kind of probabilistic training has not been reported previously.

2.3 Probabilistic weight update of the randomly selected indices

The proposed training scheme probabilistically decreases or increases the 5-bit integer weights' randomly selected indices, as mentioned in Section 2.2, without accumulating gradients in weights.

The weight update is performed as follows. First, the maximum value of the gradient map of weights is derived, which is referred to as Max(gradient(W)). Second, the indices to update in the gradient map of the weights randomly are selected as described in Section 2.2. The gradient of weights of the selected index is (i_s) $g_k | W[i_s]$. Third, a random integer number, RAN_NUM, is generated from 1 to 255. Fourth, the weights are updated by the probabilistic condition, which is summarized in Fig. 8. The gradient update is performed using Eq. (3) and is greater than RAN_NUM.

$$\frac{abs(g_k | W[i_s])}{Max(abs(g_k | W))} \times 255 \tag{5}$$

Eq. (5) is explained as follows. In the case of the gradient descent technique, the variation of weights in one iteration is proportional to the magnitude of gradients of the weights. Therefore, the author also allows the update probability to be proportional to the magnitude of the gradient (i.e., **abs (Gradient(W)[i_s])**, as shown in Eq. (5).

The distribution of weights in each layer differs according to the layers. Therefore, the weights of a layer with a larger standard deviation can be updated more frequently than that with a smaller standard deviation.

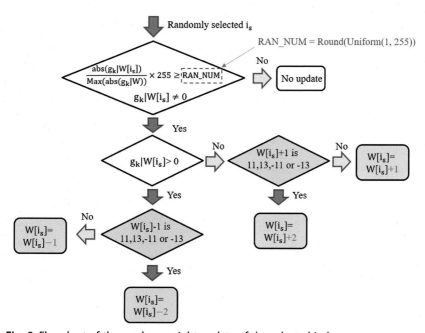

Fig. 8 Flowchart of the random weight update of the selected index.

To compensate for the uneven update frequency according to layers, the proposed scheme normalizes the weights of each layer by dividing **abs (Gradient(W)[i$_s$]** by **Max(abs(Gradient(W)))** and multiplying it by 255.

If the normalized gradient of the selected index is greater than RAN_NUM, the proposed training scheme determines whether to increase or decrease the weight of the selected index according to the sign of the weight. In the case of the gradient descent scheme, a positive gradient decreases the weight and a negative gradient increases the weight. The proposed training scheme also increases the weight when the sign of the gradient of the selected index is negative, and decreases the weights when the sign of the gradient of the selected index is positive under the condition that Eq. (5) is greater than RAN_NUM, where the weight is increased or decreased by 1 or 2 such that the updated weight does not include integers of ± 11 and ± 13.

3. Consideration of hardware implementation of the probabilistic weight update

The proposed weight-update scheme is implemented using digital hardware. Fig. 9 shows an example of the weight update using digital neuron hardware. Fig. 9A illustrates the derivation of **Max(abs(gradient(W))**. In this example, it is assumed that the digital neuron calculates one gradient of a weight every clock. The hardware stores the **Max(abs(gradient (W)))** of in the Max register and compares it with the magnitude of the gradient of the next clock calculated by the digital neuron. The greater value of this comparison is stored in the Max register. The circuit for generating **Max (abs(gradient(W)))** is only a 2's-complement circuit, a binary adder, and registers. Fig. 9B shows the weight update based on the random index and random number. The indices of weights to be updated are determined randomly. The random index generation does not require the generation of a true random number. A pseudo-random generator circuit can perform random index generation. The pseudo-random generator circuit can be implemented by only the recursive structure of shift registers [34].

In the example of Fig. 9B, the gradients of the randomly selected indices 1, 6, 33 are −12, 0, and 45, and the calculated normalized gradient of Eq. (5) of the indices were 68, 0, and 255, respectively. In this example,

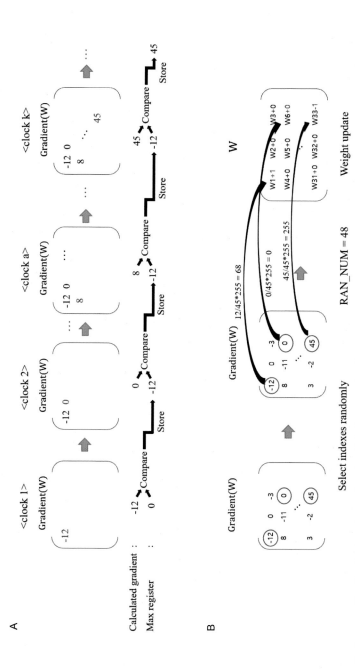

Fig. 9 An example of weight update using the digital neuron hardware. (A) Derivation of Max(abs(gradient(W))). (B) Weight update based on random index and random number of 48.

RAN_NUM is generated at 48 and **Max(abs(gradient(W)))** is 45. The sign of the gradient -12 of index 1 is negative, and the normalized gradient of index 1 is greater than 48. Therefore, the weight of index 1 is increased. The gradient of index 6 is not both negative and positive. Therefore, the weight of index 6 remained unchanged. The sign of gradient 45 of index 33 is negative, and the normalized gradient of index 33 is greater than 48. Therefore, the weight of index 33 is decreased. The increase and decrease in weight are implemented by look-up table circuits.

The gradient descent scheme requires additions for all weights. However, the proposed training scheme requires only weight updates of the selected indices based on look-up table circuits. This does not require binary adders for the weight update, and the number of look-up-table-based calculations is much less than that of the gradient descent scheme. Therefore, it can simplify the circuit complexity.

4. Simulation results of the proposed scheme

In this section, we simulated a Python code for the proposed training scheme with inference of the MNIST database, where the employed network is LeNet-5 [29]. The Python code is executed based on the FP32. However, the proposed training scheme propagates 8-bit integers to both forward and backward. Therefore, the author made and added custom functions for clipping with upper limit and rounding for integerization in the graph of LeNet-5.

The proposed training scheme, in backward propagation, employs activations with 5-bit integers except ± 11 and, ± 13 which are converted from activations with 8-bit integers. To perform a typical convolution, the *conv2d* function provided by the Tensor flow library can be used. However, the *conv2d* function cannot receive the activations with 5-bit integers except ± 11 and ± 13 in the input port. Therefore, the convolutional function is modified for employing activations with 5-bit except ± 11 and ± 13 in backward propagation. For the modification, the author used py_func function that allows users to make custom functions. In the py_func function, the 8-bit activation matrix is divided by 16, and the matrix is rounded to integerization. After that, the proposed scheme forcibly allows the function to set ± 11 and ± 13 to ± 10 and ± 12, respectively.

The number of weights in Layer 1 is $5 \times 5 \times 6 = 150$. The proposed scheme sets the code to randomly select one index among the 150 gradients in layer 1 for updating weights in layer 1 in one iteration of training. This selection ratio (1:150) was applied to the other layers. For example, the number of weights in layer 2 is $5 \times 5 \times 6 \times 16 = 2,400$, which is 16 times the number of weights in layer 1. Therefore, the code is set to randomly select 16 indices among the 2400 gradients in layer 2 for updating weights in layer 2 during one training iteration.

The scalabilities for scaling down from layer 1 to layer 5 are set to 64, 128, 256, 128, and 128, respectively, which are identical to those of Chapter 1. Fig. 10 shows the MNIST inference accuracy of the proposed training scheme versus epoch, where the batch size is 1000. The maximum accuracy was 98.15% at the 462nd epoch. The maximum accuracy of LeNet-5 with FP32 precision was 98.96% in Chapter 1. The degradation in the proposed integer backpropagation was only 0.81% compared to FP32.

In this section, COIL-20 and COIL-100 databases are also trained with the proposed training scheme with LeNet-5 with Python simulation. Inference accuracies of COIL-20 and COIL-100 reached 100% and above 95%, respectively.

Fig. 11 shows the distributions of weights of the proposed training scheme when the MNIST database is trained with LeNet-5. The distributions

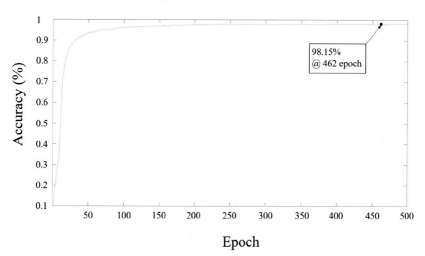

Fig. 10 MNIST inference accuracy of the proposed training scheme versus epoch.

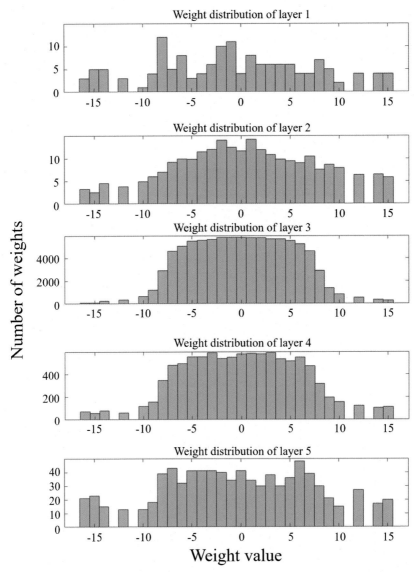

Fig. 11 Distributions of weights of the proposed training scheme, when MNIST database is trained with LeNet-5.

of from layer 2 to layer 5 are similar to the weight distributions of FP32 illustrated in Fig. 2, where the magnitude of the distribution is the largest close to zero and decreases with increasing magnitude of weights. In addition, Fig. 12 shows that the weights are trained with 5-bit integers except for ± 11 and ± 13.

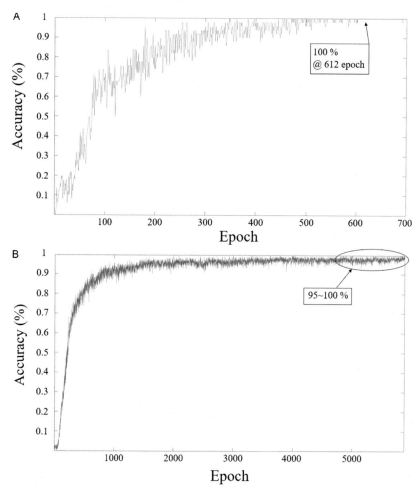

Fig. 12 Inference accuracy of the proposed training scheme versus epoch. (A) COIL-20 and (B) COIL-100.

5. Discussions

We investigated four structural advantages of the proposed training scheme from a hardware perspective, which helps to reduce the power consumption and circuit area, as summarized in Fig. 13. First, the proposed training scheme stores activations after converting to 5-bit integers, except for ± 11 and ± 13 for backward propagation. This allows us to reduce the storage requirement for storing activations in backward propagation steps to be reduced by approximately 85% compared to FP32.

Second, although the training schemes in [34,35] employ the quantized weights in forward propagation, they incur storage requirements for the accumulation of gradients with FP32 in backward propagation. However, the proposed training scheme directly updates the weights based on the calculated 8-bit gradients. Thus, it avoids wasteful storage requirements.

Third, the gradient-descent scheme requires multiplications of the gradients and learning rate and subtractions of all weights and gradients in every training iteration. In contrast, the proposed training scheme does not require any binary adder circuits or multiplier circuits for updating weights because the update is performed by look-up table circuits in only a few selected weights.

Fourth, the calculation in a convolution node during backward propagation in the proposed training scheme is performed by the inner product of the 5-bit and 8-bit vectors. This allows the digital neuron to perform calculations for both the forward and backward propagations, and therefore helps to reduce the power consumption.

Fifth, the proposed training scheme stores activations after converting to 5-bit integers, except for ± 11 and ± 13 for backward propagation. This allows the storage requirements for the activations to be reduced by approximately 85% compared to FP32.

	Training based on gradient descent	Proposed integer backpropagation
Computation for weight update	NUM_W $\times$ 2	NUM_W $\times \frac{1}{150}$ $\left(\text{Assuming predetermined probability} : \frac{1}{150} \right)$
Total registers to accumulate g|W	NUM_W $\times$ 32-bit	0
Gradient precision	32-bit	8-bit
Weight precision	32-bit	5-bit

*NUM_W : number of weights of all layers

Fig. 13 Summarization of structural advantages of the proposed training scheme.

6. Summary

This chapter proposed a training scheme that propagates 8-bit integer gradients backward and updates 5-bit integer weights in a probabilistic manner. The Python simulation for training and inference of the MNIST database confirmed the effectiveness of the proposed scheme. The inference accuracy of the MNIST database was 98.15% in the Python simulation. The reduced bit width and direct weight update from the gradient to quantized weight gave the hardware structural advantages in power consumption. To address the hardware accelerator for training with integer backpropagation and probabilistic weight updates, this chapter considered the hardware implementation of the probabilistic weight-update scheme. The proposed training technique achieved 98.15% recognition accuracy on the MNIST dataset.

Acknowledgments

The authors would like to thank the Institute for Information & Communications Technology Planning & Evaluation (IITP). This work was support by the Ministry of Science and ICT (MSIT), Korea, under the ICT Consilience Creative Program (IITP-2019-2017-0-01015), and partially by (No. 2020-0-00056, to create AI systems that act appropriately and effectively in novel situations that occur in open worlds) supervised by IITP.

Key terminology and definitions

Edge device An edge device is a type of device that connects a cloud server with a wireless or wired the Internet connection. It usually provides interconnectivity and traffic translation between different networks or network boundaries.

Hardware accelerator A hardware accelerator is a hardware system specially constructed to perform certain functions more efficiently than possible on software running on a general-purpose processing unit, such as a CPU or GPU. An artificial intelligence accelerator is a type of specialized hardware accelerator designed to accelerate artificial-intelligence-based applications, such as deep neural networks and machine learning.

Inference and training accelerator An optimized device hardware accelerator for inferencing and training artificial intelligence applications for embedded or edge devices.

Quantization Quantization is the process of constraining data representation from that based on real numbers to that based on a discrete set, such as integers. In the context of deep learning, quantization denotes the process of approximating the variables or parameters of a neural network represented by floating-point numbers using low bit-width binary numbers or integers.

Digital neuron A digital neuron is a hardware inference accelerator proposed by Park and Kim for convolutional deep neural networks with integer inputs and integer weights for embedded systems. The fundamental concept is to reduce the circuit area and power consumption by manipulating the dot products between input features and weight vectors using barrel shifters and parallel adders. This reduction allows a greater number of computational engines to be mounted on the inference accelerator, resulting in higher throughput than achieved by existing HW accelerators.

Neural tile (NT) Neural tile (NT) is a basic unit of hardware inference accelerator composed of eight neural elements of Digital Neuron.

References

[1] Y. Chen, Y. Xie, L. Song, F. Chen, T. Tang, A survey of accelerator architectures for deep neural networks, Engineering 6 (2020) 264–274.

[2] L. Deng, G. Li, S. Han, L. Shi, Y. Xie, Model compression and hardware acceleration for neural networks: a comprehensive survey, Proc. IEEE 108 (4) (2020) 485–532.

[3] Wei, Bingzhen, Xu Sun, Xuancheng Ren, and Jingjing Xu. (2017) "Minimal Effort Back Propagation for Convolutional Neural Networks." arXiv preprint arXiv:1709.05804.

[4] C.H. Lu, Y.C. Wu, C.H. Yang, A 2.25 TOPS/W fully-integrated deep CNN learning processor with on-chip training, in: 2019 IEEE Asian Solid-State Circuits Conference (A-SSCC), IEEE, 2019, pp. 65–68.

[5] S. Choi, J. Sim, M. Kang, Y. Choi, H. Kim, L.S. Kim, An energy-efficient deep convolutional neural network training accelerator for in situ personalization on smart devices, IEEE J. Solid State Circuits 55 (10) (2020) 2691–2702.

[6] H. Joonas, Consumer Big Data and the Influences of General Data Protection Regulation, M.S. thesis, Lut School of Business and Management, Lappeenranta University of Technology, Finland, 2018.

[7] T. Flew, F. Martin, N. Suzor, Internet regulation as media policy: rethinking the question of digital communication platform governance, J. Digital Media Policy 10 (1) (2019) 33–55.

[8] S. K. Venkataramanaiah, Y. Ma, S. Yin, E. Nurvithadhi, A. Dasu, Y. Cao, J. Seo, August 2019. Automatic Compiler Based FPGA Accelerator for CNN Training, arXiv:1908.06724 (accessed October 13, 2020).

[9] H. Nakahara, Y. Sada, M. Shimoda, K. Sayama, A. Jinguji, S. Sato, FPGA-based training accelerator utilizing sparseness of convolutional neural network, in: Proc. 29th International Conference on Field Programmable Logic and Applications (FPL 2019), Barcelona, Spain, 2019.

[10] B. Fleischer, et al., A scalable multi-TeraOPS deep learning processor core for AI training and inference, in: Proceedings of the IEEE Symposium on VLSI Circuits, Honolulu, HI, USA, 2018.

[11] H. Sim, J. Choi, J. Lee, SparTANN: sparse training accelerator for neural networks with threshold-based sparsification, in: Proceedings of ACM/IEEE International Symposium on Low Power Electronics and Design (ISLPED 2020), Boston, USA, 2020.

[12] J. Li, et al., TNPU: an efficient accelerator architecture for training convolutional neural networks, in: Proceedings 24th Asia and South Pacific Design Automation Conference (ASPDAC 2019), Tokyo, Japan, 2019.

[13] L. Song, X. Qian, H. Li, Y. Chen, Pipelayer: a pipelined reram-based accelerator for deep learning, in: 2017 IEEE International Symposium on High Performance Computer Architecture (HPCA), IEEE, 2017, pp. 541–552.

[14] X. Sun, X. Ren, S. Ma, H. Wang, June 2017. meProp: Sparsified Back Propagation for Accelerated Deep Learning With Reduced Overfitting, arXiv:1706.06197 (accessed October 13, 2020).

[15] B. Wei, X. Sun, X. Ren, J. Xu, September 2017. Minimal Effort Back Propagation for Convolution Neural Networks, arXiv:1709.05804 (accessed October 13, 2020).

[16] S. Zhou, Y. Wu, Z. Ni, X. Zhou, H. Wen, and Y. Zou, "DoReFa-Net: Training Low Bitwidth Convolutional Neural Networks With Low Bitwidth Gradients," arXiv:1606.06160, n.d.

[17] S. Wu, G. Li, F. Chen, and L. Shi, "Training and Inference With Integers in Deep Neural Networks," arXiv:1802.04680v1, n.d.

[18] L. Lai, N. Suda, V. Chandra, Deep convolutional neural network inference with floating-point weights and fixed-point activations, in: Proceedings of the 34rd International Conference on Machine Learning (ICML 2017), Sydney, Australia, 2017.

[19] M. Peemen, A.A. Setio, B. Mesman, H. Corporaal, Memory-centric accelerator design for convolutional neural networks, in: Proceedings of the 2013 IEEE 31th International Conference on Computer Design (ICCD), Asheville, NC, USA, 2013, pp. 13–19.

[20] C. Zhang, et al., Optimizing FPGA-based accelerator design for deep convolutional neural networks, in: Proceedings of the 2015 ACM/SIGDA International Symposium on Field-Programmable Gate Arrays, Monterey, CA, USA, 2015, pp. 161–170.

[21] Y. Chen, T. Krishna, J. Emer, V. Sze, Eyeriss: an energy efficient reconfigurable accelerator for deep convolutional neural networks, in: Proceedings of the IEEE International Solid-State Circuits Conference (ISSCC 2016), San Francisco, USA, 2016, pp. 262–263.

[22] M. Motamedi, P. Gysel, V. Akella, S. Ghiasi, Design space exploration of FPGA-based deep convolutional neural networks, in: Proceedings of the 2016 21st Asia and South Pacific Design Automation Conference, Macau, China, 2016, pp. 575–580.

[23] R. Andri, L. Cavigelli, D. Rossi, L. Benini, YodaNN: an architecture for ultralow power binary-weight CNN acceleration, IEEE Trans. Comput. Aided Des. Integr. Circuits Syst. 37 (1) (2018) 48–60.

[24] M. Courbariaux, Y. Bengio, J.-P. David, BinaryConnect: training deep neural networks with binary weights during propagations, in: Proceeding of the Advances in Neural Information Processing Systems (NIPS 2015), Montreal, Montreal, Canada, 2015.

[25] H. Park, S. Kim, Hardware accelerator systems for artificial intelligence and machine learning, in: Advances in Computers, vol. 122, Elsevier, 2020. Chapter 1.

[26] Hyunbin Park, Dohyun Kim, and Shiho Kim. "Digital Neuron: A Hardware Inference Accelerator for Convolutional Deep Neural Networks." arXiv preprint arXiv:1812.07517 (2018).

[27] O. David, N. Netanyahu, DeepPainter: painter classification using deep convolutional autoencoders, in: Proceedings of the 2016 International conference on Artificial Neural Networks, Barcelona, Spain, 2016, pp. 20–28.

[28] S. Liang, et al., FP-BNN: binarized neural network on FPGA, Neurocomputing 275 (31) (2018) 1072–1086.

[29] Y. LeCun, L. Bottou, Y. Bengio, P. Haffner, Gradient-based learning applied to document recognition, Proc. IEEE 86 (11) (1998) 2278–2324.

[30] N. Qian, On the momentum term in gradient descent learning algorithms, Neural Netw. 12 (1) (1999) 145–151.

[31] J. Duchi, E. Hazan, Y. Singer, Adaptive subgradient methods for online learning and stochastic optimization, J. Mach. Learn. Res. 12 (2011) 2121–2159.

[32] M. Zeiler, December 2012. ADADELTA: An Adaptive Learning Rate Method, arXiv:1212.5701 (accessed October 13, 2020).

[33] D. Kingma, J. Ba, December 2014. Adam: A Method for Stochastic Optimization, arXiv:1412.6980 (accessed October 13, 2020).

[34] D. Anguita, S. Rovetta, R. Zunino, Compact, digital pseudo-random number generator, Electron. Lett. 31 (12) (1995) 956–958.

[35] M. Rastegari, V. Ordonez, J. Redmon, A. Farhadi, XNOR-Net: ImageNet classification using binary convolutional neural networks, in: Proceeding of the 2016 14th European Conference on Computer Vision (ECCV 2016), Amsterdam, Netherlands, 2016, pp. 525–542.

About the authors

Hyunbin Park received his B.S. degree in Electrical and Electronic Engineering from Yonsei University, Seoul, South Korea, in 2013. He received his Ph.D. degree from a M.S.–Ph.D. joint course at the School of Integrated Technology, Yonsei University, in 2019. When he was a graduate student, his primary research topics included the system design and system architecture of neural processing units (NPUs) with integral weights and activations and training techniques for deep learning-based methods for integer gradients. Between March 2019 and June 2019, he was employed as a Postdoctoral Researcher at the Yonsei Institute of Convergence Technology (YICT), Yonsei University. He is currently involved in the analysis and benchmarking of NPUs for mobile devices at Mobile Communications Business, Samsung Electronics Co., Ltd.

Shiho Kim is a professor at the School of Integrated Technology, Yonsei University, Seoul, Korea. His previous occupations include system-on-chip design engineer at LG Semicon Ltd. (currently SK Hynix), Korea, Seoul (1995–1996);Director of the Research Center for Advanced Hybrid Electric Vehicle Energy Recovery System, a government-supported IT research center; Associate Director of the Yonsei Institute of Convergence Technology (YICT), where he conducted the Korean National ICT Consilience program, which is a Korean national program for cultivating talented engineers in the field of information and communication technology (2011–2012); and Director of the Seamless Transportation Lab, Yonsei University, Korea (2011–present).

His primary research interests include the development of software and hardware technologies for intelligent vehicles, blockchain technology for intelligent transportation systems, and reinforcement learning for autonomous vehicles. He is a member of editorial boards and a reviewer for various journals and international conferences. To date, he has organized two international conferences as the Technical Chair/General Chair. He is a member of the Institute of Electronics and Information Engineers of Korea (IEIE), the Korean Society of Automotive Engineers (KSAE), the vice president of the Korean Institute of Next-Generation Computing (KINGC), and a senior member of IEEE. He is the coauthor of over 100 papers and holds more than 50 patents in the field of information and communication technology.

CHAPTER ELEVEN

Music recommender system using restricted Boltzmann machine with implicit feedback

Amitabh Biswal, Malaya Dutta Borah, and Zakir Hussain
Department of Computer Science and Engineering, National Institute of Technology Silchar, Cachar, Silchar, Assam, India

Contents

Advances in Computers, Volume 122
ISSN 0065-2458
https://doi.org/10.1016/bs.adcom.2021.01.001

Abstract

Now-a-days with explosive growth of the internet, the amount of information available overwhelms user. Collaborative filtering is the most popular method for solving this problem. Most of the research in recommender system concentrates on explicit ratings for their recommendation. The amount of implicit ratings available are much more compared to explicit ratings as implicit data are automatically generated when user interacts with the system. With the use of implicit feedback recommender system can compute preference of users without users giving rating to the system which is large is number. With the increase of computational power and popularity of deep learning algorithms, more and more research are carried out which are implementing deep learning method to provide recommendation. In this chapter we have shown a method which implements restricted Boltzmann machine for recommendation system. Here the number of times a user has listened to any music has been used as implicit feedback. We have also explained how to use contrastive divergence algorithm to train restricted Boltzmann machine and learn its parameters. Our model has shown better results by showing less error compared to existing collaborative filtering engine from Apache.

1. Introduction

Recommender system is a system which provides recommendation to user or predicts users preferences based on user's historical data that is created while interacting with the system. Because of this long tail phenomenon, users might not find the songs they need quickly. Because of this, user might stop using the system if it takes more time to find what the user needs. With the increase and availability of digital songs online, people have access to music in abundance. Because of this long tail phenomenon, users cannot find the songs they need quickly. With the help of recommender system, users can be recommended songs based on their taste, by analyzing their past music playing history. The companies are will incorporate recommender system so that user might use the system more which in turn will increase the profit margins of the companies. So the recommender systems help both the user and the companies that implement recommender system.

With the win of Netflix prize competition of 2009, where the winning algorithm outperformed Netflix's own recommendation algorithm by more than 10% and the winner used restricted Boltzmann machine (RBM) to make recommendation. More and more efforts have been made to use machine learning and deep learning in recommender system academically and corporately. Earlier there was challenge of training deep learning models for a long time because of lack of computational power. With increase in computational power, use of deep learning methods have improved significantly in many domains. The idea of deep learning is to have multiple layers where initial layers extract low level features and higher layers uses these features to extract more high level features. Because of this added complexity, RBM has outperformed most popular memory-based technique singular value decomposition (SVD).

In this chapter we will be using a deep learning method called RBM for providing recommendation to users. With the help of deep learning method we will be able to find hidden nonlinear latent factors in the available data. Latent factors are hidden features that cannot be observed directly from the data that affects user's preferences. It is a statistical model where with observed variable, some hidden latent variable are also present. Latent factors are calculated using some machine learning methods from input features given to the model. Latent factors are considered state of the art in recommender systems. Models that calculate latent factors are called latent factor models. In RBM latent factors are represented by hidden layers.

1.1 Motivation

With exponential growth of information online popularity and need of recommender system has increased substantially. With music streaming platform such as Spotify, Amazon music, Gaana becoming popular, more and more people are going online for listening to music people have access to vast library of music. With the access of high quality portable up-to-date music on the go, this becomes a problem for users to find music among huge library of music according to their taste. With the help of music recommender system, the system can give recommendation to users by analyzing user's music listening history, so that users can find new music without the hassle of spending time searching and listening to new music. Music recommender system can help companies to make more profits as more and more people will be listening to their music which in turn increase their popularity, and more people will be hooked to their music streaming platform.

Unfortunately most of the recommender systems that are designed till date, make use of explicit feedback that user provides, such as ratings and most of the user do not provide ratings to the music they are listening to. Providing rating to each and every music they listen to hinders the natural flow of listening to music as they have to put extra effort for providing ratings. Therefore implicit feedback is better as user does not have to care about the ratings. Implicit feedback is the data collected using user's usage with the music streaming service which is available in large number, i.e., number of times user listened to any song. Depending upon the domain implicit feedback will be different. The more user interacts with the system, more data will be available for the recommendation. Unlike explicit feedback where we give explicit rating to a song in implicit feedback we compare ratings relatively, e.g., if user has listened to music A 10 times and same user has listened to music B 5 times, then we deduce that the user has more preference of music A over music B.

With large amount of user data available, it becomes easier to make personal recommendation to each user. With the increase of computational power, deep learning is able to provide great accuracy when trained with large amount of data. So we will creating music recommender system by combining deep learning method called RBM and implicit feedback.

1.2 Objective

In this chapter we will be showing a recommendation system for music domain where we will be using implicit feedback such as number of times users has listened to the music and provide recommendation based on that. Most of the recommender system designed till date concentrate on the explicit rating as it is easy to perform computation on explicit rating. In this chapter we will show how music play count can be used as rating instead of explicit rating and use of deep learning method to implement recommender system. Most of the people do not rate the music they are listening to as they consider as an extra work which they do not seem to be benefiting them. Rating each and every song they listened to breaks the natural flow of music listening for users.

Music recommender system can also help a user to find artist similar to the artist they already like instead of music.

1.3 Previous work

Much research has been done on recommender systems but very few research has been done on recommender system using implicit feedback. Jawaheer et al. [1] show performance comparison between recommendation system using implicit and explicit feedback on last.fm dataset which removes the notion that implicit feedback gives less accuracy compared to explicit feedback and accuracy of the implicit feedback can be improved. In Ref. [2] RBM has been applied on video watch time as implicit feedback and showed that recommendation works better on video with watch time history than without watch time history. Here instead of explicit ratings confidence scores are calculated based on how much user has listened to the music. Confidence score is the measure that tells how confident we are that the user is going to interact with the video again. In Ref. [3] how to calculate confidence scores from implicit ratings that can be used for recommendation. It also shows how implicit feedback can be used by converting implicit feedback into confidence and preference. In paper titled "Walk the Talk - Analyzing the Relation between Implicit and Explicit Feedback for Preference Elicitation" shows that there is strong relationship between implicit feedback and ratings. It even shows that many recommender systems designed for explicit feedback can be adapted to implicit feedback [4]. In paper titled "A comparative analysis of memory-based and model-based collaborative filtering on the implementation of recommender system for E-commerce in Indonesia: A case study PT X" shows that recommendation speed and accuracy of model-based approach is better than memory-based approach [5]. Here the recommender system was created on e-commerce dataset. Here naive Bayes method was used as model-based approach and SVD was used as memory-based models. In [6] shows the use of RBM for recommender systems. In paper titled "Restricted Boltzmann machines for collaborative filtering" author has shown the application of RBM for collaborative filtering and showed that the method slightly outperforms SVD which is the most popular memory-based technique on Netflix data set [7]. In recent years because of the amount of research done in deep learning, it is getting widely popular and being implemented in various domains. So lot of research has to be done in the field of recommender system. Chiliguano et al. [8] have created hybrid recommender system using CNN to provide recommendation, in this method audio is represented in n–dimensional vector. It has outperformed traditional content-based recommender system. Kereliuk et al. [9] has used adversary to classify music according to genere using CNN. In [7] the author has shown the application of RBM for collaborative filtering. In recent years because of the amount of research done in deep learning, it is getting widely popular and being implemented in various domains. So lot of research has to be done in the field of recommender system.

2. Types of recommender systems

There are mainly three types of recommender systems. Please refer Fig. 1 for pictorial representation of types of recommender system.

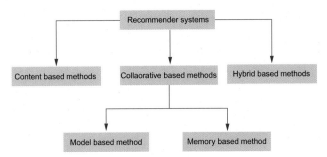

Fig. 1 Types of recommender system.

1. Content-based methods: These methods are dependent on the information of user–item interactions. In item-based methods the information about the products or also known as item metadata that user interacted or purchased are used for recommendation. In user-based methods information about the users are used for recommendation. Information can categories tags or properties of products that user purchased. The implementation of this model is simpler compared to other methods. One of implementation of this method is using tf–idf.

   Advantages of content-based methods
   - Cold start problem: It handles this problem better. If items are not rated by users still it can provide recommendation. But it performs good with new items but it does not performs well with new user.
   - Provides user independence as their own data are used to provide recommendation to them.
   - Transparency: It gives explain to the user about the recommendation they are getting.

   Limitations of content-based filtering
   - Slower: The prediction accuracy of this model is less compared to other methods. It also does not take into account what other user thinks about the item.
   - Scalability: This model becomes difficult to scale if the number of item grows.
   - Overspecialization and Serendipity: In this method it uses characteristics of the items the user has interacted or purchased then this method will only recommend products similar to the items they have already interacted with [10]. This is known as overspecialization. Serendipity means recommending some random items that user has never seen before. Serendipity is good thing as user can come across some items that user did not know he like it. A perfect

content-based filtering method will have over specialization problem. Its problem can be solved using genetic algorithm.

- Lot of knowledge of particular field is required by this technique. If a recommender system to recommend a movie is to be created then the designer of the system should know about the features of movie required to recommend the movie. If all the necessary features are not included then it can lead to poor recommendation accuracy.
- Synonyms: In content-based method, it cannot find similarity between similar words such as movie and film. Both the words represent same things but words are different which content-based method cannot detect.

2. Collaborative filtering method: Prediction is made by finding users based on the similarity of their taste. If user A have liked some items then this filtering method finds another user called B which also liked similar products and recommends products that user B has liked and user A has not interacted with or not know about that item. For example, if two user has liked similar television shows then the recommender system will recommend shows that one user has viewed which other users has not watched. Collaborative filtering can be either user–user based or item–item based. In user–user based, recommendation is provided based on search of similar users in terms of interaction with the system. As only few products are purchased or interacted by user, the method becomes very sensitive to any recorded interaction but provides more personalized recommendation. In item–item-based method similar items are recommended based on user item interaction. As item has interacted with many user the recommendation provided is less sensitive to interaction while providing less personalized recommendation. There are two types of collaborative filtering techniques:

    Memory-based approach: In memory-based approach or also known as neighborhood-based model the algorithm uses whole matrix for calculation. The whole matrix is loaded into main memory and based on that recommendation is made. So more storage is required for storing the matrix. Prediction cannot be made if the user is not present in the user–item matrix [2]. Every time prediction is made using whole matrix which slow downs the system.

    Model-based approach: In model-based approach based on given input model is created and based on that model prediction is made. Different machine learning algorithms are applied to create a model. In model-based approach some of the most popular model-based methods are clustering techniques, Bayesian technique and neural networks. But because of the neural networks ability to deal with nonlinear data and nonnumerical data, such video, text, audio have made neural network-based methods more popular than other memory-based methods. Now people are working more on deep learning model which is a neural network because of growing popularity of this method. Advantage of model-based approach is that size of trained model is smaller in size so it takes less space in memory contrary to memory-based approach which takes more memory. RBM which is a model-based approach has shown more accuracy compared to most popular memory-based methods [5, 7].

Advantages of collaborative filtering method

- Does not have over-specialization problem as it uses similarity measures which takes into other user's liking of item.
- This method provides faster recommendation and accuracy compared to content-based methods. It is also helpful in fields where not much content associated with item is there. As user uses the system, the accuracy increases as more data will be available.
- It can incorporate implicit feedback such number of times user has interacted with the item, or amount of time user has spent on a given web page, using the user's social network information or using user's browsing history.

Limitations of collaborative filtering

- Gray ship problem: Sometimes some users taste does not match with other user then it is difficult to recommend products to them. But generally they are less in number.
- Cold start problem: When new users comes to the system then there is no history of past interactions with other items then this leads to cold start problem. This problem can be solved by giving recommendation based on popularity of items or demographic information [11, 12]. Fig. 2 shows the comparison of collaborative filtering and content-based filtering.

  Types of cold start problem:
  Cold start problem can be classified into the following categories:
  - **(a)** New community: If items are available but no users are present then it is difficult to provide recommendation because of lack of interaction.
  - **(b)** New item: whenever a new item is added then but no interaction is present.
  - **(c)** New user: whenever a new users comes to the system then because of lack of history, so it is difficult to provide personalized recommendations.

COLLABORATIVE FILTERING   CONTENT-BASED FILTERING

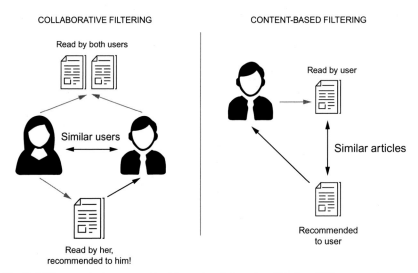

Fig. 2 Difference between content-based collaborative filtering. [13]

3. Hybrid-based method: Hybrid recommendation system is created by combining more than one recommendation algorithm. It can be designed by combining content-based and collaborative filtering algorithm. Very few hybrid recommendation systems designed till date include deep learning method because still more research is going on and people are implementing deep learning methods for collaborative filtering. As in Ref. [14] some of the hybrid techniques include:

- Weighted: This is one of most simplest form of hybrid recommender systems. In this system two scores are calculate separately and both the components are combined using linear-weighted scheme.
- Switching: In some recommendations it is difficult to combine two scores which are generated separately and selected one is implemented
- Mixed: In this hybrid system recommendations generated separately and then it is combined based on some ranking.
- Feature combination: The idea of this hybrid system is to inject data from source to another hybrid system designed for another data source.
- Feature augmentation: Feature combination and feature augmentation are very similar. Instead of using features from the participating recommender's domain, it generates a new feature for each item by using the logic of the participating domain.
- Cascade: In cascade recommender system the results generated from weak recommender system cannot overturn results generated by stronger recommender systems it can only enhance them.

- Meta–level: It uses a model learned from other recommender system as input to another.

Limitations of hybrid recommender system

Hybrid systems solves some of the limitations of above two methods but it has its own limitations. One of advantages of using deep learning method is to keep user's privacy, but by incorporating other methods in hybrid methods privacy is not maintained. Implementing hybrid recommender systems are complex and deep learning-based recommended systems can outperform the individual algorithms used in hybrid systems. Because of added complexity in hybrid system it can be expensive to implement. It needs some external information which is usually not available. Lot of research is still going on in deep learning method and more exploration is needed. Because of this hybrid recommender system involving deep learning can be developed in future.

2.1 Real world examples of recommender system

Recommender systems are utilized in many areas such as video streaming platform like YouTube and Netflix is using it to provide video recommendation. Netflix even provides personalized thumbnail based on user's taste to make particular video attractive to customers. Google uses recommendation to provide relevant ads' to its users which is their primary way of making money. Travel recommendation system is used to provide travel recommendation while, job recommender system provides job recommendation based on user profiles and skill set. Music streaming site such as Spotify provides music recommendation and based on their taste provides weekly new playlist target to each user based on their taste which is one of the popular features of Spotify. They use collaborative filtering with deep learning to provide recommendation. E-commerce giants such as Flipkart, Amazon uses recommendation system as a targeted marketing tool to recommend products similar to products we have purchased or products which we are interested in. Social networking sites such as Facebook make friends recommendation and many more applications are available in our day to day life.

2.2 Need for recommender systems

In 2000 an experiment was conducted by psychologist Sheena S. Iyengar and Mark R. Lepper of Stanford University and published the paper based on their field experiment also known as Jam experiment. They have showed that when user's are presented with more variety of choices then it is more likely that user will purchase less products compared to traditional notion that more choice will encourage more purchase by user. Recommender system solves this problem where user does not have to go through all the items to make the decision. Instead recommender system will suggest few items based on the user's taste. Companies has shown to get more profits when they provide recommendation to users and users are more likely to like a company which provides personalized recommendation to them. Companies try to retain more customer over a period of time as customers are more attracted toward personalized recommendation. Profit-based recommendation is also used by companies to increase profit.

2.3 Distinction of music recommender system from other recommender systems

First, the length of a music can be much shorter than other items such as using a product or duration of going on a trip. Second, the number of music item available is of large in number so the music are more disposable compared to other items. Short consumption time means that if some wrong recommendation is provided to user then it will not affect negatively as compared to other recommendation. This is opposite to movie recommender system where it takes more time for user to figure out that the recommendation provided to them are bad so it can affect negatively compared to music recommendation system. Third, repeated recommendation is sometimes appreciated by music listener because users like to listen to their favorite music again and again. Unlike movie recommender system where user will not like if same movie is recommended again and again. Fourth, music typically as playlists. The most important task is to recommend meaningful list of music, not just random songs is an import task of music domain.

2.4 Explicit vs implicit feedback

With users not ratings most of the songs they listen to explicit ratings are generally scarce and the amount of explicit ratings are much less compared to amount of songs users listen to. Most of the times implicit feedback are easily available compared to explicit feedback. Implicit feedback are available in large number, where user does not have to explicitly provide ratings. The more user interacts with the system more data will be available. Because of more data the recommendation of the system will be better. Implicit feedback removes the burden from users by automatically learning the features by monitoring user's action.

Challenges of collecting implicit feedback

Implicit feedback needs to be dealt separately compared to explicit feedback.

1. Number of negative feedback: If user has not interacted with the item that does not mean that user does not like that product [3]. It may also be possible that user is not aware of the product or the item may not be available in their region. Item may be any song or movie or can be any product. But in explicit rating the user specifies their liking or disliking by rating that product. So instead of using the user's liking we calculate the confidence value which tells that how confident we are that user liked the item compared to other item user has interacted with.

2. Noisy data: Implicit data tent to be noisy. Suppose a product is purchased by a user that does not mean that user liked that product, the user might have bought it for somebody else [3, 7].

3. Numerical data: Explicit ratings are generally in integers and all the ratings have same range. It is easier to perform mathematical operations on this values. Implicit feedback are such as user's click behavior or amount time user listened to the music which does not have a range. So implicit feedback needs to be carefully selected and normalized so that computation can be performed [3]. We use absolute value for explicit ratings while for implicit rating we calculate scores relative to other item's score.

2.5 Benefits of using deep learning approaches

- Practical advantages of deep learning models is that it ensures user's privacy. With today's focus on privacy in the age of internet more people are concerned about their data. Training of the model can be done locally and only the model parameters have to be sent online on centralized system for prediction. So the users data are not kept online because of that it can ensures security and privacy of users' data.

- Space efficiency: Generally the size of trained models which are kept online is much smaller compared to matrix methods used in memory-based models.

- Prediction speed: Prediction speed of deep learning models are much faster compared to other methods. Because instead of using whole database prediction is only done using learned model which greatly improves the prediction time.

- Deep learning has the ability to perform task such as such as extracting features from audio, pictures which can also be used to provide recommendation using these features. It has the ability to deal with nonnumerical data.
- Accuracy: Prediction performance of deep learning is better than non-deep learning methods.
- Elimination of feature engineering: In machine learning feature engineering is one the important task where domain knowledge is required. Benefits of deep learning is the automatic detection of features and performing computation on that.
- Elimination of costs: The use of deep learning eliminates some of the direct cost such as cost involved in labeling the data. It can also detect defects which would be difficult to detect otherwise.

2.6 Unsupervised learning

Unsupervised learning algorithm are algorithms which do not require labeled data to find hidden patterns in their input. Data collected for unsupervised learning happens in real time which is most of world's data. Especially in the field of big data where amount of data are huge most of the data's available are unstructured data's. It is used to infer a priori probability. In unsupervised learning, a system may cluster information according to their similarities or differences where the details about information is unknown. In unsupervised learning, a algorithm performs computation on unlabeled, uncategorized data and the system's algorithms perform operation on the data. The output is dependent upon the coded algorithm used. A change in algorithm will yield a difference in patterns. Unsupervised learning algorithms can process more complex tasks than supervised learning systems, depending on the particular algorithm used. However, unsupervised learning can be more unpredictable than supervised learning algorithm. It may find unforeseen and undesired categories, which are not always the target of the search. The various unsupervised learning techniques can be clustering, generative analysis models. Unsupervised learning is most useful in field of density estimation in unsupervised learning. The algorithm must sort the group based on the patterns in data. In unsupervised learning algorithm it is difficult to measure accuracy of the given result. Some of unsupervised learning algorithms are:

2.6.1 Clustering
Clustering is used for clustering items based on some of the similarity measures. Based on that similarity measures the items which are similar will be in

same cluster. The cluster may differ based on the similarity measure or distance measure used. The number of cluster can be chosen using clustering can also be used for dimensionality reduction. It is based on the idea that most of the items are redundant, so removing the whole cluster and only few elements can represent the whole cluster and that element can be the mean of the cluster. It is used to find not known pattern in a group of elements.

Types of clustering includes
- Distribution–based model
- Centroid–based methods
- Connectivity–based methods
- Subspace clustering

2.6.2 Neural networks

The artificial neural networks is created to mimic the structure of human brain. In neural networks each neuron is a function which does some computation and gives output. Deeper layer model can be constructed by connecting neurons of different layers by edges and each edge will a have some weights. By changing the weights the output can be constructed. Neural network has the ability to adapt to the change input. Neural networks have the ability to learn nonlinearity in input and output which is very important for real life scenarios where most of the relations between input and output are nonlinear. After learning this pattern, it can also make prediction for unseen data. There are some challenges with neural network is that it is considered a black box. Networks automatically decides which feature is amplified to get the output without the explanation. As more and more layers are added into the neural network to learn higher level feature, it becomes more difficult to give an explanation of the trained model. Other issue is that it still requires a large amount of data to learn patterns and also requires lot of computational power. Because of large amount of data being generated by the popularity of the internet, it has become feasible to train the model but it has given rise to another issue of user's privacy. More and more user's data are collected of the users which is of growing concern.

2.6.3 Real life applications of neural networks

Neural networks is used for fraud detection by banking enterprise. It is used to find the anomaly in purchase pattern of user compared to his normal purchase to detect fraud. Handwriting detection and language translation used in Google lens in mobile application to detect hand written notes of user and making translation to other languages. Facial and object detection used by mobile phone cameras to detect faces objects in images, traditional segmentation algorithm to separate subject from foreground, use of computational photography to enhance the photos taken through mobile cameras and edge detection in photos all are the results of advancement made using neural networks. It is also used to predict the future stock prices in stock markets. Neural network is also used in optimization problem, which involves finding solution to problems which are considered nonpolynomial problems. Virtual assistant uses voice recognition to understand human language in smart devices which is considered to be the future. In 2018 Google showed the power of its Google assistant where it booked an appointment without the intervention of human. Neural network is also use in the field of medical where it used to detect various types of cancer. It is also used in the field of network security to detect network intrusion.

2.6.4 Stochastic neural network

In stochastic neural networks, the algorithm instead of providing deterministic values to each neurons it assigns probabilities to each neuron. If each neuron passes the threshold values then only the neurons will fire. It is built by introducing random variation into the network and by giving stochastic weights. This makes it useful for optimization problem as random fluctuation help from escape from local minima which is a type of problem in neural networks. It tries to maximize the likelihood of the neurons. Boltzmann machine is the most popular form of stochastic neural network. It has found application in risk management, oncology, bioinformatics.

Advantages of unsupervised learning algorithm

Complexity: The implementation complexity is much less than supervised learning algorithm. The implementer does not have to understand the label of data inputs.

Data availability: The amount of unlabeled data available is much more than labeled data because for labeled data human intervention is required. Unlabeled data are getting generated all the time.

Some of the methods that comes under unsupervised neural network is autoencoder, RBM, generative adversarial network.

2.6.5 Energy-based models

Energy word is generally never associated with the deep learning methods as it is a concept of physics. But some deep learning architectures used the idea of energy to measure the quality of model. RBM is one such model which is based on the concept of Hop-field network which was created for storing data. In this methods energy of the models are calculated and the lesser the energy netter the system. When energy of the system reaches minimum value then model is said to be in stable. Details of energy calculation is given in implementation. The idea of using energy and probability comes from the concept of Ising model. In a stable model whenever the state of the node changes then the energy of the system decreases which is more stable. Based on this model various neural networks and deep learning networks are designed such as Hop-field network, RBM, Deep Belief network, etc.

2.6.6 Boltzmann machine

A Boltzmann machine (Fig. 3) is a group of symmetrically connected, neuron that make stochastic decisions such as to be on or off [15]. Boltzmann machine was invented by Geoffrey Hinton. It was one of the initial models that can learn internal representation of input. Boltzmann machine was create to solve two problem, one was search problem and another was learning problem [15]. Boltzmann machine contains visible neurons and hidden neurons. Visible neurons and hidden neurons are represented in Fig. 3. A Boltzmann machine can be used to learn important aspects of an unknown probability distribution based on samples from distribution. Learning in Boltzmann machine is slow. Because of unconstrained connection between neurons it was difficult to use it practically. The learning of Boltzmann machine increases with increase in the number of neurons. If connections are constrained then this method can be used. This is what RBM solves. Restriction in connections helps in faster training of model. Boltzmann machine tends to learn by making small update to the weight and minimize the energy of the model.

In above diagram by calculating the sum of its own weights and biases on edges coming from other units states of neurons are updated:

$$z_i = b_i + \sum_{i,j} s_j w_{ij} \tag{1}$$

where w_{ij} is the weights on edge between node i and j. Unit i then turns on with probability given by the logistic function.

$$prob(s_i = 1) = \frac{1}{1 + e^{-z_i}} \tag{2}$$

In this model energy defines the badness of state. Lower energy is good for the model and minimum energy shows the stability of the model.

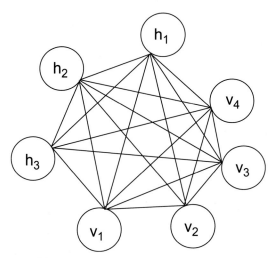

Fig. 3 Boltzmann machine.

If units are updated sequentially then the network will eventually reach a stable or equilibrium state. Learning in Boltzmann machine is given by the equation.

$$\sum_{V \in data} \frac{\delta(P(v))}{\delta w_{ij}} = \frac{\delta log(P(v))}{\delta w_{ij}} = \langle s_i s_j \rangle_{data} - \langle s_i s_j \rangle_{model} \tag{3}$$

where $\langle s_i s_j \rangle_{data}$ is the actual value while $\langle s_i s_j \rangle_{model}$ is the value predicted by model.

$$P(v) = \frac{e^{-E(v)}}{\sum_u e^{-E(u)}} \tag{4}$$

Energy of state is defined as

$$E(v) = -\sum_i s_i b_i - \sum_{i<j} s_i s_j w_{ij} \tag{5}$$

where s_i is binary state assigned to unit i, b_i is the bias.

Here the Boltzmann machine will learn by calculating the error between calculated value and observed value. Then it will change the values of slightly to get nearer to the actual result.

But generally this method will model the distribution but there is a problem when the machine is scaled up as it stops learning correctly. The time required to train this model increases exponentially with size. This model has a problem of converging to local minimum. Because of this Boltzmann machine was not used for long time. RBM overcome the limitation of Boltzmann machine by implementing restriction on the edges between neurons. Boltzmann machine can be used as associative memory.

2.6.7 Restricted Boltzmann machine

RBM is a two layer generative stochastic artificial neural network which contain one visible input layer and one hidden layer but no output layer. All visible layer is connected to all hidden layer nodes and vice versa but there is no intra layer connection between hidden layer and visible layer. They are also known as symmetric bipartite graph. Because of restriction of connection between neurons of same layer, this model is called restricted Boltzmann machine. RBM belong to energy-based models. It is based on another deep learning method called Boltzmann machine. Each node in RBM makes some computation and stochastically decides whether to pass that input or not. Boltzmann machine is a special type of Hop-field network which is based on Markov model (Fig. 4). RBM's are usually trained using contrastive divergence algorithm.

This model contains two parts, i.e., forward pass and backward pass. In forward pass the models tries to calculate probabilities at each hidden neuron and decides whether to pass that input or not, and on backward pass the model tries to create reconstruct input similar to the original input. This is also known as one Gibbs sampling step. One iteration of alternating Gibbs' sampling consists of updating all hidden node in parallel using Eq. (6) followed by updating all the visible nodes in parallel using Eq. (7). At first the reconstructed inputs will be different from original input, but by changing biases and weights, errors can be reduced. Based on this learned weights and biases, when inputs are given new outputs can be predicted or reconstructed based on the given recommender system application.

RBMs are stochastic model, i.e., instead of calculating discrete values at each node the model calculates probabilities at each nodes and then the result passes through some activation function which gives output as 1 or 0. Inputs to the RBM can be only be in binary. RBM does not contain connections between neurons of same layer. Because of this restriction, training of RBM is faster than other models as computational time decreases due to less connections between neurons. RBM is an unsupervised learning algorithm which reconstructs its input approximation from data. It does this by calculating probabilities of our input which helps in finding data points which

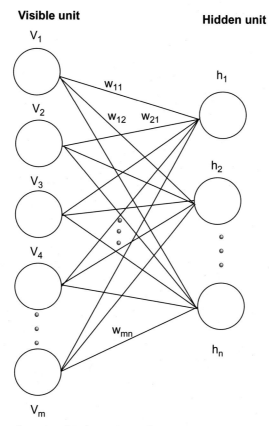

Fig. 4 Structure of restricted Boltzmann machine.

previously did not exist in our data. They try this by learning a lower representation of our data and later by trying to reconstruct the input. Availability of unlabeled data is much more compared to labeled data so use of this model makes it more compelling. To get labeled data human intervention is required where it has to be labeled by person which unlabeled data does not have.

Other applications of RBM include classification, dimensionality reduction, regression, and topic modeling.

2.6.8 Markov chain model

Markov chain model is a stochastic model which has Markov property. Markov property is satisfied when current state of the process is enough to predict the future state of the process and the prediction should be as good as making prediction by knowing their history. It is a very easy process to model random process.

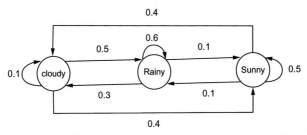

Fig. 5 Markov state diagram.

Most simple example of Markov chain is a board games played with dice. The next state of the game depends just only the current state of the game and what is the outcome of the dice. It does not depend on how the state of game reached there. Random walk and weather models are another example of Markov model.

Fig. 5 is the simplest example of the weather model. It shows the probability of transition given the current state. If current weather is cloudy then their is 50% chance of getting a rain and 40% chance of being sunny. So the probability of being sunny given that its cloudy is only dependent on the previous state which is cloudy. It does not take into account, which state was before that.

Formal definition of discrete chain Markov chain can represented as

$$P_r\left(X_{n+1} = x | X_1 = x_1, X_2 = x_2, \ldots, X_n = x_n\right) = P_r\left(X_{n+1} = x | X_n = x_n\right)$$

In the above equation if for $n \geq 0$ if value of $P_x(X)$ is similar then this equation is called homogeneous equation and the matrix becomes the transition matrix.

3. Problem statement

Using RBM which is a generative stochastic neural network for providing recommendation using number of times user listened to music as an implicit feedback.

4. Explanation of RBM

Consider we have M songs rated by N users and ratings have integer value from 1 to l then we have M visible node for M songs rated by one user. Then ratings are normalized from 0 to 1. Each user profile will become a training case for RBM. As shown in Fig. 6 the input to the normalized rating

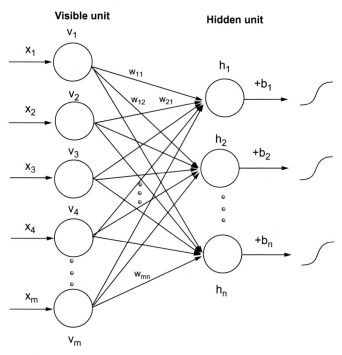

Fig. 6 Forward pass in restricted Boltzmann machine.

of each song is shown. Here number of hidden nodes are lesser than visible nodes which will force hidden nodes to learn some prominent features of input nodes. Here weights are present on edges. The visible layer has bias a_i while hidden layer has a bias b_j. The bias ensures that some of the neurons take part in the calculation if the output of that neuron becomes zero.

In forward pass (Fig. 6) all the inputs V will be multiplied with weights W and hidden layer bias is added to it which can be written as $w_{ij}v_i + b_j$ as shown in Fig. 6. Each hidden node will get input from all the visible nodes multiplied by their respective weights and then summation of it. Then this summation is given to activation function to get output for hidden nodes. Sigmoid function will be used as an activation function here but other activation function such as Rectified linear unit or hyperbolic tangent function can also be used.

$$p(h_j = 1|v) = \sigma(b_j + w_{ij}v_i) \tag{6}$$

Values of i and j is from 1 to m and 1 to n respectively.

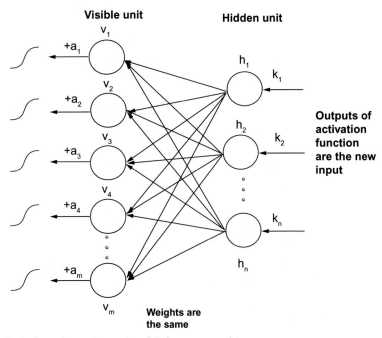

Fig. 7 Backward pass in restricted Boltzmann machine.

In backward pass (Fig. 7) the value of hidden unit becomes input to the visible nodes as shown in Fig. 7. Here each visible neuron passes through activation function and gets the value of either 0 or 1 as the model works better on binary values. Based on the above equation probability of hidden nodes are calculated which is given by the equations

$$p(v_i = 1|h) = \sigma(a_i + w_{ij}h_j) \tag{7}$$

This equation calculates the probability of hidden layer getting activated. The output of the activation function will be either 0 or 1.

Activation function is required to force output to be binary as using actual probabilities will create a serious bottleneck [16].

The RBM calculates probability to every pair of (v, h) given by equation below.

$$p(v, h) = \frac{1}{Z}(e^{-E(v,h)}) \tag{8}$$

E is the energy function. Here Z is the partition function which is used for normalizing the probability where

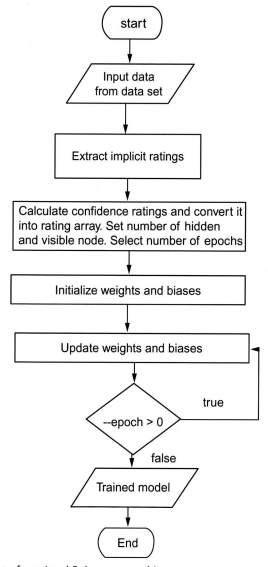

Fig. 8 Flowchart of restricted Boltzmann machine.

$$Z = \sum_{v,\,h} e^{(-E(x))} \tag{9}$$

where v, h are number of visible and hidden neurons and values are calculated for all possible pairs of visible and hidden neurons.

Fig. 8 shows the flow chart of the used method.

RBM are energy-based models which calculates energy of the system by combination of all of the hidden units and visible units. This energy function is calculated after each iteration and when the value of energy function is minimum then the model is considered trained or in stable state.

$$E(v,h) = -\sum a_i v_i - \sum b_j h_j - \sum v_i h_j w_{ij} \tag{10}$$

where i is subscript for the number of visible unit, j is the number of hidden unit. $i = 1$ to m and $j = 1$ to n.

Deriving Eq. (8) with respect to weight

$$\sum_{V \in data} \frac{\delta(p(v))}{\delta w_{ij}} = \frac{\delta log(p(v))}{\delta w_{ij}} = \langle s_i s_j \rangle_{data} - \langle s_i s_j \rangle_{model} \tag{11}$$

So for performing stochastic steepest ascent in the log probability of the training data

$$\Delta w = \alpha(\langle v_i h_j \rangle_{data} - \langle v_i h_j \rangle_{model}) \tag{12}$$

$$\Delta a = \alpha(\langle v_i \rangle_{data} - \langle v_i \rangle_{model}) \tag{13}$$

$$\Delta b = \alpha(\langle h_j \rangle_{data} - \langle h_j \rangle_{model}) \tag{14}$$

where $\langle s_i s_j \rangle_{data}$ is the outer product of values in first phase and $\langle s_i s_j \rangle_{model}$ is the outer product of values calculated in reconstruction phase. Eqs. (12), Eq. (13) and Eq. (14) is obtained from Eq. (10) [16]. Eqs. (12)–(14) will be used to update weights and biases while training the model.

5. Proposed architecture

The proposed architecture has been shown in Fig. 9. The components of the architecture have been discussed below. Information source: In this part of the process the data are collected of the users which is using the online music system for listening to the music such as userID, songID, number of times a user has played the music. We are using number of times music played as a implicit feedback. The data to be collected differ from the domain to domain. Ratings extraction: Here all the relevant ratings are extracted from the information source. We are only using one rating that is music play count. Depending on the implicit feedback to be used data is extracted from the available dataset. In e-commerce websites click through rate, amount of time spend on the page reading, etc. is used so those data need to be extracted. In this method we are only extracting number of times music

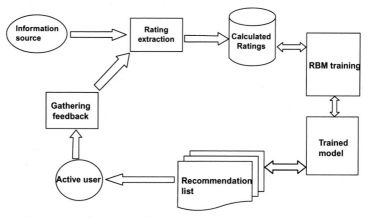

Fig. 9 Architecture of the proposed system.

played by the user. Calculated ratings: Here the implicit feedback is taken and it is converted into confidence scores which is given by

$$1 + \alpha p \tag{15}$$

where α is the importance of implicit feedback on the calculated rating. For our experiment the value of α we have taken is 1 and has shown good results. Then the confidence score will be normalized to the range of 0 to 1 to be used as ratings. To normalize the confidence score rating all the values are divided by maximum confidence score rating. As the model only performs on values from 0 and 1 the ratings needs to be normalized. Then the ratings are given as input to the visible layer. RBM training: If many users have similar confidence score for the same song then their two RBM's must share same weights for that visible unit of that song and the hidden units. The hidden units will be different for each and every user for their RBM model as hidden units. Hidden units represents the hidden latent factors.

In this part of the process the data are collected of the users who are using the music system for listening to the music such as user ID, song ID, number of times a user has played the music. We are using number of times music played as an implicit feedback. The data to be collected are used as rating, which will differ from the domain to domain. In this method we are only extracting number of times music played by the user which is our implicit feedback. Here the confidence score calculated are normalized to the range between 0 and 1 by dividing it by maximum value. Then the normalized ratings are given as input to the input layer of the model. If many users have similar input rating for the same song, then their two RBM's will share same weights for that visible unit of that song and the hidden units. The generated value of hidden units will be different for each and every user for their RBM model. Hidden units represent the hidden latent factors. Suppose we have the ratings for M songs from N number of users then we will be having M visible units for RBM and some K number of hidden units. In our experiment the number of unique songs is 384,546 so there are 384,546 visible nodes in our model. The number of hidden nodes taken is 50. Based on the input values given to the input nodes, values of hidden nodes are calculated using Eq. (6). Again based on the calculated values of hidden nodes, values of visible nodes are again calculated in reconstruction phase. The training algorithm given below which is known as contrastive divergence algorithm. After the training of the model is done each visible nodes is assigned some probability. As more user interacts with the system, the feedback can be used to improve the recommender system as time goes on.

Explanation of steps of using RBM with implicit feedback implementation

Execute the below process for set number of epochs. The below process is also called contrastive divergence algorithm [2, 7, 17].

1. Initialize w_{ij}, a_j, b_j for $i = 1$ to m and $j = 1$ to n.
2. Repeat the below processes for given number of epochs.
    (a) Select a training set S of some fixed batch size.
    (b) Set $\Delta w_{ij} = 0$, $\Delta a_i = 0 \Delta b_j = 0$.
    (c) each $v = (v_1, v_2, ..., v_n)$ in
        i. Set $v(0) = v$.
        ii. for $i = 1, ..., n$ do sample v_i and h_j
        iii. With learning parameter on each training observation, update the gradients
        iv. Calculate Δw_{ij}, Δa_i, Δb_j
        v. Update $w_{ij} = w_{ij} + \Delta w_{ij}$, $a_i = a_i + \Delta a_i$, $b_j = b_j + \Delta b_j$

Then Energy E is calculated for the system to calculate minimum energy to check and visualize the energy of the system. Change in weight in the above algorithm is calculated using Eq. (12). To calculate the errors of the reconstructed input gradients matrix or gradient vectors are calculated without which learning cannot be done. In the above algorithm initially value of w_{ij}, a_i, b_j is randomly assigned. Based on these values new value of w_{ij}, a_i, and b_j are calculated. In above equation α is the learning rate. The value of learning rate should be set carefully. If learning rate is higher than algorithm's accuracy can reduce and if learning rate is lower than it may take very long time to converge because of there is no connection between hidden units of RBM, the algorithm will get unbiased sample of the calculated data.

5.1 Contrastive divergence algorithm

Contrastive divergence algorithm is used to train RBM. Contrastive divergence algorithm greatly reduces the variance of estimates used for learning. This paper shows the convergence of this contrastive divergence algorithm [7]. Contrastive divergence is used for training undirected graphical models which are based on approximations of gradients. Prior to contrastive divergence algorithm training restricted Boltzmann machine took more time to converge but contrastive divergence algorithm has showed less time to converge. When inputs are given all the visible node values are calculated and again based on hidden node values all the visible node values are reconstructed. This is considered one Gibbs step. As shown in Fig. 10, x_0 is the visible layer of input values while x_1 is the values of visible layer after reconstruction phase or one Gibbs step. Then based on the difference of these two states values of weights and biases are updated. After k number of steps the visible layer values are represented by state x_k. Since nodes are conditionally independent, Gibbs sampling can be done which is a Markov chain Monte Carlo algorithm where all the nodes are update one at a time. Because of this Boltzmann machine practically takes huge amount of time as the number of nodes increases. But in Refs. [18, 19] the authors have shown training a probabilistic models which is used for training RBM. Unlike previous training method for Boltzmann machine here all the nodes are updated at the same time. Based on input values all the hidden node values were updated at the same time and based on the calculated values of hidden nodes all visible nodes values are updated at once. This algorithm adds bias but the added bias is very less. Theoretically if algorithm is run for infinite number of steps then it will converge to optimal situation.

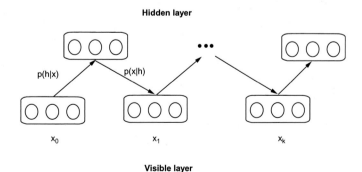

Fig. 10 Contrastive divergence algorithm.

Practically it has shown great results for k number of steps [7]. In contrastive divergence instead of approximating the second term of log–likelihood, a Gibbs chain is run for some number of steps. First v_0 is initialized then based on that $x_1, x_2, \ldots x_k$ is calculated for k steps where x is the state of visible unit. In each steps values of $p(v|h)$ and values of $p(h|v)$ is calculated.

5.2 Prediction of recommender system

Whenever user interacts with the system then to predict the output of new song then calculated rating V prediction can be made in linear time. We first draw random binary hidden units with $p(h_j = 1|v)$ for all j [20].

$$p(h_j = 1|v) = \sigma \sum_{i=1}^{n} w_{ij} \sum_{t=1}^{g_i} v_i + c_j \tag{16}$$

Then the probability of the visible unit of item i being 1 is given by

$$p(v_i = 1|h) = \sigma \sum_{j=1}^{m} w_{ij} h_j + b_i \tag{17}$$

When RBM recommender system is trained, it provides recommendation to new user if the user have few interaction with the system. When new user comes then we do not need to train the model. Prediction is made using the already calculated values of $p(h|v)$ and $p(v|h)$ given by Eqs. (16) and (17). Then recommendation output can be calculated and based on these values the movie with highest probability are recommended based on the number of recommendation to be provided. This is unlike matrix factorization method which is most popular memory-based recommender system in which the prediction cannot be made unless user is already in the training set. If user is not in training set then latent factors cannot be learned. The prediction made using this model is much quicker than matrix factorization method because matrix factorization methods keeps whole matrix in memory. Trained model is kept online while other users ratings can be kept offline which increases the security and privacy of the user data. The calculated parameters can be sent and kept online to make predictions. The recommendation can be made when users have interacted with the system. Then active users that are interacting with the system or getting recommendation, they are generating more data by their activities. This is why implicit feedback is better than explicit feedback. The more the user uses the system the more data it generates which is better for recommendation.

6. Minibatch size used for training and selection of weights and biases

Calculating gradients on whole single training set is possible but because of huge size of training set, it may slow down the training process. So instead of using whole training set, it can be divided into minibatches. Because of this it becomes advantageous to use parallel processing power of GPUs [16]. Initially the value of weights should be chosen in random from zero mean Gaussian with standard deviation 0.01 [16]. Using larger value will speed up learning but the final model will be slightly worse. Initial weights should not make values of hidden nodes closer to 0 or 1 [16]. It is usually helpful to initialize the bias of visible unit i to $\log[p_i/(1 - p_i)]$ where p_i is the proportion of training vectors in which unit i is on.

7. Types of activation function that can be used in this model

- Softmax and multinomial units: For a binary unit the probability of turning on or off is given by the logistic sigmoid function on input [16].

$$p(x) = sigm(x) = \frac{e^x}{e^x + e^0} \tag{18}$$

Here sigm is sigmoid activation function. Here the energy computed by the unit is $-x$ if is one else 0. This function can be used when we need to constraint the values between 1 and 0. This is similar to binary values, instead of discrete values probabilities are used.
- Rectified linear unit: Rectified linear also gives the output value as 0 or 1 but it gives value of 0 for all negative values which makes it difficult to fit the model. If this model is used in visible and hidden units then the value of learning rate should be small or it will make the system unstable [21].

$$\sum_{i=1}^{N} var(x - i + 0.5) \approx log(1 + e^x) \tag{19}$$

Here *var* represents variance

- Binomial units: A simple way of producing a unit of noisy numerical values between 0 and N is to create N different copies of a binary unit and assign them the same biases and weights. Since all copies receive the same total input, they all have the same probability p, of turning on and this only has to be computed once. This activation function is less stable than rectified linear unit.
- Gaussian units: When the visible and hidden units are Gaussian then this activation function can be used.

$$E(v, h) = \frac{\sum_{i \in V_i}(v_i - a_i)^2}{2} \tag{20}$$

8. Evaluation metrics that can be used to measure for music recommendation

- Accuracy: RMSE measures the difference between the actual value and the predicted value. It calculates using squared error method. The smaller the value of RMSE the better the result is considered. It then takes the average values calculated over all users to measure accuracy. Then it will sort the values of output in descending order and give top-N recommendation by selecting the songs with higher probability and suggest songs [22]. As the difference of values are squared, larger error will have large effect on the values.
- Diversity: This is to measure how diverse are the songs recommender by the system. The diversity is defined as the average randomness of the genre among user:

$$diversity = \left(\frac{1}{u}\right) \sum_{u=1}^{U} \sum_{g=1}^{G} h_u(g) log(h_u(g)) \tag{21}$$

where $h_u(g)$ denotes the number of songs for the g-th genre and G is the total number of genres.
- Novelty: This metric is to measure how many recommended songs are new.

$$\frac{RA_u / UA_u}{RA_u} \tag{22}$$

Here U denotes the number of users.

- Freshness: In this measurement we take into account how recent are the songs recommended to user's. In this songs released date are taken into account.

$$freshness = \left(\frac{1}{U}\right)\left(\frac{1}{N}\right) \sum_{u=1}^{U} \sum_{i=1}^{N} RD_{ui} \qquad (23)$$

where $n =$ top N songs and U is the number of users. $RD_{ui} =$ is the time stamp of the ith song

- Popularity: In this popularity of songs recommender are measured. Here popularity is measured by taking average among all songs recommended.

$$popularity = \left(\frac{1}{U}\right)\left(\frac{1}{N}\right) \sum_{u=1}^{U} \sum_{i=1}^{N} PS_{ui} \qquad (24)$$

$PS_{UI} =$ is the popularity of the songs

- Precision: This measures how many correct songs have been recommended among all the song recommended. This is the ratio of number of correctly recommended items and total recommended items.

9. Experimental setup

In our experimental setup shown in Table 1, we have used "The Echo Nest Taste Profile Subset" dataset [23]. The dataset contains user's unique id, song id, and user's music play count for each song. The dataset was split into 80% training set used for training and 20% testing set for calculating the error in recommendation to compare it with other model. Before using the dataset for model training, dataset was preprocessed to remove any outliers, null values and user id which contains long id's were renamed for simplicity. We have used sigmoid function as activation function but other types of activation function can also be used.

Table 1 Experimental setup.

Dataset	The echo nest taste profile subset
Language	Python
Libraries	Scikit-learn, pandas, numpy
Coding environment	Google Colab

Table 2 Hardware configuration of testing environment.

RAM	25 GB
Processing power	2.3 GHz
Disk space	68.4 GB

The model was trained on Google Colab environment which is Google's cloud server for executing Python code and training machine learning models. Google Colab provides free environment for training the model where similar physical system with similar system configuration will be expensive. Please refer Table 2 for hardware configuration of testing environment.

The initial weights used for the model was very small chosen from zero-mean Gaussian with a standard deviation of 0.01 [16]. The initial values of visible and hidden bias was taken as 0.01. Care should be taken to ensure that the initial weight values do not allow typical visible vectors to drive the hidden unit probabilities very close to 1 or 0 as this significantly slows the learning.

10. Result

This RBM model was compared against collaborative filtering engine from Apache which uses artist playback as implicit feedback to provide recommendation given in paper titled "Comparison of Implicit and Explicit Feedback From An Online Music Recommendation Service" [24]. The metric for comparison is normalized Root Mean Square Method (NRMSE). RMSE measures the standard deviation of the prediction value. Normalization is done as the range of artist play count and music play count are different. RMSE provides a way to compare results of different range of output. It would be unfair to compare RMSE values of both the values. So NRMSE value is used that is normalized to the range between 0 and 1. The results have been shown in Table 3.

The dataset used for testing is sparse as it represents real world dataset. In this experiment we have shown that the RBM model with music play count as implicit feedback has better accuracy compared to Collaborative filtering engine by Apache Mahout Project with artist play count as implicit feedback.

Table 3 Result.

Technique	Error (NRMSE)
Restricted Boltzmann machine (RBM)	0.0075
Collaborative filtering engine by Apache	0.0200

11. Conclusion

This chapter introduces a music recommender system using RBM that uses implicit feedback as input contrary to the explicit rating for recommendation that most of people concentrate on. This chapter describes the structure of RBM, how to convert implicit feedback and represented to be given input to RBM and how recommendation is provided. The tests were successfully performed on RBM and Collaborative filtering engine by Apache Mahout Project. RBM with music play count has shown less error compared to collaborative filtering engine by Apache Mahout project using artist play count as implicit feedback. The NRMSE we obtained in our model is 0.0075 is better compared Collaborative filtering engine's NRMSE value which is 0.0200. This experiment also shows how to use the implicit feedback in RBM for music domain as input for recommender systems.

12. Future works

In future, this work can be extended to include more types of implicit feedback to increase the accuracy of the model. Other type implicit feedback which can be quantified to rating can be used in this model in future. This method can be tested on more datasets to understand the performance of this model. Other error metrics can also be used to calculate the error.

Reference

[1] G. Jawaheer, M. Szomszor, P. Kostkova, Comparison of implicit and explicit feedback from an online music recommendation service, in: Proceedings of the 1st International Workshop on Information Heterogeneity and Fusion in Recommender Systems, ACM, New York, NY, USA, 2010, pp. 47–51.
[2] F. Yang, Y. Lu, Restricted Boltzmann machines for recommender systems with implicit feedback, in: IEEE International Conference on Big Data (Big Data), Seattle, WA, USA, 2018, pp. 4109–4113.

[3] Y. Hu, Y. Koren, C. Volinsky, Collaborative Filtering for Implicit Feedback Datasets, in: 2008 Eighth IEEE International Conference on Data Mining, Pisa, 2008, pp. 263–272.

[4] D. Parra, X. Amatriaain, Walk the talk–analyzing the relation between implicit and explicit feedback for preference elicitation, in: 19th International Conference on User Modelling, Adaptation and Personalization (UMAP), 2011, pp. 255–268.

[5] P.H. Aditya, I. Budi, Q. Munajat, A comparative analysis of memory-based and model-based collaborative filtering on the implementation of recommender system for E-commerce in Indonesia: a case study PT X, in: 2016 International Conference on Advanced Computer Science and Information Systems (ICACSIS), Malang, 2016, pp. 303–308.

[6] K. Georgiev, P. Nakov, A non-IID framework for collaborative filtering with restricted Boltzmann machines, in: Proceedings of the 30th International Conference on Machine Learning, Georgia, USA, vol. 28, 2013.

[7] R. Salakhutdinov, A. Mnih, G. Hinton, Restricted Boltzmann machines for collaborative filtering, in: Proceedings of the 24th International Conference on Machine Learning, Corvalis, Oregon, USA, June, 2007, pp. 791–798.

[8] P. Chiliguano, G. Fazekas, Hybrid music recommender using content-based and social information, in: 2016 IEEE International Conference on Acoustics, Speech and Signal Processing (ICASSP). IEEE, 2016.

[9] C. Kereliuk, B.L. Sturm, J. Larsen, Deep learning and music adversaries, IEEE Trans. Multimedia 17 (11) (2015) 2059–2071.

[10] L. Iaquinta, M. de Gemmis, P. Lops, G. Semeraro, M. Filannino, P. Molino, Introducing Serendipity in a Content-Based Recommender System, in: 2008 Eighth International Conference on Hybrid Intelligent Systems, Barcelona, 2008, pp. 168–173.

[11] X.N. Lam, T. Vu, T.D. Le, A.D. Duong, Addressing cold-start problem in recommendation systems, in: Proceedings of the 2nd international conference on Ubiquitous information management and communication (ICUIMC '08), ACM, New York, NY, USA, 2008, pp. 208–211.

[12] A.I. Schein, A. Popescul, L.H. Ungar, D.M. Pennock, Methods and metrics for cold-start recommendations, Proceedings of the 25th Annual International ACM SIGIR Conference on Research and Development in Information Retrieval (SIGIR '02), Association for Computing Machinery, New York, NY, USA, 2002, pp. 253–260. https://doi.org/10.1145/564376.564421.

[13] M.H. Mohamed, M.H. Khafagy, M.H. Ibrahim, Recommender System Challenges and Solutions Survey, in: 2019 International Conference on innovative trends in Computer Science and Engineering, Aswan, Egypt, 2–4 February, 2019, pp. 149–155.

[14] R. Burke, Hybrid web recommender systems, in: P. Brusilovsky, A. Kobsa, W. Nejdl (Eds.), The Adaptive Web, Lecture Notes in Computer Science, vol. 4321, Springer, Berlin, Heidelber, 2007.

[15] G.E. Hinton, Boltzmann machine, Scholarpedia 2 (5) (2007).

[16] G.E. Hinton, A practical guide to training restricted Boltzmann machines, in: Neural networks: tricks of the trade, Springer, Berlin, Heidelberg, 2012, pp. 599–619.

[17] E. Geoffrey, G.E. Hinton, Training products of experts by minimizing contrastive divergence, Neural Comput. 14 (8) (2002) 1771–1800.

[18] G.E. Hinton, Training products of experts by minimizing contrastive divergence, Neural Comput. 14 (8) (2002) 1771–1800.

[19] G.E. Hinton, Products of Experts, Vol 1, Proceedings of the Ninth International Conference on Artificial Neural Networks, 1999, pp. 1–6.

[20] A.L. Yuille, The convergence of contrastive divergences, in: Advances in Neural Information Processing Systems, Los Angeles, 2005, pp. 1593–1600.

[21] V. Nair, G.E. Hinton, Rectified linear units improve restricted Boltzmann machines, in: Proceedings of the 27th International Conference on International Conference on Machine Learning, Haifa, Israel, June, 2010.

[22] H. Chou, Y. Yang, Y. Lin, Evaluating music recommendation in a real-world setting: on data splitting and evaluation metrics, in: 2015 IEEE International Conference on Multimedia and Expo (ICME), Turin, 2015, pp. 1–6.

[23] B. McFee, T. Bertin-Mahieux, D. Ellis, G. Lanckriet, The million song dataset challenge, in: Proceedings of the 12th International Conference on Music Information Retrieval (ISMIR), 2011. http://millionsongdataset.com/tasteprofile.

[24] G. Jawaheer, M. Szomszor, P. Kostkova, Comparison of implicit and explicit feed-back from an online music recommendation service, in: Proceedings of the 1st International Workshop on Information Heterogeneity and Fusion in Recommender Systems, ACM, New York, NY, USA, 2010, pp. 47–51.

About the authors

Amitabh Biswal is a M.Tech Student of NIT Silchar. Research interests are Data Mining, Machine Learning.

Dr. Malaya Dutta Borah is working as Assistant Professor in the dept. of Computer Science and Engineering at NIT Silchar. Her research interest is Data Mining, Block-chain Technology and Cloud Computing. For more information you can visit her official webpage: http://cs.nits.ac.in/malaya/

Zakir Hussain is a PhD Scholar at NIT Silchar. His research interest includes Machine Learning in Healthcare, Data Mining, and Predictive Analytics.

Printed in the United States
by Baker & Taylor Publisher Services